Transport and Energy Research

Transport and Energy Research

A Behavioral Perspective

Edited by

JUNYI ZHANG
Hiroshima University, Higashi Hiroshima, Japan

ELSEVIER

Elsevier
Radarweg 29, PO Box 211, 1000 AE Amsterdam, Netherlands
The Boulevard, Langford Lane, Kidlington, Oxford OX5 1GB, United Kingdom
50 Hampshire Street, 5th Floor, Cambridge, MA 02139, United States

British Library Cataloguing-in-Publication Data
A catalogue record for this book is available from the British Library

Library of Congress Cataloging-in-Publication Data
A catalog record for this book is available from the Library of Congress

ISBN: 978-0-12-815965-1

For Information on all Elsevier publications
visit our website at https://www.elsevier.com/books-and-journals

Publisher: Katie Hammon
Acquisition Editor: Raquel Zanol
Editorial Project Manager: Aleksandra Packowska
Production Project Manager: Nirmala Arumugam
Cover Designer: Christian J. Bilbow

Typeset by MPS Limited, Chennai, India

Contents

List of contributors

Christine E. Kormos
Sustainable Transportation Action Research Team, Simon Fraser University, Burnaby, BC, Canada

Ying Jiang
Hiroshima University, Hiroshima, Japan

Fanglei Jin
Beijing Jiaotong University, Beijing, P.R. China

Xiaoyi Li
Beijing Institute of Technology, Beijing, P.R. China

Hitomi Nakanishi
University of Canberra, Canberra, ACT, Australia

Wei-Shiuen Ng
International Transport Forum, Organisation for Economic Co-operation and Development (OECD), Paris, France

Long Pan
Beijing Jiaotong University, Beijing, P.R. China

Chunfu Shao
Beijing Jiaotong University, Beijing, P.R. China

Reuven Sussman
Behavior and Human Dimensions Program, American Council for an Energy-Efficient Economy, Washington, DC, United States

Li Qin Tan
University of Victoria, Victoria, BC, Canada

Harry Timmermans
Eindhoven University of Technology, Eindhoven, The Netherlands

Meimei Xue
Beijing Institute of Technology, Beijing, P.R. China

Dujuan Yang
Eindhoven University of Technology, Eindhoven, The Netherlands

Enjian Yao
Beijing Jiaotong University, Beijing, P.R. China

Biying Yu
Beijing Institute of Technology, Beijing, P.R. China

Junyi Zhang
Hiroshima University, Higashi Hiroshima, Japan

Linghan Zhang
Hiroshima University, Higashi Hiroshima, Japan

Rui Zhang
Beijing Jiaotong University, Beijing, P.R. China

Runsen Zhang
Hiroshima University, Higashi Hiroshima, Japan

Tuqiang Zhou
Hiroshima University, Higashi Hiroshima, Japan

Weiyan Zong
Hiroshima University, Higashi Hiroshima, Japan

Foreword by Prof. David Banister

It has proved to be extremely difficult to break the link between the use of carbon-based energy and the growth in the demand for transport. Efficiency gains over the last 30 years have to be balanced against the continuous growth in both passenger and freight transport. Transport is still dependent on oil for over 95% of its energy, and it currently accounts for about two-thirds of all oil consumption. The primary objective for the future transport and energy policy must be to reduce oil dependence and to switch to alternative renewable energy sources, but such a potentially simple objective is embedded within a hugely complex reality.

It is not just a matter of heavy investment in technological solutions that may only result in more travel, but in understanding the reasons why both people and goods are transported. This means that it is essential to engage with all stakeholders, so that the most efficient (noncarbon-based) energy sources are used. It also means that the full range of complementary measures are used to achieve targets set, to ensure rebound effects are minimized, and to make sure that the benefits of low-carbon transport are locked-in and not diluted over time. To achieve a noncarbon-based transport system means that a range of complimentary measures need to be combined as a set of strong and consistent policies. Currently, too much emphasis is given to finding a "single" solution rather than accepting that real change can only be achieved through policy packaging and the involvement of all stakeholders as part of the debate over the future of transport and energy.

Such a holistic approach would operate at all scales (international, national, regional, city, and local), and it would draw on positive experiences from around the world. It would explore different analytical approaches, different types of data, and different model specifications, accepting the limitations of the methods being used and promoting methodological innovation. It would engage with the full range of stakeholders, including politicians, the transport and energy industries (e.g., oil, electricity, aviation, shipping, and car manufacturers), businesses, transport operators, experts, and transport users, so that they are all made aware of the continued carbon dependence of transport. The debate would explore the full range of options available, look at the means by which they can be combined into effective policy packages, and examine the sequencing of interventions, so that real progress can be monitored. Such an approach provides the means by which effective implementation can take place, with all stakeholders understanding the issues involved, and so that they can ""buy into" the proposed policy packages.

This book adopts such an approach, exploring the nature and scale of the transport and energy problem, focusing on the recent changes with the increasing dominance of the Asian countries. These countries have formed the largest energy consumption region in the world (since 1992), and it has now become dominant with a 2.7 times increase (1992–2017) as compared with a situation of little change in the two other main transport energy consuming regions (Europe and North America). It also brings together the very different approaches adopted in different parts of the world, covering the technologies used, the analytical approaches, data sources and availability, behavioral issues, and policy choices. The focus on the Asian countries allows a more detailed exploration of some of the new mobility options, including the mass introduction of electric vehicles, shared mobility, and the social issues arising from (rural) migration. There are clear lessons here for Europe where more traditional institutional frameworks are still dominant, and where the focus has been on compact cities, density and mixed use development, pricing, physical restraint, and a range of other measures to restrict car use in cities.

There are good opportunities to reduce the carbon and energy intensity of transport in the cities and regions. Long-distance travel (passenger and freight) presents much greater difficulties, as there are many fewer alternatives, perhaps with the exception of the substitution of high-speed rail for air for journeys up to about 800 km. It is in tourism and visiting friends and relatives where the greatest growth in long-distance travel is taking place, and these two activities account for over 80% of all air travel. There are also important multiplier effects as many destinations are dependent on attracting international visitors. In the end, decisions on long-distance energy intensive travel are a societal issue about how priorities are decided, together with individuals making moral decisions about the frequency and importance of their own long-distance travel. This is where behavior understandings of both societal and individual priorities are essential.

This book addresses these complex issues, the trade-offs, and the decisions or choices that have to be made, as well as the outcomes that have to be monitored and evaluated, so that effectiveness can be measured. The conclusion reached is clear and unequivocal, namely, that behavioral change is a necessary condition for the achievement of sustainable transport, but that researchers and policy makers have not truly understood the complexity and diversity of either the issues faced by the transport and energy providers or by the users of transport services. The problems being confronted are not only transport-based, as more travel is often the consequence of macroeconomic forces (e.g., consequences of globalization, long supply chains, and the growth of wider social networks) or decisions made in other sectors of the economy (e.g., education, health, and housing location).

If you have an interest in a truly international perspective that combines a diversity of behavioral thinking, different methods of analysis, a range of policy options, and the involvement of all stakeholders in trying to find the means by which the transport and energy debate can be addressed and resolved, then I urge you to read this book—you will not be disappointed.

Prof. David Banister
Professor Emeritus of Transport Studies and Senior Research
Fellow at St Anne's College, School of Geography and the Environment,
University of Oxford, Oxford, United Kingdom
July 2019

Foreword by Prof. James L. Sweeney

A broad strategy has emerged to address the important problem of global climate change. Emissions can be reduced and eventually eliminated by a combination of changes on the supply and demand sides of energy markets. First, aggressive energy efficiency efforts can reduce the consumption of all energy, thereby directly reducing the greenhouse gas emissions from end-user combustion of fossil fuels and indirectly reducing greenhouse gas emissions from fossil fuels used to generate electricity. Second, end uses of energy can be switched from direct combustion of fossil fuels to the use of either electricity, hydrogen, or biofuels, all of which can be produced by zero-carbon or low-carbon means. Third, all electricity can be produced through clean energy sources, such a wind, solar, geothermal, hydropower, or nuclear. Hydrogen can be produced through hydrolysis, which can use clean electricity as the energy source to split water into hydrogen and oxygen, or through steam reforming of methanol, with carbon capture and storage. Biofuels production can be based on cellulosic processes, rather than on carbon-intensive corn-based ethanol. These three collections of changes—energy efficiency; fuel switching from direct use of fossil fuels; and clean electricity, hydrogen, and biofuels production—can be the basis of a global strategy for deep decarbonization.

Transport—either for personal mobility or for movement of freight—is a source of greenhouse gas emissions. As recently as 2013, 15% of the greenhouse gases in the world were associated with transportation. This fraction is even higher in some countries, for example, the United States, in which 29% of the greenhouse gases in 2017 were associated with transportation.

And transport—either for personal mobility or for movement of freight—is an important area in which the first two parts of this broad strategy can be applied to reduce greenhouse gas emissions.

Over the last several decades, energy efficiency has been by far the greatest driver of greenhouse gas reductions in transport. Light duty cars and trucks have become far more fuel efficient over the last 50 years. In the United States, the energy use per passenger mile for commercial airlines has decreased by 75% over the same time period, with about half the reduction based on more energy-efficient airplanes, and the other about half based on behavioral and management actions aimed at filling empty seats on flights. Energy efficiency remains an important, if not the most important, pathway toward decarbonizing transport in future years.

Energy efficiency improvement has been in the past, and can be expected to be in the future, based on a combination of behavioral motivations, technology development, economics, and policy choices. These improvements are based on a myriad of changes and can best be understood by examining the various sectors, end uses, and locations. This book addresses, at a deep research level, many of the behavioral and economic forces shaping energy use for commuting to work, for bringing children to school, for tourism, for freight movement, and for other uses.

Fuel switching for transportation is in its infancy, but it has begun. Many models of passenger cars fueled by electricity or by hydrogen are on the market. Heavy duty trucks fueled by hydrogen are available, as are buses. And there has been experimentation with electricity or hydrogen-fueled airplanes. Drop-in biomass fuels have been developed for jet fuels and have been tested in military applications. To date, only a very small fraction of transport is fueled by electricity, hydrogen, or biofuels, so there is a massive opportunity for future changes.

Such a future transformation to electricity or hydrogen of course depends on technology. But it also depends on a set of behavioral issues. Electric vehicle charging currently takes more time than refueling a car with gasoline. But high-speed chargers are being installed by private firms, selling charging services. If high-speed charging is absent, what trade-offs are people willing to take? How can new systems make these trade-offs more acceptable? How may fleet managers reorganize their operations to take advantage of hydrogen fuels? This book also addresses, at a deep research level, some of these issues.

While overall strategies are easy to describe, in practice, they will require many policy, economic, and technology changes—as well as behavioral changes. This book covers many different approaches that have been used around the world and that can be expanded over time for a deep decarbonization of the transport system. The emphasis is on the behavioral issues, because technology and policy are insufficient if humans do not adopt the low-carbon technologies and the policies are ignored, or feasible policies are not promulgated.

Behavioral issues are ubiquitous, and many are discussed in depth in this book. Where people live, in relationship to their work, social, and play locations, shapes vehicle miles. Where people live is influenced by city/rural structure, by their tastes and preferences, and by the mobility options. Public transportation can be made available, but its effectiveness for reducing greenhouse gas emissions depends on how often people choose public transportation in place of private cars. And those choices are influenced by income/wealth, by perceived value of time, by perceptions of safety, and by social norms. Freight transportation use of energy is influenced by human choices of whether and what they buy online or in traditional stores, also a behavioral issue, as well as their preferences for speed of delivery.

It is relatively easy to outline abstract changes that would improve energy efficiency and would lead to decarbonizing fuel shifting. But the changes must be made by human beings, either individually or within organizations, such as companies or governments. Hence, the need to understand the behavioral issues underlying adoption of those technologies and practices which can enhance these desirable changes. This book takes important steps in advancing the knowledge of these behavioral issues. At its heart, it is a compendium of research articles that both summarize the state of knowledge and advance that state. It is a book for those who want to be serious about understanding the state of research and who are ready to dive into cutting-edge research on behavioral issues of transport, energy, and greenhouse gas emissions.

Prof. James L. Sweeney
Professor, Management Science and Engineering, Stanford University, Stanford, CA, United States;
Former Director, Precourt Energy Efficiency Center, Stanford University, Stanford, CA, United States
August 2019

Foreword by Prof. Noboru Harata

Today, it is widely accepted that human activities are affecting the balance of greenhouse gases in the atmosphere, thereby contributing to climate change. While research on global energy and environmental issues has progressed over the past few decades, this research has been dominated by macro-level modeling approaches, while behaviorally oriented modeling has been neglected. Given that the sum of human activities is made up of individual actions, ignoring behavioral factors and decision-making mechanisms will lead to seriously biased estimation of policy scenarios that involve behavioral interventions.

This book is therefore very timely. To the best of my knowledge, this is the first book bringing together behavioral research on transport and energy. It provides an overview of up-to-date transport and energy research from a behavioral perspective, covering a wide range of geographic areas and major transport modes. This book also gives readers a chance to learn what should be done next in terms of both research and practice. For example, it presents a systematic review of effects of behaviorally oriented green policies on household transport and energy consumption in Europe, which will be of interest to other countries that are trying to persuade households to reduce energy use. Similarly, readers can learn from case studies on household energy consumption in Asia (in-home and out-of-home energy consumption), United States (travel choices), and Australia (school travel).

Importantly, this book also discusses the implications of new developments in the transport sector on energy use, including shared mobility services and electric cars in China, the biggest energy-consuming country in the world. It looks at two interrelated and rapidly growing service sectors, tourism and aviation, which have been underresearched from a behavioral perspective. Social and demographic influences on transport and energy use are also examined, such as the impact of rural migrant workers in urban China and changing preferences of younger generation in Japan.

The book underlines the importance of behavioral interventions for reducing energy consumption. Methodologically, life-oriented behavioral modeling approaches are highlighted in the context of household decisions. This life-oriented approach is extremely important for better capturing both direct and indirect rebound effects related to energy-saving technologies and policies. Various choice models are also used for capturing energy-consumption behaviors.

I agree with the editor's belief that behavioral interventions should be a "must" in transport and energy-policy menus. This book is a valuable resource for policy-makers, researchers, engineers, and other individuals interested in reducing energy consumption in various sectors and building a more sustainable future.

Prof. Noboru Harata
Professor of Department of Urban Engineering, Graduate School of Engineering,
The University of Tokyo, Tokyo, Japan
August 2019

CHAPTER 1

Introduction of research on transport and energy

Junyi Zhang
Hiroshima University, Higashi Hiroshima, Japan

1.1 Background

According to Global Energy Statistical Yearbook 2018,[1] total energy consumption in the world was 13,730 Mtoe (million tonnes of oil equivalent) in 2017, and since 1992, Asia has consumed the largest amount of energy in the world, which has been growing continuously and increased 2.73 times from 1990 to 2017. Northern America and Europe have shown a relatively stable trend of energy consumption (from 1990 to 2017, it just increased 1.17 times in Northern America and 1.04 times in Europe). Especially, China accounted for 22.9% of total energy consumption in the world and 54.7% of total energy consumption in Asia.

As shown in Table 1.1 (made based on World Energy Statistics, 2018[2]), the industry sector over the world consumed 76.9% of the final coal demand, 97.9% of coal gas, 37.3% of natural gas, 41.4% of electricity, and 47.9% of heats, which topped all sectors, respectively. The transport sector mainly consumed oil, which accounted for 64.6% of the final oil demand, while the residential sector mainly consumed heat (35.0% of the final heat demand), natural gas (29.9% of the final natural gas demand), and electricity (27.2% of the final electricity demand). Enormous efforts have been made; however, CO_2 emissions from the transport sector increased by 11% from 2010 to 2016, which is higher than those from the sectors of industry (by 8%) and buildings (by 3%, including residence).[3]

[1] <https://yearbook.enerdata.net/total-energy/world-consumption-statistics.html> (accessed 10.05.19.).

[2] <https://www.oecd-ilibrary.org/docserver/world_energy_stats-2018-en.pdf?expires = 1557319177&id = id&accname = ocid49007842&checksum = 1D814FB615-B1C08997D020B7811BDFA8> (accessed 08.05.19.).

[3] These percentages were calculated based on data from <https://www.iea.org/tcep> (accessed 08.05.19.).

Transport and Energy Research
DOI: https://doi.org/10.1016/B978-0-12-815965-1.00001-6

Table 1.1 Shares of major energy demands in global final consumption by sectors in 2016.

Major sectors	Coal (%)	Coal gases (%)	Natural gas (%)	Oil (%)	Electricity (%)	Heat (%)
Industry	76.9	97.9	37.3	8.3	41.6	47.9
Transport	0.0	0.0	7.1	64.6	1.7	0.0
Residence	8.1	1.3	29.9	5.0	27.2	35.0
Share	85.0	99.2	74.4	77.9	70.6	82.9

Source: Calculated based on World Energy Statistics, 2018.

In the transport sector, road transport, water transport, aviation transport, and rail transport accounted for 75.6%, 10.9% (8.9% in the international water transport), 11.9% (7.3% in the international aviation), and 1.2% of total oil consumption, respectively.

Of total energy consumption in the world, passenger and freight transport accounts for about 25% (Creutzig et al., 2015; IEA, 2015). This suggests that it is important to decarbonize the transport sector via various transport and energy policies.

1.2 Importance of behavioral research

Even though it is more difficult to decarbonize the transport sector than other sectors (e.g., Pietzcker et al., 2014), it is necessary to make a transition from the use of fossil fuels to renewable energies in the future. With such an energy transition, emissions from the transport sector could be reduced. For example, García-Olivares et al. (2018) estimated that a 100% renewable transport in the world would consume about 18% less energy, by assuming that the present transport services were unchanged. García-Olivares et al. further revealed that a 100% renewable transport would reduce 69% of energy consumption in road transport; however, energy consumption in the shipping and air sectors would increase by 163% and 149%, respectively, because of the need to produce natural gas from electricity. The latter part of the previous findings may suggest that it is necessary to realize an optimal mix of energy sources. In reality, there are various barriers and uncertainties in promoting the use of renewable energies, even though enforcement of regulations about energy use may work to some extent. Whether the abovementioned potential energy reductions can be actually realized in the real world or not will have to rely on voluntary choices of transportation service providers and users, via transportation users' travel mode choices and transportation service providers' choices of transportation vehicles powered by different types of energies.

Thus it is a great challenge to encourage the use of renewable energies for both transportation service providers and users. This may depend on whether and how much transportation users change their lifestyles and service providers change their business styles. Using a data from 1960s to 1970s, Schipper and Lichtenberg (1976) revealed, "Swedes use less than two-thirds as much energy per capita as Americans, at the same standard of living." They further concluded that the use of smaller automobiles and more use of mass transit were among major factors causing the reduced energy use in Sweden. Considering that transport had not been an energy issue at all for many countries until the 1980s (Boyle, 1990), the study by Schipper and Lichtenberg (1976) really pioneered the transport and energy research by showing that changing lifestyles (even though limited to travel behaviors) could lower energy consumption.

There are also more relevant choices of transportation service providers and users, which may affect energy consumption. For example, living closer to a subway may encourage less use of a car, leading to less energy consumption from the car. Thus residential location and travel choices are associated with energy consumption and the resulting emissions. Related to this, public transportation service providers may choose whether to promote transit-oriented development or not; however, this choice usually involves many stakeholders, making the decision complicated and time consuming.

Road pricing to traffic going into the city center during peak hours may encourage more through traffic to use a ring road and/or shift departure time to an off-peak period, which can also reduce energy consumption. Thus departure time and route choices (by both passenger cars and trucks) are not irrelevant to energy consumption. People may choose a job with flexible working hours, which may contribute to mitigating traffic congestion during peak hours and consequently reducing energy consumption and emissions caused by traffic congestion.

Logistics companies and their customers need to make various choices [e.g., transport modes and routes (e.g., airline, waterway, railway, expressway or ordinary road), goods processing, locations and types of warehouses or logistics centers for goods transfer] for minimizing the leading time and cost of goods transport, caused by goods pickup, loading/unloading, storage and/or processing at warehouses, and delivery to end users, and so on. Such choices also involve various stakeholders and may be further sensitive to the types of goods transported. At every step of the whole logistics process, energy is usually consumed, indicating that choices of logistics are also connected with energy consumption.

Energy consumption in the passenger transport and residential sectors commonly involves complicated household decision-making issues. For example, energy consumption involves an expenditure decision. In such a case a household needs to decide how to allocate its limited income to different purposes, including energy consumption from using vehicles and various in-home energy-consuming appliances (Yu and Zhang, 2015). Because of the limited household income, spending more money on energy consumption means a decline in available household income used for other purposes. Accordingly, decisions on various expenditures are usually not independent on each other (Yu et al., 2011). In addition, energy consumption is related to residential behavior (e.g., residential location, housing size and type) (Yu et al., 2012a) and time use behavior (e.g., time spend on in-home and out-of-home activities) (Yu et al., 2012b). On the top of the abovementioned behavioral decisions, people's long-term life choices may be further influential, such as migration, employment mobility, and family formation (Zhang, 2017). On the other hand, life choices are associated with lifestyles. Lifestyle is crucial to a better understanding of household energy consumption (Bin and Dowlatabadi, 2005; Anable et al., 2012; Yu and Zhang, 2017). After making an extensive literature review in association with the life-oriented approach (Zhang, 2017), Van Acker (2017) concluded that lifestyle could be defined as the way by which an individual indicates his or her social position through consumption and leisure behaviors as well as other behaviors, and it is not stable over time.

Pietzcker et al. (2014) projected that transport decarbonization would lag 10–30 years behind other sectors; however, they further pointed out that their transport models only introduced monetary incentives to explain transport demand and neglected behavioral factors, suggesting that introducing behavioral factors may derive a different picture of transport decarbonization. More importantly, behaviorally oriented transport and energy research is useful to properly capture both direct and indirect rebound effects of energy-saving technologies. Such rebound effects can be observed not only in the domestic and transport sectors (Yu et al., 2013, 2015, 2016) but also in other sectors (Amjadi et al., 2018; Jin and Kim, 2019; Liu et al., 2019; Vélez-Henao et al., 2019).

In summary, it is necessary to decarbonize the transport sector for ensuring a sustainable future of our planet, and it is important to promote transport and energy research for altering behaviors of various stakeholders in a way of reducing energy consumption.

1.3 Outline of chapters

This book aims to present a comprehensive picture about behavioral research connecting transport and energy. In total, there are 16 chapters, including this introductory chapter. All sectors related to transport and energy are targeted. Major countries in the world are focused on, even though Europe, the United States, Japan, China, and several other Asian countries are mainly investigated. Behavioral decisions and factors are emphasized for a better understanding of transport and energy decisions by various stakeholders. Behavioral interventions are recommended as a key policy instrument for reducing energy consumption in a sustainable way.

To better position behavioral research in the literature from a global perspective, Chapter 2, The role of the transport sector in energy transition and climate change mitigation: insights from an integrated assessment model, explores the role of the transport sector in energy transition and climate change mitigation. The study was based on an integrated assessment model consisting of transport, economic, and climate submodels, where the transport submodel contains behavioral and technological factors of the transport sector. An empirical analysis showed that promoting electric vehicles (EVs) and car sharing could effectively help the decarbonization in the transport sector. The analysis supports not only technological transformations but also social innovations for energy transition.

For decarbonizing the transport sector, various policies have been implemented; however, studies on the effects have been scattered in the literature. By taking Europe as an example, Chapter 3, Effects of green policies on household transport and energy consumption in Europe: a literature review, reviews and evaluates the effects of green policies aiming at encouraging households to reduce household CO_2 emissions from both transport and domestic sectors. Main focus is on the long-term effects of the green policies on behavioral change, reduction in both conventional energy usage and renewable energy production. The previous policy review has its significant global implications.

As stated in the beginning of this chapter, Asia is the largest energy consumption region in the world. Because of the rapid urbanization in Asia, various problematic development issues have been observed, including the excessive use of energy resources (Zhang and Feng, 2018); however, behavioral insights about household energy consumption in Asian cities are very limited. To provide additional behavioral insights by reflecting household energy consumption decision-making mechanisms, Chapter 4, Energy consumption in the transport and domestic sectors: a household-level comparison between capital cities of Japan, China, and Indonesia, compares both in-home and out-of-home energy consumption in three major capital cities of Japan, China, and Indonesia. A multiple discrete-continuous choice model is built, and influential factors to energy consumption were extensively investigated. This study presents not only a unique Asian context but also a useful modeling tool to represent cross-sectoral behavioral decisions of households.

Needless to say, transport energy consumption mainly comes from travel choices; and therefore a better understanding of travel choices is needed, by providing insights on mode choice, travel patterns and even time of travel, as well as future changes in behavior under different policy scenarios. Such efforts are crucial to realize sustainable transportation by lowering energy use and emission levels. Chapter 5, Travel choices, preferences, and energy implications in the United States, analyzes travel choices and preferences and discusses energy implications. This chapter first illustrates a nation-wide picture of travel behavior and energy use as well as the potential transformation of the transport sector in the United States. Then, it further investigates heterogeneous travel choices of trip makers based on focus group surveys in California. Capturing such heterogeneities is useful to better predict future travel demand, energy use, and CO_2 emissions under different transport policies.

Travel to school is an important travel behavior that has long-term impacts on energy consumption. It involves intrahousehold interactions; however, there are a lot of unknowns. Further, there has been little focus in the literature on the energy use associated with children's car travel to school. Chapter 6, Children's travel behavior and implication to transport energy consumption of household: a case study of three Australian cities, challenges this difficult topic based on a qualitative analysis approach, which has been largely neglected in the transportation and energy literature. Data from qualitative interviews to parents from Sydney, Melbourne, and Canberra is used. It is found that time pressure is the main reason why many parents drove their children to schools on the way to their work. Energy policy implications are extensively discussed, together with important future research issues.

The abovementioned intrahousehold interactions are also popular to other travel-related decisions. Different from the qualitative approach adopted in Chapter 6, Children's travel behavior and implication to transport energy consumption of household: a case study of three Australian cities, Chapter 7, Residential location and commuting mode choices: intrahousehold interaction modeling and its implications for energy policy, presents a new joint model of household residential location and commuting mode choices. This research is related to transit-oriented development. Here, attribute-based and outcome-based intrahousehold interaction models are developed, where the former model uses the concept of a generalized mean, and the latter model adopts a weighted probability approach. A stated preference (SP) data in the context of a Japanese city is used. Empirical analysis results suggest that transport and energy policymaking should pay more attention to different roles of household members.

The importance and necessity of deploying EVs for decarbonizing the transport sector has been clarified in Chapter 2, The role of the transport sector in energy transition and climate change mitigation: insights from an integrated assessment model. Targeting China, the largest energy consumption country in the world, Chapter 8, Battery electric vehicles in China: ownership and usage, extensively analyzes ownership and usage of EVs as well as charging choice. Current situations of EVs in China are reviewed. Attitude, intention, SP, and choices are emphasized. The limited cruising range of EVs, charging demand, and the location of chargers are also incorporated into the analyses. The modeling methods used in this study are promising for future transport and energy research, while various methodological challenges are also discussed.

To reduce carbon emissions from the transport sector, it is necessary to reduce the dependence on cars. Among various transport policies, car sharing is a promising option, as highlighted in Chapter 2, The role of the transport sector in energy transition and climate change mitigation: insights from an integrated assessment model. Chapter 9, Impacts of shared mobility on energy consumption and emissions in China, evaluates impacts of shared mobility (Zhuanche, Kuaiche, and Shunfengche) on energy consumption and emissions in China. For exploring the sustainable development of car sharing in China, a Big Data of trips from the famous

Didi Chuxing company is used. A life cycle analysis is conducted to measure the direct and indirect impacts of sharing mobility on the energy consumption and emissions. The analysis confirms the reduction of the direct energy consumption, CO_2 emissions, and NO_x emissions, caused by shared mobility. Policy challenges are also identified.

Energy consumption increases with urbanization, which is featured by rural-to-urban and urban-to-urban migration. Especially in developing countries, new migrants from rural areas need time to adapt to urban life, due to various difficulties and concerns in life. However, the impacts of such life factors on energy consumption have remain unclear, because of limited studies. Chapter 10, Rural migrant workers' energy consumption in passenger transport and domestic sectors of China: case study from a social perspective, examines influential factors to Chinese rural migrant workers' energy consumption, including living conditions, residential behaviors, ownership of both in-home and out-of-home energy-consuming end uses, and the built environment. Such a social perspective is emphasized because of the existence of the China-unique household registration system (hukou). A structural equation model is estimated based data from three Chinese cities. It is found that effects of living conditions are limited, in comparison with other factors. Policy implications and future research issues are discussed.

Car-dependent lifestyles have been widely pointed out to be a key factor contributing to huge amounts of energy consumption and the resulting emissions from the passenger transport sector. In recent years, a peak-car phenomenon has been observed in some developed countries and explained from various angles, where a declining trend of young people's vehicle ownership and usage has been revealed (Bastian and Börjesson, 2015; Metz, 2015; Van Wee, 2015; Bastian et al., 2016; Focas and Christidis, 2017; Lee-Gosselin, 2017; Stapleton et al., 2017; Bussière et al., 2019). Car ownership and usage is a major part of household expenditures, which may be affected by other types of expenditures, as suggested by the life-oriented approach (Zhang, 2017). However, no study can be found in the literature from an expenditure perspective. Chapter 11, Japanese young people's car ownership and usage: research from an expenditure perspective, revisits research on young people's car ownership and usage from an expenditure perspective in the context of the whole Japan. A new type multiple discrete-continuous choice model based on the concept of multilinear utility is built to accommodate interactions between energy and transport expenditures as well as other expenditures. Major factors affecting energy and transport expenditures are captured in a fully behaviorally oriented way. An empirical analysis based a national expenditure data of Japan in 2009 confirmed the applicability of the new model, revealed the roles of various interactions and clarified influential factors.

Behavioral research associated with energy consumption can also be observed with respect to aviation. The aviation industry is criticized as one of the fastest growing industries in emitting greenhouse gases, which accelerates the process of global warming. The current trend of increasing demand of air travel has made it difficult to have a significant effect in the short term with existing policies and regulations as well as technological improvement. Therefore the emissions reduction now is heavily relying on both air travelers' and airline industries' behavioral change. For this, Chapter 12, Behavioral research on transport and energy in the context of aviation, overviews the energy-related behavioral research on aviation, from perspectives of air travelers, airlines, and aviation logistics. The air traveler perspective emphasizes travelers' attitudes towards flying and climate change, and willingness to pay (WTP) for carbon offset, at the individual level. From a macroscopic viewpoint, airlines' fuel usage and carbon emissions, choice of aircraft size, environmental friendliness and energy efficiency are reviewed. Furthermore, the behavioral models for both air travelers and airlines are summarized. Finally, relevant behavioral research issues are discussed.

Related to the aviation sector, the growth of tourism sector has also caused the increase of energy consumption and concerns about its sustainability. Chapter 13, A systematic review on tourism energy consumption, sustainable tourism, and destination development: a behavioral perspective, examines the relationships between tourism energy consumption, sustainable tourism, and destination management based on literature review. A systematic framework is proposed for identifying major research gaps, by combining energy consumption studies and stakeholder theory within sustainable goals. Existing behavioral studies on tourism energy from a perspective of stakeholders are summarized. The links between tourism energy consumption, sustainable tourism and destination are also discussed. Finally, the study summarizes various research issues.

To reduce energy consumption, it is crucial to encourage the various stakeholders to voluntarily change their behaviors. In this regard, Chapter 14, Behavioral interventions for sustainable transportation: an overview of programs, and guide for practitioners, reframes transportation sustainability as a human behavior issue and explains how behavioral sciences can be used to encourage sustainable transportation decisions. It summarizes theories of behavior change and major behavioral interventions in practice. It is found that most interventions to date focus on personal transportation behaviors and rarely target professional transportation. Importantly, this chapter presents readers with an effective transportation behavior change program using the DRIVE process (defining the problem, researching, intervening, verifying impact, and evaluating next steps).

Transport is an essential part of human life. Encouraging behavioral changes toward a less car-dependent and energy-saving lifestyle need to understand whether people can make a better life without a car or not. In this regard, such a life-oriented approach is important; however, impacts of a large set of life choices on energy consumption have remained unknown. Chapter 15, Life-oriented household energy consumption research, makes an initial attempt to make use of the life-oriented approach to better understand household energy consumption by using a household-level questionnaire data in Japan. A random forest approach is applied to quantify the influences of 87 variables of life choices covering 10 life domains on household consumption of electricity, gas, kerosene, gasoline, and diesel. Various insights are derived for supporting energy-saving policies.

In the last chapter, Chapter 16, To connect the past with the future, major findings and future research issues derived from Chapters 2–15 as well as promising policies are summarized in a consistent way and a set of behavioral interventions are proposed as a framework for supporting future energy-saving policymaking.

1.4 Final remarks

There is a variety of studies on energy consumption in the transport sector, most of which, however, have focused on technologies and energy efficiency. There is a wide consensus that more efficient technologies do not necessarily lead to a reduction of energy consumption, partially due to rebound effects. However, very limited behavioral research can be found in the literature about transport issues associated with energy consumption. This is the first book about transport and energy research from a behavioral perspective in an interdisciplinary and systematic way.

This book presents a series of behaviorally oriented transport and energy research, throughout extensive literature review and case studies, where various useful and insightful findings are summarized and discussed. In each chapter a rich set of future research directions are also provided. This book focuses on the behaviors of not only transport and energy service users but also transport and energy service providers, policy makers, organizations, company managers, and other stakeholders who are involved in and/or affected by transport and energy policymaking. Behavioral research is useful to assist the development of transport technologies and services via a proper reflection of users' needs and behavioral patterns. Such a behavioral orientation also emphasizes that transport and energy policymaking needs to reflect various stakeholders' behaviors and opinions as well as factors behind the behaviors and opinions, based on interdisciplinary research methodologies and cross-sectoral policymaking theories. Such an emphasis is also useful to achieve long-lasting consensus building.

This book can serve as a textbook for graduate students in the fields of transport, energy, environment, planning, public policy, behavioral studies, and so on to have knowledge about how to understand and reduce energy consumption from various behavioral aspects based on various interdisciplinary approaches. It can also be a handbook for researchers to overview existing transport and energy studies as well as relevant future research issues. As for behavioral topics, not only travel behavior and energy consumption but also their associated life choices (e.g., residential behavior, migration, expenditure, tourism and leisure, time use, and health) are included. Transport systems cover not only traditional systems (e.g., air, railway, and road) but also new mobility systems (e.g., EVs and shared mobility). Both behavior and behavioral change theories are investigated. Not only microlevel models (behaviorally oriented models and data mining approaches as well as statistical models) but also macrolevel integrated models with behavioral factors are adopted. Behavioral issues on transport and energy in Europe, the United States, Japan, China, and other Asian countries are further extensively examined.

This book can provide policy makers with comprehensive insights into making more effective policies over the whole process of policymaking via behavioral interventions than the traditional approaches that only emphasize technologies and energy efficiency. First, it can help policy makers and planners identify the issues motivating policy decisions and collect relevant data by focusing on various stakeholders' decision-making mechanisms related to the identified issues. Second, policy goals can be set in a more socially acceptable way, and policy options can be generated in a more practically feasible way by reflecting various stakeholders' behaviors and opinions. Third, future prediction and evaluation can be made in a much more accurate way because introducing behavioral factors makes the assumptions for future prediction and evaluation more realistic. Fourth, policy decisions and implementation could achieve long-lasting consensus building in a well-prepared way by reflecting various stakeholders' behaviors and opinions from the beginning of the whole policy-making process. Dynamic and complicated decision-making mechanisms involved in transport and energy consumption behaviors suggest that transport and energy policies should be better monitored over time.

References

Amjadi, G., Lundgren, T., Persson, L., 2018. The rebound effect in Swedish heavy industry. Energy Econ. 71, 140–148.
Anable, J., Brand, C., Tran, M., Eyre, N., 2012. Modelling transport energy demand: a socio-technical approach. Energy Policy 41, 125–138.
Bastian, A., Börjesson, M., 2015. Peak car? Drivers of the recent decline in Swedish car use. Transp. Policy 42, 94–102.
Bastian, A., Börjesson, M., Eliasson, J., 2016. Explaining "peak car" with economic variables. Transp. Res., A 88, 236–250.
Bin, S., Dowlatabadi, H., 2005. Consumer lifestyle approach to US energy use and the related CO_2 emissions. Energy Policy 33, 197–208.
Boyle, S., 1990. Transport and energy policies—only connect. Energy Policy 18 (1), 34–41.
Bussière, Y.D., Madre, J.-L., Tapia-Villarreal, I., 2019. Will peak car observed in the North occur in the South? A demographic approach with case studies of Montreal, Lille, Juarez and Puebla. Econ. Anal. Policy 61, 39–54.
Creutzig, F., Jochem, P., Edelenbosch, O.Y., Mattauch, L., van Vuuren, D.P., McCollum, D., et al., 2015. Transport: a roadblock to climate change mitigation? Science 350, 911–912.
Focas, C., Christidis, P., 2017. Peak car in Europe? Transp. Res. Procedia 25, 531–550.
García-Olivares, A., Solé, J., Osychenko, O., 2018. Transportation in a 100% renewable energy system. Energy Convers. Manage. 158, 266–285.
IEA, 2015. CO_2 Emissions From Fuel Combustion Highlights 2015. International Energy Agency, Paris.
Jin, T., Kim, J., 2019. A new approach for assessing the macroeconomic growth energy rebound effect. Appl. Energy 239, 192–200.
Lee-Gosselin, M.E.H., 2017. Beyond "Peak Car": a reflection on the evolution of public sentiment about the role of cars in cities. IATSS Res. 40 (2), 85–87.
Liu, H., Du, K., Li, J., 2019. An improved approach to estimate direct rebound effect by incorporating energy efficiency: a revisit of China's industrial energy demand. Energy Econ. 80, 720–730.
Metz, D., 2015. Peak car in the big city: reducing London's transport greenhouse gas emissions. Case Stud. Transp. Policy 3 (4), 367–371.
Pietzcker, R.C., Longden, T., Chen, W.Y., Fu, S., Kriegler, E., Kyle, P., et al., 2014. Long-term transport energy demand and climate policy: Alternative visions on transport decarbonization in energy-economy models. Energy 64, 95–108.

Schipper, L., Lichtenberg, A.J., 1976. Efficient energy use and well-being: the Swedish example. Science 194 (4269), 1001–1013.

Stapleton, L., Sorrell, S., Schwanen, T., 2017. Peak car and increasing rebound: a closer look at car travel trends in Great Britain. Transp. Res. D 53, 217–233.

Van Acker, V., 2017. Lifestyles and life choices. In: Zhang, J. (Ed.), Life-Oriented Behavioral Research for Urban Policy. Springer, pp. 79–96.

Van Wee, B., 2015. Peak car: the first signs of a shift towards ICT-based activities replacing travel? A discussion paper. Transp. Policy 42, 1–3.

Vélez-Henao, J.-A., Vivanco, D.F., Hernández-Riveros, J.-A., 2019. Technological change and the rebound effect in the STIRPAT model: a critical view. Energy Policy 129, 1372–1381.

Yu, B., Zhang, J., 2015. Modeling household energy consumption behavior: a comparative analysis. Transp. Res., D 39, 126–140.

Yu, B., Zhang, J., 2017. Household energy consumption behavior. In: Zhang, J. (Ed.), Life-Oriented Behavioral Research for Urban Policy. Springer, pp. 123–148.

Yu, B., Zhang, J., Fujiwara, A., 2011. Representing household energy consumption behavior in Beijing with the consideration of in-home and out-of-home interdependence. Energy Policy 39 (7), 4168–4177.

Yu, B., Zhang, J., Fujiwara, A., 2012a. Analysis of the residential location choice and household energy consumption behavior by incorporating multiple self-selection effects. Energy Policy 46, 319–334.

Yu, B., Zhang, J., Fujiwara, A., 2012b. A household time use and energy consumption model with multiple behavioral interactions and zero-consumption. Environ. Plann. B 40 (2), 330–349.

Yu, B., Zhang, J., Fujiwara, A., 2013. Rebound effects caused by the improvement of vehicle energy efficiency: an analysis based on an SP-off-RP survey. Transp. Res., D 24, 62–68.

Yu, B., Tian, Y., Zhang, J., 2015. A dynamic active energy demand management system for evaluating the effect of policy scheme on household energy consumption behavior. Energy 91, 491–506.

Yu, B., Zhang, J., Fujiwara, A., 2016. Who rebounds in the private transport sector? A comparative analysis between Beijing and Tokyo. Environ. Plann. B 43 (3), 561–579.

Zhang, J., 2017. Life-Oriented Behavioral Research for Urban Policy. Springer.

Zhang, J., Feng, C.-M., 2018. Routledge Handbook of Transport in Asia.

CHAPTER 2

The role of the transport sector in energy transition and climate change mitigation: insights from an integrated assessment model

Runsen Zhang*
Hiroshima University, Higashi Hiroshima, Japan

2.1 Introduction

The human-induced global climate change is serious not only due to the magnitude of the damages but also the potential for their irreversibility (Solomon et al., 2009). Carbon dioxide (CO_2) makes up the vast majority of current greenhouse gas (GHG) emissions, which contribute to global warming (Lashof and Ahuja, 1990). As the source of most CO_2 emissions, more than 80% of the energy consumed in modern society comes from fossil fuels. Global energy consumption is rapidly increasing due to higher living standards and population explosion worldwide (Bose, 2010). The transportation of people and goods is responsible for a quarter of global energy consumption (Creutzig et al., 2015; IEA, 2009, 2015). Due to the rapid motorization worldwide and overwhelming dependence on fossil fuels, the decarbonization of the transport sector has proved more challenging than other sectors (Loo and Banister, 2016; Pietzcker et al., 2014; Yin et al., 2015). Without active and powerful policy intervention, future transport-related energy consumption and CO_2 emissions will maintain rapid growth, and the transport sector will still remain heavily reliant on fossil fuels. The adoption of advanced vehicle technologies and fuels and various options for behavioral changes can achieve the significant decarbonization of transport sector (McCollum and Yang, 2009). Some research suggested that reduced activity growth and modal shift toward low-carbon modes only have a limited contribution to reducing emissions, but technological improvement is likely to be more effective for emission

* Corresponding author.

Transport and Energy Research
DOI: https://doi.org/10.1016/B978-0-12-815965-1.00002-8

reduction (Edelenbosch et al., 2017; Melo, 2016). In particular, electric vehicles (EVs) have been considered as clean technology alternatives provided that electricity gets decarbonized in the long run (Dhar et al., 2017). On the contrary, as shown in empirical studies, there is no compelling evidence to support the alternative energy sources as a low-carbon option (Neves et al., 2018); moreover, energy consumption and emissions from transport are not only determined by technical efficiency but also influenced by lifestyles and sociocultural factors (Anable et al., 2012). Dramatic progress of technology cannot achieve the stringent low-carbon goals in the absence of considerable behavioral transformations such as decreased mileage, increased occupancy, increased rail use, and increased use of public transport (Bristow et al., 2008; Rayner et al., 2008).

The transport sector has synergistic relationship with other sectors and affects the energy system and economy through multiple dimensions. The electricity supply and transport electrification have cross-sectoral trade-offs, and it is necessary to keep consistency between power sector and transport policies (Kannan and Hirschberg, 2016). The transformation of the power generation and transport systems need the successful coordination of the interrelated market actors, consumers, and policy makers (Siskos et al., 2018). Regarding the economic effects, the decarbonization of transport sector might exert negative impacts on the GDP compared to business as usual (Karkatsoulis et al., 2017), and even the renewable energy use is likely to obstruct economic growth (Neves et al., 2018). Because the interactive mechanism among transport sector, energy system, and economy is complex, there is limited information on how transport policies would contribute to energy transition, climate change, and economic development. Therefore it is needed for policymaking to model the cross-sectoral interactions and to detect how the policies of transport decarbonization would modify the energy and emission profiles as well as the economic development.

Integrated assessment models (IAMs), which couple models of energy system technologies with economic and climate models, have been widely used to evaluate the technological and economic feasibility of achieving specific climate change mitigation goals. IAMs are designed to involve socioeconomic factors as well as natural sciences components to achieve a better understanding of how human activities and social development affect the natural world, in particular, climate system. A key characteristic of IAMs is that a broad category of research approaches and disciplines related to climate change are combined to model human society alongside parts of the Earth system. Although transport sector has been taken into account in IAMs (Daly et al., 2014; Girod et al., 2012, 2013a,b; Karkatsoulis et al., 2017; Kyle and Kim, 2011; McCollum et al., 2016; Waisman et al., 2013; Zhang et al., 2016), it is usually defined at a highly aggregated level without behavioral and technological factors. Due to the high aggregation level of transport representation, transport demand projection is simply formulated as a sort of industrial activity, but the modal shift and technological selection are not incorporated explicitly, implying that current IAMs cannot be used to assess the potential of energy transition and emission reduction due to transport behavioral and technological options. A few of IAMs offer more elaborate representation of transport sector that incorporates mode choice and individual vehicle technologies, but it is still challenging to capture the interaction between behavioral and technological determinants of transport sector, energy system, climate change, and macroeconomy.

To create a better understanding of whether transport sector contributes to energy transition and climate change mitigation, the main purpose of this chapter is to investigate how transport behavioral and technological factors impact the energy use and emission trajectories, and which transport policies and strategies help achieve the goal of low-carbon future, by an IAM that couples a global transport model for detailed transport representation. The integrated modeling methodology not only enables the existing IAMs to consider detailed factors in transport sector such as mode choice, preference, load factor, and technological development, to structure scenarios for transport policy assessment but also offers a balanced perspective to the impacts on energy system and economic system.

2.2 Model

2.2.1 Model structure

To investigate the role of transport sector in energy transition and climate change mitigation, here a global transport model is employed to provide an elaborate transport sector representation in IAMs. The transport model offers spatially flexible and temporally dynamic simulations of transport demand, energy use, and emissions in consideration of detailed behavioral and technological descriptions such as transport cost, vehicle speed, time, modal choice, and preferences. The transport model is developed in the tradition of a recursive-type transport choice model to project the global transport demand and energy use (Zhang et al., 2018a,b). The transport model consists of different tiers including trip distances, travel modes, vehicle sizes, and technologies. For the first tier, passenger and freight transport volumes are distinguished between short and long distances. At the second tier, passenger and freight transport modes compete for short- and long-distance travel, respectively. Here, short-distance passenger transport modes include car, bus, and two wheeler, while passenger rail, domestic, and international passenger air are available for long-distance passenger transport. Long-distance freight transport modes include small truck, large truck, freight rail, domestic shipping, international shipping, domestic freight air, and pipeline, while short-distance freight transport mode only includes small and large trucks. For the next tiers, mode-specific transport demand would be divided into different vehicle sizes and technologies. Then, transport-related energy consumption and CO_2 emissions can be computed based on technology-specific transport demand.

The transport model is developed for 17 regions, including Brazil; Canada; China; the European Union; Former Soviet Union; India; Japan; Middle East; North Africa; Oceania; rest of Africa, Asia, Europe, and South America; Southeast Asia; Turkey; and the United States. The regional transport demand is determined by the GDP, industrial value added, population, and generalized transport cost for passenger and freight, respectively. Discrete-choice model is used to split the future transport demand by distances, mode structures, and mixture of vehicle sizes and technologies. The shares of distances, modes, sizes, and technologies are calculated using a modified logit model (Kyle and Kim, 2011; Mishra et al., 2013) based on the generalized transport cost that includes fuel cost, device cost, time cost, infrastructure cost, and carbon tax. Technological improvements in advanced vehicles have been incorporated into the process of technological selection. Technological selection parameters for advanced vehicle technologies will increase gradually, accompanied by the implementation of transport policies.

The transport model is coupled with an IAM to capture the interactive mechanism between the transport sector, energy system, emissions, macroeconomy, and climate change (Fig. 2.1). The frameworks of computable general equilibrium (CGE) model and the Model for the Assessment of Greenhouse Gas–Induced Climate Change (MAGICC) will be employed for integrated assessment modeling. The CGE model is classified as a multiregional and multisectoral CGE model to detect the

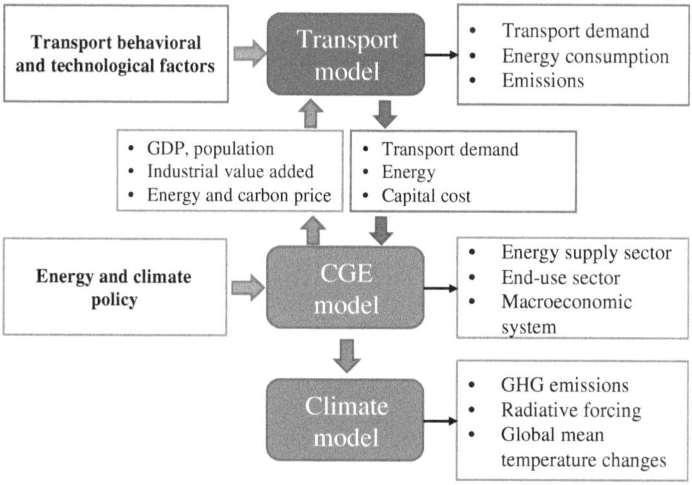

Figure 2.1 Model structure.

future climate change mitigation and its impact on economic conditions, which covers all economic goods while considering production factor interactions (Fujimori et al., 2012, 2014). The CGE model passes the macroeconomic variables to transport model for projecting the transport demand as well as energy consumption. Correspondingly, outcomes from transport model such as transport demand, transport-related energy consumption, and capital cost for transport device are passed to CGE model for reestimating the parameters. Next, global GHGs and other air pollutant emissions from CGE model will be input into a climate model, which generates climate outcomes such as radiative forcing and global mean temperature changes. The mitigation cost such as carbon price and GDP loss can be calculated by CGE model according to the emission constraint given by a Dynamic-Integrated Climate-Economy (DICE)-type intertemporal model. An iterative algorithm is used to obtain the convergence of this integrated transport–CGE–climate model, which can help enrich the transport representation in IAMs via providing technology details or behavior factors. This loop of feedback stops when the transport-related energy consumptions in CGE model and transport model are equal.

2.2.2 Data

The global transport demand and energy data for parameter estimation and calibration are collected from the database of the Asia-Pacific Integrated Model that is a large-scale model developed to reduce GHG emissions and to avoid the climate change impacts. Socioeconomic data such as GDP and population are obtained by the shared socioeconomic pathways (SSP) database (IIASA, 2015). Transport demand for different modes, sizes, and technologies, technology-specific energy intensity, and vehicle device cost is acquired from a bottom-up technological end-use model (Akashi and Hanaoka, 2012; Akashi et al., 2012). The load factor of each passenger transport mode and the door-to-door speed used to estimate the mode-specific time cost are taken from the global change assessment model (GCAM) (Mishra et al., 2013). Parameters for transport demand projection and shares of distances, sizes, modes, and technologies are calibrated based on the base year data in 2005. The study phase of the integrated model is from 2005 to 2100 with 1-year interval.

2.2.3 Scenario settings

To explore the contribution of transport sector to low-carbon transition and climate change mitigation, a set of scenarios was structured according to the socioeconomic pathways, transport behavioral and technological factors, and climate change mitigation efforts. SSP2 estimates are employed as default values for the socioeconomic settings such as GDP and population. Transport policy scenarios were designed based on the technological transformation and lifestyle changes. Regarding transport technological innovation, it is assumed that EVs will be economically competitive and strongly advocated by the government, and therefore internal combustion engine (ICE) vehicles will be phased out gradually. In this *EV* scenario, 100% of EV market share will be achieved around the world by 2050 due to high preference on EVs (car, bus, two wheeler, and small truck), while moderate preference is considered in the *Default* scenarios. With respect to the innovations in lifestyle, car sharing (CS) is considered one representative lifestyle change. In the *CS* scenario the behavioral modification is represented as the increased load factor with the promotion of shared mobility. It is supposed that carpooling and CS will help reduce the car energy efficiency per person-kilometer by increasing car occupancy rates. The patterns of future travel demand would be also altered because the greater load factor might change the transport cost per person-kilometer. The *CS* scenario assumes that the load factor of car will increase to 3 by 2100.

Climate change mitigation efforts correspond to a climate stabilization target versus no climate action, which can be denoted by *BaU* and *2D*,

Table 2.1 Scenario settings.

Scenario	Description
BaU_Default	Neither transport policy nor carbon pricing
BaU_EV	100% EV market share by 2050 without carbon pricing
BaU_CS	Car load factor will increase to 3 by 2100 without carbon pricing
2D_Default	No transport policy but carbon pricing is imposed for 2°C target
2D_EV	100% EV market share by 2050 with carbon pricing for 2°C target
2D_CS	Car load factor will increase to 3 by 2100 with carbon pricing for 2°C target

respectively. In the *BaU* scenario, no climate mitigation action is assumed, while the *2D* scenario puts a price on carbon. The *2D* scenario is created to reach the 2°C target, with a global mean temperature peaking at 1.8°C in 2100. The radiative forcing level associated with the *2D* scenario is approximately 2.8 W/m^2 in 2100. Then, six scenarios can be created combing dimensions of transport policies and climate change mitigation efforts (Table 2.1). Such scenario settings considering different model assumptions of EV adoption, load factor, and carbon pricing can be structured to analyze cross-sectoral relations and trade-off while assessing mitigation pathways with technological innovations and lifestyle changes in transport sector.

2.3 Results

2.3.1 Energy use

The energy results can be explained by examining the trajectories of energy use in transport sector during 2005–2100 across scenarios. Fig. 2.2 shows the energy use by fuel for six scenarios considering both climate efforts and transport policy incentives. In the *BaU_Default* scenario without carbon pricing or transport policies, energy use increases from 81 to 201 EJ during 2005–2100, with oil continuing to power the transport sector as a dominant contributor, though electricity, natural gas, and biomass are playing increasing roles. The energy consumption required by transport decreases from 201 EJ (*BaU_Default*) to 168 (*BaU_EV*) and 151 EJ (*BaU_CS*) in 2100, indicating both of technological and lifestyle transformations can reduce the energy consumption in

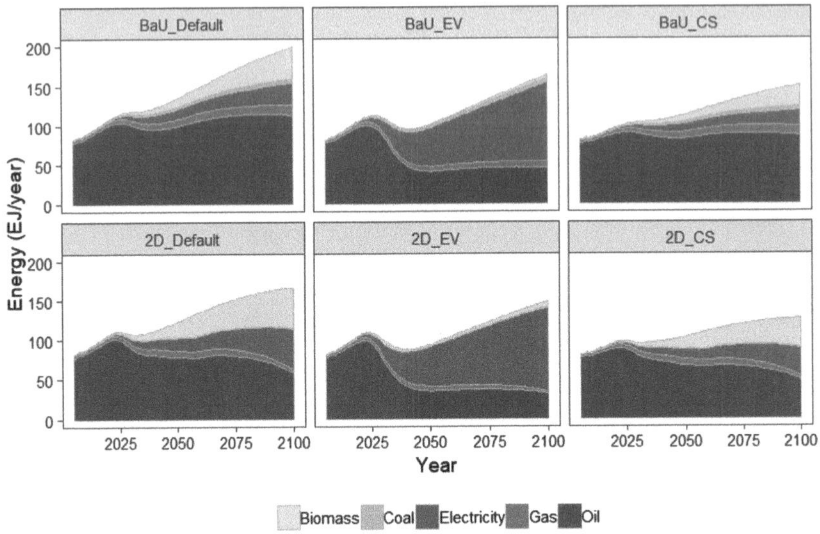

Figure 2.2 Global energy use from transport by fuel type (2005–2100).

transport sector. Moreover, the comparisons between results of *BaU* and *2D* suggest that the imposition of a carbon price can significantly reduce the oil consumption and stimulate the usage of electricity and biomass. Scenario simulations of energy use with carbon pricing also show that *2D_EV* and *2D_CS* produce less energy consumption than *2D_Default*, implying the effectiveness of transport policies is robust, regardless of whether the climate change mitigation efforts such as carbon pricing are made or not.

Transport sector consumes higher electricity when the targets for the implementation of electric road transport would be achieved by 2050 in the scenario of *BaU_EV* and *2D_EV*, while the global consumption of oil and biofuels is lower with the deployment of EVs, implying that transport electrification is able to reduce the oil dependency and moderate the imbalance between the supply and demand of oil. Instead, oil still accounts for considerable proportions in other scenarios without stringent penetration of EVs, although CS can reduce the oil consumption to some extent. Combining the carbon pricing and stringent penetration of EVs projects a significant transition of energy mix over the next few decades. Oil is no longer in demand for road transport with vehicle technological innovation, and such energy transition can be achieved more quickly with climate change mitigation efforts.

2.3.2 Emissions

Fig. 2.3 shows the CO_2 emissions by transport modes from 2005 to 2100. In the *BaU_Default* scenario, transport-related CO_2 emissions increase from 5.8 Gt in 2005 to 9.5 Gt in 2100. This increase is primarily attributed to the increased use of liquid fossil fuel by transport sector. Under the scenario of *2D_Default* that stabilizes radiative forcing for the 2°C goal, CO_2 emissions eventually decrease toward 4.6 Gt in 2100, because the use of liquid fossil fuel has been replaced by gas, electricity, and biomass. Without vehicle technological innovations, car and trucks are major contributors to CO_2 emissions, while emissions from road transport including car, bus, two wheeler, and small truck drop down to zero by 2050 with the stringent penetration of EVs. In the scenarios with 100% market share of EVs, transport modes such as large trucks, aviation, and navigation are expected to emit most of emissions in the future, which are at present difficult to be electrified without breakthrough efforts and technological changes. However, CS can reduce the emissions to some extent to 7.4 (*BaU_CS*) and 3.9 Gt (*2D_CS*) in 2100 with and without carbon pricing, respectively. In spite of the promotion of CS, road transport such as car and trucks still account for large proportions in contributing CO_2 emissions, implying that lifestyle changes alone are tough to achieve the target of net zero emissions in transport sector.

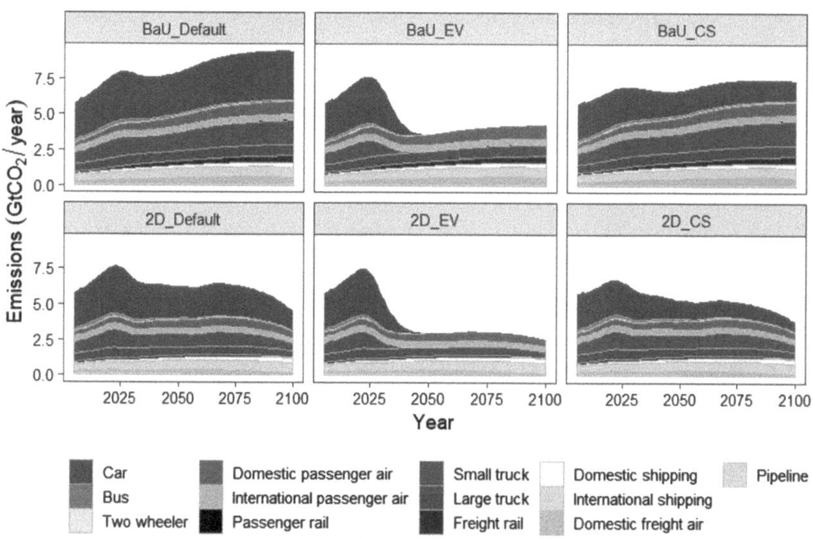

Figure 2.3 Global CO_2 emissions from transport by mode (2005–2100).

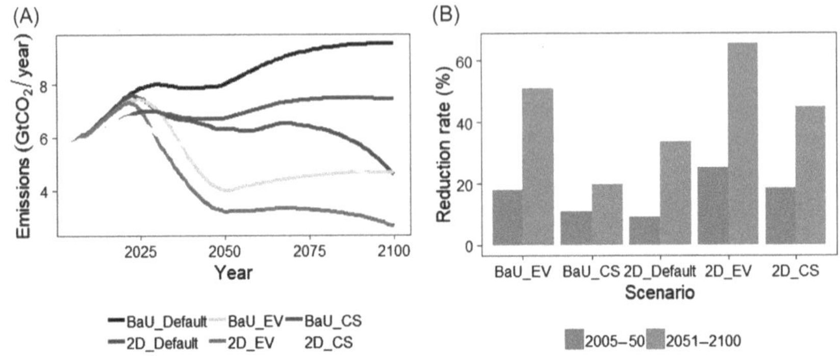

Figure 2.4 (A) CO_2 emission trajectories of transport sector and (B) reduction potential of cumulative emissions.

Fig. 2.4A presents the emission trajectories of transport sector during 2005–2100. The stringent penetration of EVs without carbon pricing (*BaU_EV*) is more effective to reduce the emissions than carbon pricing scenario (*2D_Default*), because the road transport cannot have zero emissions by the only implementation of carbon pricing. CS will contribute less to emission reduction in transport sector than EV policy, regardless of whether carbon price is imposed or not, because the emission trajectories of *BaU_CS* and *2D_CS* are significantly higher than those of *BaU_EV* and *2D_EV*, respectively. Fig. 2.4B shows a more detailed analysis of the influences of each mitigation scenario on cumulative emissions during 2005–50 and 2051–2100. Stringent transport electrification together with carbon pricing (*2D_EV*) has the highest mitigation potential for transport sector, as 66% and 25% of CO_2 emissions can be reduced during 2005–50 and 2051–2100, respectively, compared with the benchmark scenario in the absence of transport policies and climate change mitigation efforts, whereas the lowest reduction rate is attributed to *2D_Default* during 2005–50 and *BaU_CS* during 2051–2100. The reduction rates of cumulative emission imply that the contribution of lifestyle innovations on low-carbon transition is not as remarkable as technological breakthrough and transformation.

2.3.3 Economic effects

The long-term economic costs and benefits of transport scenarios can be evaluated by employing the global transport model coupled with the economic model, because the integrated model describes the feedback and interplay between transport sector and economy. The indicator of GDP loss, calculated by CGE model according to the emission constraint, can be used to analyze how the imposition of the carbon price in mitigation

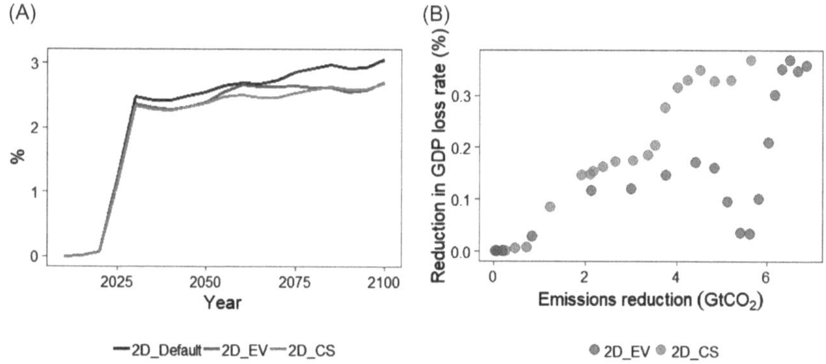

Figure 2.5 Mitigation cost metrics: (A) GDP loss of scenarios with carbon pricing and (B) correlation between reductions in emissions and GDP loss rate.

scenario will reduce GDP as compared with a *BaU* scenario in which no climate action is assumed. A measure of the economic impacts of transport scenarios is to detect how the cost of climate change mitigation would be modified by the transport policies, which can be indicated by GDP loss rate to achieve emission reduction consistent with the stabilization objective of 2°C target.

Fig. 2.5A shows that GDP loss in 2100 decreases from 3.1% to 2.7% with undertaking transport policies, indicating that the economic cost associated with pricing carbon can be thereby mitigated significantly because the goal of emission reduction can be achieved more easily by the penetration of EVs and the promotion of shared mobility, rather than putting a heavy price on carbon emissions. To further clearly investigate the different effects on mitigation cost of transport policies, the correlations between the reduction in emissions and the GDP loss rate are employed, as shown in Fig. 2.5B. The values of the overwhelming majority of scattered data points in the *CS* scenario are greater than those in the *EV* scenario, indicating that the contribution to reducing GDP loss generated by the decrease in the same quantity of emissions is relatively more significant in the *CS* scenario. Interestingly, the degree of contribution of transport sector to reducing economic cost is more effective when implementing share mobility rather than EV adoption.

2.4 Discussion and conclusion

A transport model is employed to provide elaborate behavioral and technological descriptions in the transport sector, and to detect the impacts of transport behavioral and technological factors on energy use and GHG emissions. Conventional IAMs cope with the transport sector at an aggregate level by using a production function or price elasticity but cannot handle the behavioral and technological details such as travelers' mode choice, vehicle preference, technological development, load factor, and speed. The integration of transport model, economic model, and climate model enables the IAMs to incorporate explicitly the detailed representations of transport sector to create transport scenarios for policy evaluation. Moreover, model integration can capture the interplay between transport sector and economic system. Not only can such integrated methodology assess the impacts on energy consumption and climate change, but it is also possible to investigate how transport policies would influence the macroeconomy, since the outcomes of transport model would be passed to the economic model.

Scenario simulations reveal that the behavioral and technological factors such as higher preferences on EVs and greater load factor due to CS would have a significant influence on energy transition and emission reduction. The penetration of EVs has more effective effects because the emissions from road transport would decrease to zero provided that the goal of 100% market share of EVs would be achieved by 2050, while shared mobility only exerts a moderate influence through decreasing the energy efficiency per person-kilometer owing to greater load factor. On the other hand, another reason of lower effectiveness of shared mobility is attributed to the modal shift from public transport to personal travel mode, because personal vehicle sharing service might lead to the declining cost of car. When the ICE vehicles have not been phased out, more road trips by car are bound to produce more fossil fuel consumption and GHG emissions.

The potential of emission reduction from transport sector would increase by implementing transport policies together with carbon pricing, although the imposition of carbon tax is bound to bring economic losses. Detecting the economic impacts of transport policies is one advantage of the integrated model, since the feedbacks from transport model are passed to economic model to examine the interaction between transport sector and macroeconomic system. Mitigation cost measures represent the economical attractiveness of EVs and shared mobility as mitigation tools, because they reduce the loss rates of economic growth due to imposition of carbon tax for achieving climate targets. Interestingly, the reduction in mitigation cost generated by shared mobility is greater than that produced by EV adoption, implying the CS services deserve more attention for achieving stringent climate targets. In spite of more powerful emissions-reducing effects of EV policy, electrified road transport might result in an increase in emissions if the electricity is produced by fossil fuels. Compared with shared mobility, additional demand for electricity in EV scenario leads to higher economic losses because a heavier carbon price needs to be imposed on the energy sector for electricity generation to power EVs.

Implications can be obtained based on the integrated modeling and scenario simulations. First, in addition to the mechanism how the transport behavioral and technological factors would modify the energy use and emission profile, the impacts on the macroeconomic dynamics need to be considered when evaluating the feasibility and cost-effectiveness of policy incentives in transport sector. Second, the results of economic cost should not be interpreted to underestimate the role of EVs as a promising solution toward green future, but to highlight the interaction between the implementation of EVs and the macroeconomy to formulate more harmonized policy designs. Although most countries do not have concrete plan for an action against fossil fuel—powered vehicles, there is no doubt that EVs will play a vital role in energy transition and low-carbon transport development. Third, lifestyle changes such as CS need attention as a supplementary policy instrument. Harmonious technological and lifestyle innovations need to be taken into account for constructing a decarbonized transport system, since they have their own advantages and disadvantages on reduction in GHG emissions and economic losses.

The global transport model copes with the transport behavioral mechanisms mainly by the discrete-choice model. Generalized transport cost is the critical decision variable in the choice model. Although the calculation of generalized transport cost involves the fuel price, energy efficiency, vehicle speed, transport infrastructure, and carbon pricing, transport cost cannot reflect the noneconomic factors that influence the travel behaviors. Furthermore, the global model is developed based on the highly aggregated regional data rather than the personal data such as age, gender, income, and education, so it is incapable of analyzing the behavioral factors and mechanisms at the individual level. The integration of multiscale data deserves consideration for the future research. In addition, behaviors in other sectors, such as consumption behaviors of households and industrial behaviors of firms, are handled by a global economic model in the form of CGE model. Since the residential, industrial, and transport behaviors are a part of people's life choice, it is necessary to promote studies on the transport behavior together with the behavioral factors in other sectors for cross-sectoral policymaking.

Since climate change is the biggest global environmental problem, IAMs draw knowledge and strengths from various disciplines related to climate change and have developed into a mature interdisciplinary research area that involves plentiful models. It is still challenging to integrate disciplinary components with each other based on a unified modeling platform, even if the certain relationships linked to climate change can be mathematically represented by plausible sophisticated approaches. Typical transport models such as four-step model, which is the primary tool for projecting future transport demand and performance of a transport system, are usually developed at a city scale or neighborhood scale. The application of this modeling approach requires a detailed database of travel behaviors, household demographics, and other socioeconomic factors in each traffic analysis zone. In particular, origin–destination matrix and traffic network provide essential basic data for trip distribution and route assignment, which are tough to be incorporated in the conventional IAMs because of the aggregate nature of IAMs for climate change. Another major oversight when integrating the transport sector with IAMs is the absence of any interplay between transport model and land use model. Not only do transport behaviors respond to land use patterns, but they also shape it as well. Such feedbacks have been described in land use–transport interaction model at city scale but have not been considered in the transport sector representation of IAMs.

Acknowledgment

The author acknowledges support from the Japan Society for the Promotion of Science (JSPS) KAKENHI grant number JP19K20507.

References

Akashi, O., Hanaoka, T., 2012. Technological feasibility and costs of achieving a 50 % reduction of global GHG emissions by 2050: mid- and long-term perspectives. Sustain. Sci. 7 (2), 139–156.

Akashi, O., Hijioka, Y., Masui, T., Hanaoka, T., Kainuma, M., 2012. GHG emission scenarios in Asia and the world: the key technologies for significant reduction. Energy Econ. 34, S346–S358.

Anable, J., Brand, C., Tran, M., Eyre, N., 2012. Modelling transport energy demand: a socio-technical approach. Energy Policy 41, 125–138.

Bose, B.K., 2010. Global warming: energy, environmental pollution, and the impact of power electronics. IEEE Ind. Electron. Mag. 4 (1), 6–17.
Bristow, A.L., Tight, M., Pridmore, A., May, A.D., 2008. Developing pathways to low carbon land-based passenger transport in Great Britain by 2050. Energy Policy 36 (9), 3427–3435.
Creutzig, F., Jochem, P., Edelenbosch, O.Y., Mattauch, L., van Vuuren, D.P., McCollum, D., et al., 2015. Transport: a roadblock to climate change mitigation? Science 350 (6263), 911–912.
Daly, H.E., Ramea, K., Chiodi, A., Yeh, S., Gargiulo, M., Gallachoir, B.O., 2014. Incorporating travel behaviour and travel time into TIMES energy system models. Appl. Energy 135, 429–439.
Dhar, S., Pathak, M., Shukla, P.R., 2017. Electric vehicles and India's low carbon passenger transport: a long-term co-benefits assessment. J. Clean. Prod. 146, 139–148.
Edelenbosch, O.Y., McCollum, D.L., van Vuuren, D.P., Bertram, C., Carrara, S., Daly, H., et al., 2017. Decomposing passenger transport futures: comparing results of global integrated assessment models. Transp. Res. D: Transp. Environ. 55, 281–293.
Fujimori, S., Masui, T., Matsuoka, Y., 2012. AIM/CGE [Basic] Manual (Discussion paper series). National Institute for Environmental Studies, Tsukuba.
Fujimori, S., Masui, T., Matsuoka, Y., 2014. Development of a global computable general equilibrium model coupled with detailed energy end-use technology. Appl. Energy 128, 296–306.
Girod, B., van Vuuren, D.P., Deetman, S., 2012. Global travel within the 2 degrees C climate target. Energy Policy 45, 152–166.
Girod, B., van Vuuren, D.P., de Vries, B., 2013a. Influence of travel behavior on global CO_2 emissions. Transp. Res. A: Policy Pract. 50, 183–197.
Girod, B., van Vuuren, D.P., Grahn, M., Kitous, A., Kim, S.H., Kyle, P., 2013b. Climate impact of transportation A model comparison. Clim. Change 118 (3–4), 595–608.
IEA, 2009. Transport, Energy and CO_2: Moving Toward Sustainability. International Energy Agency, Paris.
IEA, 2015. CO_2 Emissions from Fuel Combustion Highlights 2015. International Energy Agency, Paris.
IIASA (Ed.), 2015. SSP Database.
Kannan, R., Hirschberg, S., 2016. Interplay between electricity and transport sectors—integrating the Swiss car fleet and electricity system. Transp. Res. A: Policy Pract. 94, 514–531.
Karkatsoulis, P., Siskos, P., Paroussos, L., Capros, P., 2017. Simulating deep CO_2 emission reduction in transport in a general equilibrium framework: the GEM-E3T model. Transp. Res. D: Transp. Environ. 55, 343–358.
Kyle, P., Kim, S.H., 2011. Long-term implications of alternative light-duty vehicle technologies for global greenhouse gas emissions and primary energy demands. Energy Policy 39 (5), 3012–3024.
Lashof, D.A., Ahuja, D.R., 1990. Relative contributions of greenhouse gas emissions to global warming. Nature 344 (6266), 529.
Loo, B.P.Y., Banister, D., 2016. Decoupling transport from economic growth: extending the debate to include environmental and social externalities. J. Transp. Geogr. 57, 134–144.
McCollum, D., Yang, C., 2009. Achieving deep reductions in US transport greenhouse gas emissions: scenario analysis and policy implications. Energy Policy 37 (12), 5580–5596.
McCollum, D.L., Wilson, C., Pettifor, H., Ramea, K., Krey, V., Riahi, K., et al., 2016. Improving the behavioral realism of global integrated assessment models: an application to consumers' vehicle choices. Transp. Res. D: Transp. Environ. 55, 322–342.

Melo, P.C., 2016. Driving down road transport CO_2 emissions in Scotland. Int. J. Sustain. Transp. 10 (10), 906–916.
Mishra, S.G., Kyle, P., Teter, J., 2013. Transportation Module of Global Change Assessment Model (GCAM): Model Documentation. Institute of Transportation Studies, University of California, Davis, CA.
Neves, S.A., Marques, A.C., Fuinhas, J.A., 2018. Could alternative energy sources in the transport sector decarbonise the economy without compromising economic growth? Environ. Dev. Sustain. 20, 23–40.
Pietzcker, R.C., Longden, T., Chen, W.Y., Fu, S., Kriegler, E., Kyle, P., et al., 2014. Long-term transport energy demand and climate policy: alternative visions on transport decarbonization in energy-economy models. Energy 64, 95–108.
Rayner, T., Russel, D., Lorenzoni, I., 2008. 'It's demand, stupid': the failure and future of integrating climate change concerns into UK transport policy. Br. Polit. 3 (3), 373–389.
Siskos, P., Zazias, G., Petropoulos, A., Evangelopoulou, S., Capros, P., 2018. Implications of delaying transport decarbonisation in the EU: a systems analysis using the PRIMES model. Energy Policy 121, 48–60.
Solomon, S., Plattner, G.-K., Knutti, R., Friedlingstein, P., 2009. Irreversible climate change due to carbon dioxide emissions. Proc. Natl. Acad. Sci. U.S.A. 106 (6), 1704–1709.
Waisman, H.D., Guivarch, C., Lecocq, F., 2013. The transportation sector and low-carbon growth pathways: modelling urban, infrastructure, and spatial determinants of mobility. Clim. Policy 13, 106–129.
Yin, X., Chen, W.Y., Eom, J.Y., Clarke, L.E., Kim, S.H., Patel, P.L., et al., 2015. China's transportation energy consumption and CO_2 emissions from a global perspective. Energy Policy 82, 233–248.
Zhang, H.J., Chen, W.Y., Huang, W.L., 2016. TIMES modelling of transport sector in China and USA: comparisons from a decarbonization perspective. Appl. Energy 162, 1505–1514.
Zhang, R., Fujimori, S., Dai, H., Hanaoka, T., 2018a. Contribution of the transport sector to climate change mitigation: insights from a global passenger transport model coupled with a computable general equilibrium model. Appl. Energy 211, 76–88.
Zhang, R., Fujimori, S., Hanaoka, T., 2018b. The contribution of transport policies to the mitigation potential and cost of 2° C and 1.5° C goals. Environ. Res. Lett. 13 (5), 054008.

CHAPTER 3

Effects of green policies on household transport and energy consumption in Europe: a literature review

Dujuan Yang* **and Harry Timmermans**
Eindhoven University of Technology, Eindhoven, The Netherlands

3.1 Introduction

The Earth's climate is changing and the average global temperature is rising which has unprecedented consequences around the world. These abnormal weather conditions are having various impacts on the economy, environment, health, and our daily lives. Although some debates deny the influence of human behavior, the majority of world's climate scientists seem to agree that human activities and greenhouse gases, such as the burning of fossil fuels and deforestation, have detrimental effects on the climate (Grimm et al., 2008; Solomon et al., 2009). To control the effects of human behavior on global warming, international action is needed since climate change has no national boundaries. The first major international agreement—The United Nations Framework Convention on Climate Change—was established in 1992. In 1997 the convention was supplemented by the Kyoto Protocol to reduce the emission of greenhouse gases (CO_2). The protocol entered into force in 2005. It represents the first step toward reversing the global trend of rising emissions. Later, several rounds of negotiations were organized aiming at reaching a new comprehensive climate agreement. Finally, in December 2015, more than 195 countries adopted the first-ever, legally binding global climate deal—The Paris Agreement. This deal will help to make an end to the fossil fuel era, as the world rapidly replaces coal, oil, and gas with clean energy sources (Climate Council, 2015).

* Corresponding author

Transport and Energy Research
DOI: https://doi.org/10.1016/B978-0-12-815965-1.00003-X

The European Union (EU) has a long history in taking actions to reduce greenhouse gas emissions and be involved in pushing the international action. The need to control CO_2 emissions and keep global warming under control has been argued for a long time, back to 1990 when the EU committed to stabilizing its CO_2 emissions at that year's level. Meanwhile, the union has set up a whole series of policy measures through the European climate change program. The EU has set the energy targets for 2020 and passed binding legislation to ensure these are achieved. Moreover, the EU adopted the 2030 climate and energy framework and fixed ambitious emission targets for 2050 (European Commission, 2017). The 2020 directive sets national binding targets for all EU countries with the overall aim of making renewable energy sources account by 2020 for 20% of EU energy and for 10% of energy specifically in the transport sector (European Commission, 2017). This directive also includes two other climate goals, namely, a 20% reduction in EU greenhouse gas emissions (compared to 1990 levels) and 20% improvement in the EU's energy efficiency. These goals are average goals for the entire union. To reach the national renewable energy targets, specific targets have been set for each country, depending on the current state of each country and the possibilities to become fully independent of fossil energy sources. Malta has the lowest target (10% renewable energy) compared with Sweden's target (49%—renewable energy), which is the highest of all EU countries. The 2030 and 2050 targets are succeeding the 2020 targets, striving for even higher goals. In 2030 the EU strives for at least 40% greenhouse gas emission cuts and a 27% incensement in both energy efficiency and renewable energy use. In 2050, the EU has committed to cut its emissions by 80%–95% compared to the 1990 levels (Ecofys, 2010; European Commission, 2013).

Besides this directive and framework, member states in the EU have taken specific national actions, such as green policies. The green policies have focused on the fossil fuels that are the main source of CO_2 emissions and applied a broad-based approach to encourage behavioral adjustments throughout the whole economy. These policies are targeting three sectors, which are the energy production sector, consumption of energy for the production sector, and the final household energy consumption sector. Six common policy instruments are applied in the EU: regulations, tax instruments, emission trading systems, research and development, and public investment (INST-EC no. 12 paper). Among these instruments, regulation, tax instruments, and emission trading systems are recognized as the more effective approaches (Broberg and Kažukauskas, 2014; Gillingham and Palmer, 2014; Laes et al., 2018). Among these three, tax instruments were widely applied to the third sector—household energy consumption—by EU member states. However, for the design and reform of green policy, there is no one-size-fits-all approach. Each member state needs to find its best approach to address the specific needs, challenges, and priorities. Some of them have taken a more active role than others in implementing green policies. Particularly, the United Kingdom, the Netherlands, and Italy implemented more than 50 policies since 1990.

EU green policies, affecting household energy consumption, can be divided into two categories which are efficiency and curtailment behavior (Gardner and Stern, 2002). Efficiency behavior refers to one-shot behavior and entails the purchase of energy-efficient equipment, such as house renovation (insulation) or purchasing electric vehicles. Curtailment behavior refers to repetitive efforts to reduce energy usage, such as driving less private vehicles, consuming less electricity, and lowering room temperatures in winter. The empirical studies reviewed in this chapter consider both perspectives.

The purpose of this chapter is threefold. First, we will summarize EU green policies which have a direct and indirect relationship with household energy consumption. The aim is to examine the overall strategic plan of the EU and to understand variations between countries. Second, empirical studies related to EU household energy consumption, especially energy conservation behavior, will be reviewed on the effectiveness of policies and interventions to promote household energy conservation. The aim is to find consistent findings with respect to the effectiveness of these policies or interventions. Third, the strengths and shortcomings of the policies and interventions are discussed. Suggestions are provided for promoting energy conservation behavior to achieve 2050 EU energy goals.

The remainder of this chapter is organized as follows. In Section 3.2 the methodology underlying the literature review is presented. In Section 3.3, we will summarize and discuss released green policies related to household energy consumption since 1990 by EU member states. Then, we will focus on household travel behavior and home energy consumption behavior separately to discuss to what extent the environmental policies have resulted in behavioral change, reduction in conventional energy usage, and/or an increase of renewable energy production. This chapter is concluded with a discussion and recommendation.

3.2 Methodology

To provide a consistent review of both EU green policies and their effects on household energy consumption, an interactive approach was applied. It includes two main steps. The first step focuses on screening literature on green policies in the EU and identifies relevant green policies related to households' energy consumption. Based on the green policies identified at the first step, the literature that focuses on the study of policy effects on household energy conservation behavior was selected. Meanwhile, references were checked to catch any missing green policies from the first step. The search stopped once not new household-related green policy in the EU was found.

Various public and government databases were consulted for green policies in the EU, including the website of the European Commission, EU, Energy Community, European Network of Transmission System Operators for Electricity, International Energy Agency, ENGIE, International Renewable Energy Agency, Wind Europe, World Energy Council, and REN21. Two selection criteria were used to identify green policies. First, the policy has to be implemented between 1990 and 2015 in any of the EU countries. Second, the group of the green policy concerned households. The selected policies were then classified according to the INST-EC Discussion Paper No. 12, in which an in-depth discussion of the various tools available for EU policymakers to encourage the structural change toward a low-carbon economy is provided. The main categories related to household energy consumption are regulations, tax instruments, and public investment.

In addition, the database from both Elsevier and Google Scholar were scanned. The keywords used include "energy policy," "household," "travel behavior," "electric vehicles," "solar panel," "heat pump," "house renovation," "fuel tax," "incentives/subsidies," "energy consumption," "energy conservation behavior," and "Europe." We selected the peer-reviewed articles that were published between 2000 and 2018 from high-impact factor journals. Book chapters and EU reports were selected as well. In order to be selected for the final review, the study has to either include one or more green policies or needs to have a design allowing the assessment of the effects of implementing a green policy on household energy consumption/conservation behavior. The study has to be conducted with a decent number of respondents according to methodologies applied in European countries. The main reason for this criterion is the differential effect a green policy may have depending on the target group. Alternatively, it has to be a well-cited literature review paper which mainly compared literature from Europe. Finally, 30 papers were selected. These are summarized and discussed in Section 3.4.

3.3 European Union green policies related to households

It has found that markets alone are inadequate in addressing climate change. Green energy policy options are adopted to address different kinds of barriers existing in technology adaptation and household energy consumption behavior. Among the green policy instruments, regulations and tax instruments are the two most widely applied instruments in the household sector. However, there is a blurred boundary between these two policies. Based on ILO, EU, IILS (2013), the green policy instruments related to households could be further categorized into five categories according to their implementation format: regulatory instruments, incentives/subsidies, fee/charge, education and outreach, and tax. Moreover, considering the final target of these polices, these green policies could be further divided into two categories which are travel-related policies and home appliance—related policies. The green energy policies of 25 countries in the EU are summarized in detail in Appendix 1.

Fig. 3.1 presents an overview of the green policies implementation situation across EU countries. It can be seen that some countries are much more active in formulating green policies to trigger behavior adaptation or stimulate green technologies adaptation than others. As shown in the figure, Belgium has implemented 17 green policies for the household section, while Estonia and Malta have implemented only 3 green policies targeting households. On average, nine green policies have been implemented per country. When we divide the green policies further into transport and home-related policies, it shows that more green policies aim at transport than at home. Overall, there are 134 green policies for transport and 89 for home. Moreover, there are eight policies practically targeting carbon dioxin emission, which concern both transport and

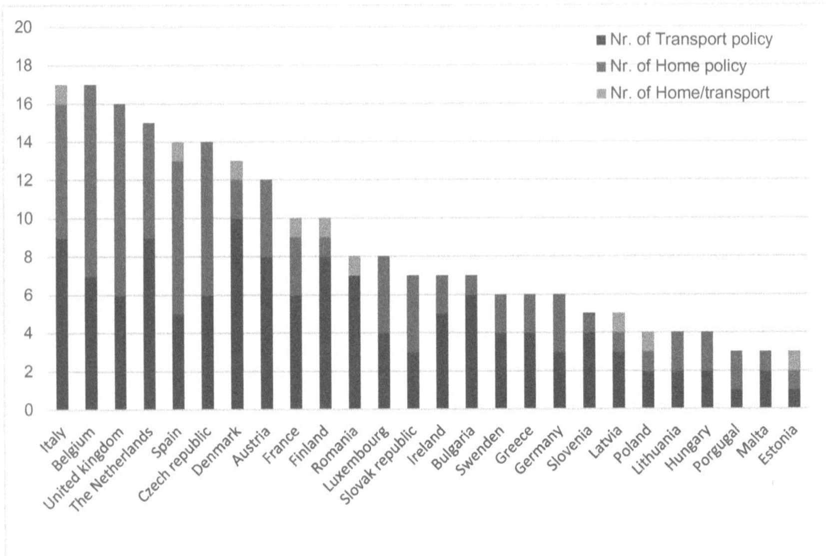

Figure 3.1 Number of green policies for households in the EU countries. *EU*, European Union.

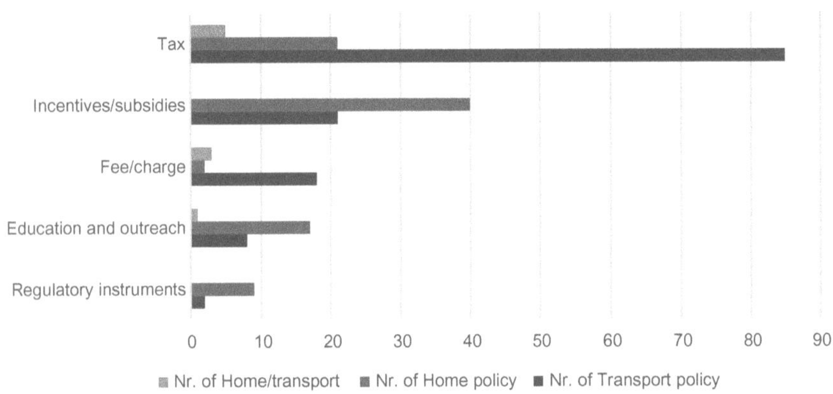

Figure 3.2 Distribution of types of policy instruments for household green policy.

home. An interesting pattern is that countries which have implemented more green policies for households also implemented more green policies targeted at the for home subsection than transport subsection.

Fig. 3.2 presents the different types of green policies adopted by EU countries. It shows the popularity of a market-based approach, reflected in tax, incentives/subsidies, and fee/charge. Education and outreach as a soft instrument is well used for home-related policies especially for educating about energy-efficient appliances and energy labels. Regulation is the least adopted instrument for promoting household energy conservation behavior. To be more specific, tax is mainly applied to the consumption of natural gas, coal, electricity, petrol, diesel, and other mineral oils. Tax also has been applied for the use of motor vehicles, insurance of motor vehicles, and registration or purchase of motor vehicles. Moreover, many countries used the tax instrument for controlling emission levels, including CO_2, SO_2, and heavy metal. It has been especially used for promoting low-emission vehicles by adding extra tax for heavy polluting vehicles and relaxation policies to favor environment-friendly vehicles with lower CO_2 emissions and better fuel economy. For in-home appliances, tax instruments have been used for promoting the purchase of high-efficiency home appliances/equipment, such as fridges, freezers, and lighting.

Incentives/Subsidies used in the EU countries particularly aim at increasing energy efficiency in environmental vehicles (plug-in hybrid electric vehicles (PHEVs) and plug-in electric vehicles (PEVs)), friendly house equipment, and buildings. Fifteen countries have provided incentives/subsidies for purchasing PEVs and PHEVs. The incentives and subsidies vary from country to country and up to 30% of the price of a new PEV (9190 euros). Besides the beneficial environmental vehicles, to reduce traffic emission levels, incentives/subsidies for using public transportation and incentives to encourage the sharing of private means of transport among users have been implemented in Belgium and Italy. Incentives/Subsidies have been widely used in the home subsection and for three main purposes. First, they have been adopted for stimulating renewable energy investments by households, such as solar photovoltaic electricity. The second purpose is to promote energy efficiency measures in residential buildings including the refurbishment of existing residential buildings and constructing new energy efficiency residential buildings. The incentives/subsidies could be direct grants, a tax reduction of loans, reduced interest for loans, and greater flexibility for the payment of new houses. Last, some green incentives/subsidies specifically target environment-friendly house equipment and purchasing eco-friendly goods and furniture. Poland implemented a green policy (incentives/subsidies) partially targeting the rationalization of heat consumption behavior in households.

The fee/charge instrument has been mainly applied for the household transport subsection. It has been used to reduce the use of fossil fuels and air pollution (CO_2, SO_2, heavy metal, or other particle emissions), the use of vehicles in general or on certain roads. It also has been applied to control parking time in limited parking zones. In Latvia, it has been used to promote EVs by substantially reducing the annual technical inspection fee. For the home subsection, it only has been used for electricity usage.

The EU governments have launched educational programs to provide information concerning energy labeling of various consumer products, energy performance certificates (EPCs), and energy labels for new passenger cars (energy consumption and CO_2 emissions). Education and outreach as a green policy instrument plays a soft role in promoting household energy conservation behavior. However, it is a quite well-used instrument which has been adopted by 26 countries. Austria provided voluntary information activities in the field of energy-efficient home electronics. France developed a website to help people making sustainable choices and reduce their CO_2 emissions. The United Kingdom provided a "one-stop shop" service to help households make their homes greener. Worth to mention, both Belgium and the Netherlands have tried to enforce the rational use of energy and the reduction of CO_2 emissions. Belgium used "Eco-score" to help the consumer choose cars that have a lesser impact on the environment, while the Netherlands promoted optimal driving behavior, called eco-driving, and encouraged modal shift from passenger vehicles to other forms of transport.

Regulation is the least used green policy instrument for the household section in the EU. It has been used to address various issues, including promoting the use of renewable energy sources, secondary legislation on the methodology for purchasing electricity, vehicle certification, building energy certification/performance standards, and progressive replacement of electricity meters.

To further highlight the main policy targets in the household section, Table 3.1 summarizes five main targets of the green policies across EU countries. These five targets are not independent. They are linked with each other but focus on different perspectives. Reducing fossil fuel usage focuses mainly on household curtailment behavior including reducing household use of conventional cars, electricity, and gas usage by introducing a tax or fee/charge. Controlling emission levels of CO_2, SO_2, and heavy metal has been applied to both household efficiency behavior and curtailment behavior. The purpose is to increase household attention to their energy emissions and adjust their energy consumption to reduce emission levels. The remaining three targets all focus on one-shot behavior and entailed the purchase of energy-efficient equipment. The purpose

Table 3.1 Main green policy targets (household section).

Policy target	Instrument type	Country
Reducing fossil fuel usage	Tax	Austria, Bulgaria, Belgium, Czech Republic, Denmark, Estonia, Finland, France, Germany, Greece, Ireland, Italy, Latvia, Lithuania, Luxembourg, Poland, Portugal, Romania, Slovak Republic, Slovenia, Spain, Sweden, The Netherlands
	Incentives/Subsidies	Austria
	Fee/Charge	Bulgaria, Denmark, Germany, Italy, Lithuania, Malta
Controlling emission levels of CO_2, SO_2, and heavy metals	Tax	Estonia, Finland, France, Germany, Ireland, Romania, Spain, United Kingdom
	Fee/Charge	Latvia, Poland, Romania
	Education and outreach	Belgium, Finland, France, Italy, Slovak Republic, Sweden, United Kingdom, The Netherlands
	Regulatory instruments	United Kingdom
Promoting PHEVs and PEVs	Tax	Bulgaria, Czech Republic, Finland, France, Greece, Hungary, Ireland, Italy, Luxembourg, Malta, Poland, Portugal, Romania, Slovak Republic, Slovenia; Sweden, United Kingdom
	Incentives/Subsidies	Austria, Belgium, Denmark, France, Germany, Luxembourg, Romania, Slovenia, Spain, Sweden, United Kingdom, The Netherlands
	Fee/Charge	Malta
	Education and outreach	Belgium and Denmark
	Regulatory instruments	Portugal

(*Continued*)

Table 3.1 (Continued)

Policy target	Instrument type	Country
Promoting efficient home appliances	Tax	Italy and Portugal
	Education and outreach	Austria, Belgium, Hungary, Portugal, Sweden, United Kingdom
Promoting energy efficiency measures (including renewable energy sources)	Tax	United Kingdom
	Incentives/Subsidies	Austria, Bulgaria, Belgium, Czech Republic, Finland, France, Germany, Greece, Hungary, Ireland, Italy, Lithuania, Luxembourg, Spain, United Kingdom
	Education and outreach	Denmark, Luxembourg, Spain, Sweden, United Kingdom, The Netherlands
	Regulatory instruments	Czech Republic, Portugal, Spain, Sweden, United Kingdom

of separating them is due to the fact that these green policies were supposed to trigger specific household energy conservation behavior.

Moreover, the combination of different policy targets might have synergy results for directing household energy conservation behavior, such as purchasing an electric vehicle (EV) could be triggered by the green policies targeting on promoting PHEVs and PEVs and the policies targeting on reducing fossil fuel usage or/and controlling emission level of CO_2, SO_2, and heavy metal. The purchasing behavior of solar panels could be triggered by the green policies targeting on promoting PEVs and PHEVs, or verse vice, since there might be a synergy between transport and home electricity consumption in future smart homes. However, few studies systematically examined the effects of the combination of green policies.

3.4 Selected empirical studies of household energy conservation behavior in the European context

3.4.1 Reducing fossil fuel usage

Most studies did not distinguish between energy tax, energy fee/charge, and energy price, assuming that people treat costs indifferently. Because we were unable to find a peer-reviewed paper regarding fossil fuel incentives/subsidies for the household sector in the English language, we selected five papers which investigated energy price/energy tax elasticity for electricity and gas in Norway, Spain, Germany, and the Netherlands.

A summary of reported price and income elasticities is presented in Table 3.2. Most studies found that households are elastic to both heating and electricity price, although there is also some evidence of price inelasticity in the short run (Hanemann et al., 2013; Meier and Rehdanz, 2010). However, these elasticities are hardly comparable since different estimation methods, theoretical models, assumptions, exogenous variables, data types, and measurements were used in these studies. Therefore we discuss their findings country by country and highlight the behavioral mechanism.

One Norwegian study focused on heating (Vaage, 2000), the other on electricity (Halvorsen and Larsen, 2001). Both studies used historical revealed data from the Norway governmental agency with quite detailed information but different years. Vaage (2000) used an extensive survey data on household energy consumption, organized by the Norwegian Central Bureau of Statistics in the late spring of 1980. The survey covers 2289 households who were equally distributed throughout the country. Halvorsen and Larsen (2001) used the annual Norwegian Survey of Consumer Expenditure (SCE) for the period 1975–94. The SCE yields information on each household's electricity consumption, the purchase of new household appliances as well as the current stock, and detailed information about household characteristics. In addition, they had disaggregated information about prices and outdoor temperature. Both studies applied a two-step model. Since Vaage (2000) used cross-sectional micro-data, he only estimated one time choice of different heating systems. He found a relatively high-price elasticity comparing to the literature. Halvorsen and Larsen (2001) used panel data to estimate long-term residential electricity demand in terms of the household's investment in new appliances. They found no significant difference between long-term and short-term elasticity since there is almost no alternative regarding electricity for households. Both studies have limitations regarding the data. First, the data is old. The behavior of households in the 1980s and earlier 1990s might be different from the current generation of households. Second, for the study on heating, there is a high degree of combined heating technology. But there are no objective measures of quality and capital costs. The data cannot be used to model technology choice

Table 3.2 Summary of electricity and gas elasticities in the European Union.

Paper	Country	Energy good	Price elasticity	Income elasticity
Vaage (2000)[a]	Norway	Average for heating	(−1.29, −1.24)	(−0.07, 0.00)
Nesbakken (2001)	Norway	Electricity	−0.43	0.08
		Electricity, fuel oil	−0.23	0.04
		Electricity, wood	−0.22	0.04
		Electricity, fuel oil, wood	−0.17	0.03
Halvorsen and Larsen (2001)[a]	Norway	Electricity	−0.433 SR	0.06 SR
			− 0.442 LR	0.13 LR
Labandeira et al. (2006)	Spain	Electricity	−0.78	0.81
Labandeira et al. (2012)	Spain	Electricity	−0.25	–
Bernstein and Madlener (2011)	Spain	Electricity	(−0.14, 0.01) SR	(0.14, 0.30) SR
			(−0.35, −0.30) LR	(1.21, 1.24) LR
Hanemann et al. (2013)[a]	Spain	Electricity	−0.14	0.30
		Natural gas	−0.18	0.23
		Liquid fuels	−0.27	0.30
Rehdanz (2007)[a]	Germany	Fuel oil space heating	(−2.03, −1.68)	–
		Natural gas	(−0.63, −0.44)	–
Berkhout et al. (2004)[a]	Netherlands	Natural gas	−0.28	−0.27
		Electricity	−0.57	0.61

Note: *LR*, Long run; *SR*, short run.
[a]Reviewed paper in this study.

objectively. The study reflects that heating is much more complex than electricity usage. Respondents have a shared mixed heating system. With the limited data, the two-step model could not fully capture the joint decision-making mechanism of heating equipment.

In Spain, Hanemann et al. (2013) applied a similar method to estimate the choice of energy source and the amount of energy used for heating. They used the Spanish household survey for the period 2006–08 (SHS, Encuesta de Presupuestos Familiares), carried out by the Spanish National Institute for Statistics (INE). It contains 63,054 observations from households that remain in the sample at most 2 years. Compared with the study of Halvorsen and Larsen (2001), this study includes energy source data but no capital cost data for heating equipment. It is possible to estimate switching energy source for heating. However, the heating situation in Spain is different from Norway. Spain has mild winters, and there are a large number of households without residential heating (around 40%). There is no mixed heating energy source considered in one household. There is no joint decision-making for heating energy source and equipment purchase in Spanish households. They found that for all fuels, the price of the energy source has a negative and significant influence on the probability of choosing that energy source, whereas the price of the alternative sources has a positive though not always significant impact on that probability. It can be explained by the fact that many households do not have heating and the possibility of changing energy source for heating is limited. It might depend on the locations of the households and available options for heating in that area. The short-term impact of fuel price on heating consumption is not significant which indicates that heating demand in the short term is mainly driven by comfort requirements. Therefore it is also easy to understand that the income plays a significant role in Spain. The richer households have a higher probability of owning a heating system than poorer households. For the long run, the results indicate that if there is a possibility to switch to a cheaper energy source for heating, households prefer to do so.

To identify the impact of green tax on household energy expenditures of space heating and hot water supply, Rehdanz (2007) conducted a study in Germany using a socioeconomic panel survey which includes more than 12,000 households administered between 1998 and 2003. This chapter focused on green tax introduced in 1999 and 2003. The data set includes detailed information on the heating and hot water supply systems of households. However, the data does not include energy consumption but only energy expenditure. After controlling for socioeconomic characteristics, regional differences, and dwelling differences, results indicate that the doubling of gas or heating oil prices increase household expenditures by 40%–50% on average. However, it is not clear whether households decreased their energy consumption by modernization of their heating system or by adopting their comfort level or behavior. The results found that energy price increases lead to higher energy expenditures for rented accommodation than for owned accommodation. It reflects the impact of house isolation investment, and new heating systems. The difference is likely to become smaller over time due to the further modernization of the heating system in rented accommodation.

Berkhout et al. (2004) investigated the impact of Regulating Energy Tax (REB) on Dutch households demand for energy before and after the implementation. Two large Dutch panel data sets were used, which contain information on gas and electricity energy use for more than 1000 and 1500 households, respectively. The empirical analysis was performed for the years 1994–99 for electricity and 1992–99 for gas. Different from previous studies, this study collected not only the sociodemographic data and household appliances but also energy consumption data. A two-stage budgeting model was applied. The results are consistent with previous studies. Households are more sensitive to the electricity tax than the gas tax since gas is mainly used for heating in the Netherlands.

Based on these studies, we can conclude that the tax for electricity may have a short-term impact on households' electricity usage. From a long-term perspective, it may trigger investments in energy-efficient appliances. Energy tax regarding heating and hot water energy source cannot easily influence short-term households' energy consumption. However, from the long-term perspective, it can trigger renovation or efficient heating system investments.

Transport energy demand for gasoline and diesel has been closely examined by academics and politicians over the last few decades. There is an extensive literature on gasoline and diesel demand elasticity, and high-quality reviews have been published (e.g., Espey, 1998; Graham and Glaister, 2002; Ajanovic et al., 2012; Dahl, 2012). Still, comparisons are difficult due to differences in estimation methods, theoretical models, assumptions, exogenous variables, data types, and measurements. To overcome these limitations, Aklilu (2016) estimated price elasticity across 28 EU countries by applying an Autoregressive Distributed-lag bounds testing approach using data from 1978 to 2013. In this study the estimated average gasoline price elasticity among the EU-28 countries is −0.17 in the short run and varies from −0.005 in Spain to −0.58 in Romania. The estimated long-run elasticity is −0.72 on average. The short-run income elasticity of gasoline demand is estimated to be 0.46, and the long-run elasticity is 1.45. Both short-run and long-run income elasticities of gasoline and diesel demand were found to be more elastic than their price equivalents. It implies that if a policy intends to decrease fuel consumption by increasing price, then the price needs to rise faster than income.

The literature mentioned earlier assumes that there is no difference in effects between the fuel tax and fuel price. Recently, two papers particularly investigated the effects of fuel tax on consumer's fuel usage (Grigolon et al., 2018; Nikodinoska and Schröder, 2016). Grigolon et al. (2018) studied consumers' intertemporal trade-off between the initial purchases price of the car and future fuel costs and highlighted the key parameters that drive consumer responses to fuel taxes versus product taxes. They focused on long-term households investment decisions and compared the fuel tax impacts with product tax impacts. Nikodinoska and Schröder (2016) studied the environmental, distributive, and welfare effects of the car fuel tax in Germany. The car fuel tax is charged as a fixed monetary amount per liter in Germany. They focused on the short-term impact of fuel price. They used expenditure data but from different resources. Grigolon et al. (2018) used a panel dataset from the European car market. The dataset includes the sales, prices, and product characteristics for every new passenger car sold during 1998–2011 in seven European countries: Belgium, France, Germany, Italy, the Netherlands, Spain, and the United Kingdom. Nikodinoska and Schröder (2016) used the Income and Expenditure Survey and the consumer price data for various expenditure categories from the German Federal Statistical Office. Based on these two studies, we can conclude that a fuel tax is more effective in reducing fuel usage than a revenue-equivalent product tax on fuel consumption because it especially targets high mileage consumers to substitute their cars for cars with lower fuel consumption. The fuel tax impacts the long-term household car ownership decision. But in the short run, fuel demand is highly price-inelastic. In the short run, there might be a substitution effect among car fuels, electricity, and other fuels, which indicate that environmental taxes not only alter demand, they also affect households' budget (income after taxes) allocation. However, due to the data limitations, a dedicated survey design is needed to investigate the emissions-inequality and emissions-welfare trade-off.

3.4.2 Controlling emission levels of CO_2

It is difficult to observe the direct effects of policies targeting households' emission levels. There is almost no scientific literature that investigated the effects of regulation instruments and education and outreach instruments about emission levels on households' behavior. Therefore we selected three papers which capture the impact mainly from tax and fee/charge perspectives. Based on the extensive discussion among economists on public policy interventions for environmental protection, taxes and fee/charge seem the most effective instruments to achieve a given environmental target. However, most of this literature focuses at the national level (Baranzini et al., 2000). It is necessary to understand the distributional implications of carbon tax to provide acceptable policy recommendations. Tiezzi (2005) summarized several studies in the EU, conducted before 2000 (Barker and Kohler, 1998; Cornwell and Creedy, 1996; Labandeira and Labeaga, 1999; Smith, 2000; Symons et al., 2000). The results indicate that the predicted impacts are country specific and depend on the modeling framework. It can be mildly regressive or even reversed. Tiezzi (2005) also performed a complete almost ideal demand system with historical household living cost data from Italy. The study aggregated disaggregated household data into five major types of households. She found similar results as Smith (2000), Labandeira and Labeaga (1999), and Symons et al. (2000): tax mainly hits the transport fuel and the burden affected more effectively on one or two adults' households. The effects decreased for larger families due to cars' widespread use. There is a variation in welfare losses across different income groups. The effect becomes bigger with higher income groups. Besides the distribution impact, the final effectiveness should be evaluated as well. It is normally mixed with energy price elasticity. As we discussed in Section 3.4.1, people may treat costs indifferently no matter it is fuel tax or emission tax. Tiezzi (2005) found that emission tax elasticity is very high in Italy for both transport fuel (−1.282) and domestic fuel (−1.057) which differs from results found in Germany for transport fuels (0.084). It indicates that the tax policy is much more effective in Italy than in Germany. However, the results are not consistent with the fuel price elasticity results of Aklilu (2016). We have to acknowledge that the comparison of the results does not have much meaning since they applied different methods on different types of data.

Besides the calculation of price elasticity, a group of researchers has investigated people's willingness to pay (WTP) for mitigation policies with choice experiments to understand the acceptance of emission policies. Alberini et al. (2018) presented a stated choice experiment which included hypothetical scenarios, while Löschel et al. (2013) used revealed preference data and offered participants a tangible opportunity to buy European Union Allowances from the EU Emissions Trading System. In these two studies, three countries WTP were examined, which are Italy, Czech Republic, and Germany (Mannheim). The stated WTP found by Alberini et al. (2018) is consistent with the literature, which are 133 euro/ tCO_2 for the Italians and 94 euro/ tCO_2 for the Czech respondents. However, the revealed WTP in Germany is much lower than the stated WTP, which is 11.89 euro/ tCO_2. The revealed WTP is almost 10% of the stated WTP. Moreover, both studies have explored preferences for carbon tax policies. They both revealed an "inconvenient truth" from a political economy perspective that "carbon tax" is significantly less acceptable. The renewable energy policy and improving energy efficiency policy are more acceptable compared to the "carbon tax."

3.4.3 EVs promotion policies

EVs are being introduced as a solution to the problem of dependency on fossil fuels and environmental issues. As mentioned in Section 3.3, all five types of policy instruments have been used to promote PHEV and PEV in different counties. Most commonly used instruments are tax, incentives/subsidies, and fee/charge. Some countries have implemented regulation and education/outreach. However, we could hardly find any study using revealed preference data to examine the effectiveness of these policies. Extensive studies of consumer EV adoption have been conducted during the last 10 years. Many of them have applied stated choice experiments to predict future market penetration rates. Considering the focus of this chapter, we only selected studies exploring the relationships between adoption behavior and EV-promoting policies based on revealed preference data. Besides, there is a branch of literature that focused on the consequence of EVs promotion policies and life-cycle environmental impacts of EVs. They argued that EV promotion policies are a gamble with tax money. To cover the full story, we selected five papers in total which discussed both the effects of EVs promotion policies on revealed households' adaptation behavior and effects of the EVs policy in well-adopted countries (Holtsmark and Skonhoft, 2014; Langbroek et al., 2017; Ozaki and Sevastyanova, 2011; Prud'homme and Koning, 2012; Sierzchula et al., 2014).

Regarding the effectiveness of financial policy instruments, the literature suggested that the financial policy instruments have a significant influence on the PHEV/PEV adoption rate. Sierzchula et al. (2014) found that financial policy instruments are strongly and positively correlated with national EV market shares based on revealed country-level data from 30 countries. Ozaki and Sevastyanova (2011) validated the relationship by identifying the motivations to purchase the Prius. They found that the perceived financial benefits strongly affected the decision to purchase a hybrid car. The financial benefits recognized by consumers were a lower tax, congestion charge exemption, and fuel economy. However, some exceptional cases were identified by Sierzchula et al. (2014). Belgium and Denmark had very high financial incentives, but relatively low rates of adoption. Conversely, Switzerland and Sweden exhibited the opposite dynamic with low consumer subsidies but high EV uptake levels. The disproportional adoption rate can be partially explained by the finding of Ozaki and Sevastyanova (2011). They found that social norms and pressure, practical compatibility, self-expression, positive attitudes toward technology, and the role of trial experience were factors strongly influencing the PHEV/PEV purchase decision.

Although there is almost no disagreement regarding the effectiveness of these financial policy instruments on PHEV/PEV adoption rates, a branch of researchers questioned these policies concerning the promotion of PHEV/PEV on the basis of three main arguments. First, EV promotion policies are a kind of gambling considering the unclear CO_2 gain and higher cost from a life-cycle calculation (Holtsmark and Skonhoft, 2014; Prud'homme and Koning, 2012). The CO_2 gains are difficult to compare since they mainly depend on the way of generating electricity. Second, the impacts of EVs on replacing gasoline or diesel are unclear (Langbroek et al., 2017; Halvorsen and Frøyen, 2009). A large-scale deployment of EVs might increase overall car use and congestion and lead to a substantial reduction in the use of public transport and bicycles. Last but not least, these policies lead to inequality of tax benefits across income levels since richer households are more likely to purchase a PHEV/PEV.

3.4.4 Efficient home appliance policies

Two types of policy instruments have been implemented for promoting efficient home appliances, which are tax and education/outreach. To date, only two countries applied tax to promote energy-efficient fridges (Italy) and energy-efficient bulbs (Portugal). Unfortunately, we could not find any recent revealed preference study from these countries. Most countries applied education/outreach using efficiency labeling policy. Only a few studies have investigated the impact of efficiency labeling at the household level despite its widespread use (Blasch et al., 2017; Dieu-Hang et al., 2017; Mills and Schleich, 2010). Mills and Schleich (2010) explored the determinants of choice of class-A appliances using a large survey data set including more than 20,000 German households. They estimated the latent relationship between household attributes and choice of a class-A appliances. Dieu-Hang et al. (2017) investigated the adoption of energy and water-efficient equipment collecting quota samples from 11 OECD countries. Probit and Logit models were applied for choice of three water-efficient devices. Blasch et al. (2017) designed an experiment to examine the role of information, energy literacy, and investment literacy on the choice of decision-making strategies for choosing an electrical appliance. They organized a web-based survey involving 2835 respondents cooperating with 3 electricity providers in 3 major urban areas. A randomized controlled trial was used to calculate the lifetime costs of the appliances under different conditions by varying displayed information for light bulbs and refrigerators. In addition, respondents were asked about their decision-making strategy when making a choice. A series of binomial Probit models was applied to analyze the role of information display, energy, and investment literacy on the choice of electric appliances.

The literature suggests that energy/water labeling policies (including having labels, recognizing labels, understanding labels, and trust of the labels) have a significant impact on the household choice of energy/water-efficient equipment. It also indicates that consumers need prior knowledge to process label information. Consumers who possess energy-related knowledge and have high cognitive abilities were more likely to find the most cost-efficient appliance. It indicates the education/outreach plays an important role in household energy-efficient equipment choice. Moreover, as expected, green attitudes, energy saving habits, and water saving habits affect household investment in energy/water-efficient equipment. The results also indicate that house appliance choice is a midterm decision given the long average lifecycle of many house appliances, such as washing machine, dishwasher, freezers, and refrigerators. Consumers are less likely to replace their newly bought or still functional device due to high operation cost. The energy labeling information takes time to influence consumers' purchase decisions (Mills and Schleich, 2010).

3.4.5 House efficiency measures (renovation and photovoltaics (PV) panels)

The energy performance of residential buildings depends on many factors, such as the characteristics of the building, use patterns, and climate conditions. In the EU, over 40% of the residential buildings were constructed before 1960. Therefore major renovation would contribute to around 36% savings in energy consumption based on the report of Buildings Performance Institute Europe. Stimulating household investments in renovation measures including renewable energy microgeneration to reach nearly zero energy building standard is a critical point. Multiple types of green policies have been implemented in various countries to initiate or speed up progress, such as tax, incentives/subsidies, education/outreach, and regulatory instruments. Evidence of the impact of green energy policies is found in many studies. A selection of eight studies is discussed here. We first selected four studies focusing on how EPC, energy labels for buildings, Green Deal, and knowledge of climate change influence household energy renovation practices (Bartiaux et al., 2011; Christensen et al., 2014; Gram-Hanssen et al., 2007; Pettifor et al., 2015). These studies investigated the impacts of both financial policies and nonfinancial policies. Then, we moved to some particular cases of renovation including household willingness to adopt renewable energy technologies and to become prosumers by adopting PVs (Keirstead, 2007; Kotilainen and Saari, 2018; Sardianou and Genoudi, 2013; Vasseur and Kemp, 2015), and thermal renovation of existing homes (Galvin, 2012).

Gram-Hanssen et al. (2007) collected 10 qualitative interviews from Belgium and 10 from Denmark to analyze how house owners interpreted and used the knowledge from labels for the renovation of their homes. The 10 respondents from Belgium were recruited as volunteers who would like to receive energy recommendations, while in Denmark the respondents were randomly drawn from a sample of new house owners. This is because Denmark has an active energy policy on savings, whereas

Belgium has been more reluctant to promote energy savings. They compared two different systems and found residents are not only recipients of new information but also actors that interpret or reject new information based on their previous knowledge and the norms of their social network. Moreover, even though residents trust the advice, it does not mean they will follow the advice. Getting economic benefit from the house renovation cannot be prioritized without taking esthetics, identity, and convenience into consideration in their daily life. To further investigate the impact of EPC on energy renovation practices in Denmark after purchasing a house, Christensen et al. (2014) used data from a cross-national stated preference survey carried out in five EU countries in 2010. They confirmed the finding of Gram-Hanssen et al. (2007) that the Danish trust the information provided by EPC and understand the information provided. They also found that EPC is not useful for home energy renovation. The information provided by EPC is too general and trivial. It does not provide tailored recommendations considering homeowners' motivations. These two studies focused on the education/outreach instrument.

Bartiaux et al. (2011) conducted a qualitative study in Bulgaria and Latvia to investigate apartment owners' renovation motivations. It is worth mentioning that the number of studies on energy conservation behavior in East Europe is limited, at least in the English language. They conducted 16 interviews in Bulgaria and 18 in Latvia. They found that the major motivations to underlying energy-related renovations are the poor technical conditions of multiapartment buildings and poor comfort. Respondents did not take any environmental concern as a lever for undertaking renovation. There exist some ambivalence and skepticism concerning climate models.

To reduce the financial burdens of residents, in 2013 the United Kingdom installed a new energy policy, called Green Deal. It provides a financing mechanism by third parties to pay for the upfront costs of installing energy-efficient measures in the houses. Pettifor et al. (2015) published a study which examined the effectiveness of the Green Deal on strengthened intentions for renovations by private owners. They collected panel data before and after the implementation of the Green Deal. They did not find any evidence to prove that the Green Deal strengthened households' intentions for energy efficiency renovation. However, they found that there were significant changes in "awareness and knowledge" and "attitudes and beliefs" toward energy efficiency in the households who had the intention to do energy efficiency renovation compared to households who did not have any intention to renovate.

These four studies consistently suggest that energy efficiency cannot be the trigger for house renovation. The most common reason for energy efficiency renovation is discomfort. Residents trust their knowledge and social network more than a third party which provides nontailored feedback. Moreover, energy efficiency is not their primary concern when living in a house. Comfort is much more important than the energy bill within budget limitations. There is also skepticism concerning climate challenges.

Regarding renewable energy technology acceptance, both Sardianou and Genoudi (2013) and Kotilainen and Saari (2018) investigated which factors affect the acceptance. Sardianou and Genoudi (2013) focused on financial policies only and conducted a stated choice experiment design in Greece to investigate which type of respondents are more likely to adopt renewable energy technologies and how financial policies play a role in their decisions. They found that residents with a higher education level and income are more likely to accept renewable energy technologies. For the financial policies, they found that tax deduction is more effective than energy subsidies. Kotilainen and Saari (2018) compared the impact of financial policies with the impact of nonfinancial policies. Based on a sample of 197 respondents including both consumers and prosumers from 5 countries in Europe, they found that the nonfinancial policies are more effective than financial policies in steering consumers' attitudes toward using renewable energy technologies.

PV is the most popular renewable technology followed by wind and ocean energy. The adaptation of PVs in residential houses has a great impact on further decentralized energy networks and defusion of EVs (Keirstead, 2007; Vasseur and Kemp, 2015). Vasseur and Kemp (2015) found that knowledge about PV is equally important as cost. But the economics of PV is a complex issue because it can be viewed in terms of payback period or cost savings. The finding is consistent with findings of Kotilainen and Saari (2018). Once households have installed the PVs, Keirstead (2007) found that they consumed 6% less electricity. Moreover, they adjusted their electricity consumption patterns by load shifting to the peak time of PV generation. It indicates that PV system has both real-time and accumulated impacts on households.

Besides renewable energy technologies, thermal renovation plays a vital role in energy efficiency improvements. It is commonly argued that space heating can be substantially reduced. Quite a few policies have focused on this issue. Galvin (2012) examined the efficacy of a subsidy policy in German, examining a large number of existing case studies. Again, results indicated that energy efficiency only cannot trigger the replacement of thermal equipment. It involves a high investment and long-term payback period that bring high economic risk to homeowners. Knowledge is equally important as costs. Many homeowners did not trust the calculation of economic viability of the thermal renovation.

3.4.6 Rebound effects

Rebound effects play an important role in evaluating the impact of green policies. In order to discuss the effects of green policies aiming at increasing energy efficiency, rebound effects should be addressed properly. Considering there is a relatively small number of studies, it is difficult to split them into five policy targets discussed earlier. We, therefore, report general findings.

Different types of rebound effects, such as direct, indirect, and secondary, have been examined in a range of studies in the United States and Europe (Sorrell and Dimitripoulos, 2007). In this chapter, we will discuss the rebound effects from households' domestic and transport perspectives separately based on three recently published reports (Athavale and Knaus, 2017; Maxwell et al., 2011; Sorrell, 2018). Extant studies seem to agree that rebound effects exist, but the measurement of rebound effects has been subject to debate.

The direct rebound effect associated with energy efficiency interventions is explored best. For domestic sector, most studies estimated the direct rebound effect for household heating and lighting due to data availability. Results show a variation between 10% and 60%. Comfort factor plays an important role in raising temperature since it was cheaper and more efficient to heat the home after renovation. Fewer studies have measured the indirect rebound effect for the domestic sector. Sorrell (2018) summarized these studies and emphasized the remarkably wide range of indirect effects. Evidence of backfire has been recognized by a few studies (Alfredsson, 2004; Freire-González, 2017). Combining direct and indirect rebound effects, total effects vary substantially across countries in Europe due to differences in commodity prices and taxation regimes. Within a country, the rebound effects are higher for low-income groups than mid-high income groups.

For the transport sector, Maxwell et al. (2011) found 30%–80% direct rebound effects for fuel efficiency in commercial road transport. Long-term direct rebound effects are higher than short-term effects, indicating that the impact of green policies decays with time. Similar to the domestic sector, indirect effects are difficult to measure. There are a few published European studies (Alfredsson, 2004; Bjelle et al., 2018; Freire-González, 2017). Sorrell (2018) compared the results of these studies and found that the indirect rebound effects for the transport sector are larger than for the domestic sector because road fuels tend to be subject to higher levels of taxation. The combined effect of direct and indirect rebound effects for the transport sector is higher than for the domestic sector.

Besides the direct and indirect rebound effects, secondary effects should be taken into consideration. Because secondary effects cannot be measured individually, macroeconomic models are widely used. Based on Lecca et al. (2014) and Figus et al. (2017), estimated rebound effects exceed 50%. No comparison can be made between secondary effects and direct/indirect effects due to different definitions and metrics.

Thus, overall, on balance green policies seem effective although rebound effects can be quite substantial. Rebound effects should not be ignored when designing a green policy. Direct, indirect, and secondary effects tend to reinforce each other.

3.5 Discussion and recommendations

3.5.1 Discussion

A vast number of studies have provided evidence that without political interventions, energy efficiency and renewable energy targets will not be realized. This study summarized the green policies which have been implemented in the EU countries and discussed the variation among countries and policy instrument types. Moreover, the study investigated the effectiveness of these green policies on household energy consumption based on evidence, which has not been reviewed systematically. However, many studies indicate that to date green policies have largely failed in their attempt to deliver economically efficient outcomes regarding energy usage, CO_2 emission reduction, and energy-efficient measure adaptation. When taking into account the various rebound effects, policies are even less effective. To be more specific, we will discuss the success and failure of green energy policies category by category.

The green policies that are targeting the reduction of fossil fuel usage mainly involve tax policy instruments. Households seem price elastic in most countries although some inelasticity has been found, not only for short-term energy usage but also for long-term equipment purchase decisions. However, price elasticities vary by type of energy. For electricity demand, households can be motivated by tax reduction. In contrast, for heating, tax reduction did not establish fruitful results. For long-term investment decisions, this policy can motivate household owners to invest in energy-efficient heating and hot water supply systems. However, for tenants who have little control over home improvement, the tax reduction did not play any role. The energy consumption for heating and hot water mainly depends on equipment and house isolation. Households with low income have a higher chance to live in low-quality dwellings. Besides, rebound effects should be considered. There is clear evidence that there is a remarkable wide range rebound effect for household energy efficiency for space heating/cooling for European countries. The efficiency improvements associated with cost reduction for future technology development will result in increased consumption by consumers. The evolution of household income has a significant influence on residential energy demand for heating since energy accounts for a greater amount of expenditure in lower income households. The tax-based policy may cause more social issues than intended. Therefore green policies aimed at reducing heating demand would be more effective if using nonprice-based policy instruments, such as regulation, and education and outreach. Instead of a tax, subsidies are more useful for encouraging households with low income to renovate houses with high-efficiency heating equipment.

Fuel taxes are intended to alter the demand for car fuels by influencing car usage decisions and indirectly car type purchasing decisions. Based on the literature, we found that taxes have triggered the trade-off between the initial purchases price of a car and future fuel costs. The discriminatory taxes play an important role in shifting fuel type usage due to the special targets of these policies, which are the high mileage consumers. Although some researchers mentioned that fuel taxes perform much better in terms of total welfare, it affects households' budgets allocation. The rebound effects for personal transport should not be neglected, which will have significant effects especially for long-term policy effectiveness. The improvement of energy efficiency transport modes and road infrastructure results in more personal trips by car. The rebound effect may increase in the future as demand increases and income increases.

In order to reduce the rebound effects from energy efficiency improvements, income arise, and better infrastructure, households, must limit consumption of the relevant energy services. They have to find a better way to spend the cost savings and establish energy efficiency improvements across multiple areas of consumption, which needs good information. This also requires strong motivation and a very good understanding of the environmental impact of different actions. Therefore education and outreach as a policy instrument should play an important role for long-term effectiveness.

The emission levels of CO_2 are strongly linked with energy usage. The green policies targeting this category mainly are tax and education and outreach. The impact of tax/fee can be explored by the concept of WTP for CO_2. In some EU countries, such as Italy, the existing residential CO_2 reduction in terms of cost per ton is higher than the WTP. Moreover, the real monetary contribution by individuals in a well-educated and high-income country, such as Germany, is much lower than the indicated contribution in hypothetical scenarios.

Moreover, tax is a significantly less acceptable policy instrument than other instruments, even when they have the same characteristics (Brännlund and Persson, 2012; Kallbekken and Aasen, 2010; Kallbekken et al., 2011). In general, households are more likely to express support for or accept policies that have a lower chance of affecting or modifying their own behavior. Meanwhile, households' attitudes toward taxes strongly depend on their beliefs about the environmental consequences of the taxes.

Tax and incentives have been widely used for promoting PHEVs and EVs. There seems little doubt that EV policies have successfully contributed to EV penetration rates in many European countries, although markets share are still very small in most countries and there is substantial inequality in the benefits. Based on the EU's Renewable Energy Directive, a high penetration of EVs presents an opportunity with vehicle-to-grid technology. However, there are some debates regarding the EV policies, especially the policies related to the privilege to use bus lanes, free parking, and exemption from toll road charges. First, the reason that the fossilized transport sector is still dominant is due to the price of current fossil fuel energy. There is no cheaper integrated solution for personal transport than the one based on gasoline/diesel-powered cars. It is unlikely to be a technological lock-in. Second, to achieve lower CO_2 emissions, both the source of the electricity and a full life-cycle assessment including the battery need to be taken into consideration. It is important to keep in mind that two-thirds of the world's electricity supply feeding the EV batteries are currently generated from fossil sources. To achieve the worldwide CO_2 problem in a long term, policymakers need to be more cautious and aware of their national electricity production situation. Third, the privilege policies for EVs have resulted in more families holding two cars together with a switch from the use of public transport and bicycles to the use of private cars. They are compelled to own a conventional car due to the battery technology and the deficiency of the charging network for long-distance travel. Last, the current beneficial group of EV policies is mainly the small group of high-income families who have the chance to buy a second car. From a social welfare perspective, the EV policies need to be carefully examined.

The green policies applied for promoting efficient home appliances are mainly incentives/subsidies, and education and outreach. Empirical evidence suggests that nonpricing instruments are most effective. Both efficiency labeling and informational campaigns on conservation appear to be effective public policy tools to promote the adoption of energy-efficient and water-efficient appliances. Energy saving habits may increase the probability of investing in energy-efficient appliances. However, it is important to remember that consumers show bounded rationality and there is a rebound effect. They prefer to use shortcuts instead of optimizing their expenditure. Therefore to achieve a long-term effect, from both supply and demand perspectives, green policies need to play a role. From the supply perspective, right and sufficient information needs to be offered in order to enhance households' energy-related knowledge and the ability to make energy-efficient investments considering their increased comfort requirements associated with income rising. From the demand perspective, green policies should continue to educate consumers about their energy consumption and how to identify the most efficient appliances. Meanwhile, respondents have limitations to perform energy-efficient behavior consistently for all different energy consumption areas in household. Well-designed decision support websites or tools should be promoted by governments to support their decisions.

Energy efficiency measures have potential appeal to all households. They can improve indoor comfort, reduce energy bills, and reduce CO_2 emission. Multiple green policies have been implemented to influence house purchasing behavior and promote energy-efficient renovation behavior. However, evidence suggests that these policies play only a minor role in the purchasing decision due to house ownership, trust issues, and imperfect information. Moreover, the housing market is demarked by scarcity and heterogeneity, which implies that households rarely choose between similar objects. As a result, energy efficiency as a criterion of minor importance cannot play the same role for buildings as for electronic appliances even if energy efficiency is perfectly visible to the purchaser. Renovation decisions as complicated decisions usually include a series of stages. Reducing energy bills only motivates a small proportion of households (Judson and Maller, 2014; Haines and Mitchell, 2014). Energy efficiency is often incorporated into other types of home improvements, such as esthetics and comfort.

Incentives and subsidies are widely applied to stimulate renewable energy installation in many countries. Households are provided with financial benefits for renewable energy sources which include so-called microgeneration technologies, such as solar photovoltaic, microwind, thermal, and heat pumps. The green energy policies, to a certain extent, reduce households' uncertainty about future financial benefits. However, subsidies and incentives are quite costly for governments and cannot last forever. It is better to provide education or tailed information considering homeowners' motivations in the long run.

3.5.2 Recommendation for future studies

With the reducing greenhouse gas target of the EU, green policies will remain for the coming years. Various efforts were focusing on energy efficiency, promoting new transport mode, and promoting renewable energy technologies. The current relatively low energy price situation (oil, gas, coal), combined with uncertainty about technology development, and future prices, makes it challenging to achieve the target. Green policies were focused on overcoming obstacles that hinder the realization, notably in the finical way. Evidence-based results were discussed in this chapter. However due to data limitations, and the fact that various measurement models have been applied, we can hardly draw any conclusions regarding the accuracy of measurements, since there is a general lack of accurate time series data, for household expenditures before and after policy interventions. Virtually all studies have assessed the effects of a single policy. Little is known about the combined effects of different policies. Maybe some synergy can be established, but it may also be that budget constraints force households to prioritize their investment decisions. In addition, still little is known about the dynamics of adaptation and diffusion processes. Moreover, most researches have not taken into account the likely rebound effects or only addressed direct rebound effects. In future studies, it might be relevant to collect time series data for household expenditures before and after serious policy interventions. Models need to be developed that dynamically specify utility functions as a function of the innovation process. Considerations that make consumers decide not to adopt a new technology immediately after its introduction, and may disappear after some time, implying that the importance of attributes shifts over time.

In addition, there are always inherent limitations in the measurement and modeling approaches. It is worthwhile to investigate the goal of such research whether it is to develop more accurate calculations or make the best possible decisions with the available information. More in-depth insight can be gained by moving beyond the economic models and take psychological and social factors in account as these tend to be the main drivers of people's actions in this context as suggested in Löschel et al. (2013), Ozaki and Sevastyanova (2011), Pettifor et al. (2015). It is useful to investigate these factors to gain more and better insights in the reasons behind specific impacts and identify possible ways to maximize the policy effects.

References

Ajanovic, A., Dahl, C., Schipper, L., 2012. Modelling transport (energy) demand and policies—an introduction. Energy Policy 41, 3–16.

Aklilu, A.Z., 2016. Sveriges Lantbruksuniversitet, Institutionen För Ekonomi Gasoline and Diesel Demand Elasticities: A Consistent Estimate across the EU-28. <https://pub.epsilon.slu.se/13860/1/zeleke_a_161205.pdf>.

Alberini, A., Bigano, A., Ščasný, M., Zvěřinová, I., 2018. Preferences for energy efficiency vs. renewables: what is the willingness to pay to reduce CO_2 emissions? Ecol. Econ. 144, 171–185.

Alfredsson, E.C., 2004. 'Green' consumption—no solution for climate change. Energy 29, 513–524.

Athavale, S., Knaus, K., 2017. Energy Efficiency Policies and the Rebound Effect Review of International Instruments and Recommendations for Austria Literature Review. <http://www.energyagency.at> (accessed 29.04.19.).

Baranzini, A., Goldemberg, J., Speck, S., 2000. A future for carbon taxes. Ecol. Econ. 32, 395–412.

Barker, T., Kohler, J., 1998. Equity and ecotax reform in the EU: achieving a 10 per cent reduction in CO_2 emissions using excise duties. Fiscal Stud. 19 (4), 375–402.

Bartiaux, F., Gosselain, V., Vassileva, D., Stamova, G., Ozolinz, L., 2011. Knowledge on climate change and energy saving renovations by apartment owners in Bulgaria and Latvia. A qualitative study. Int. Rev. Soc. Res. 1 (3), 11–31.

Berkhout, P.H.G., Ferrer-i-Carbonell, A., Muskens, J.C., 2004. The ex post impact of an energy tax on household energy demand. Energ. Econ. 26 (3), 297–317.

Bernstein, R., Madlener, R., 2011. Responsiveness of residential electricity demand in OECD countries: a panel cointegration and causality analysis. In: FCN Working Paper 8/2011.

Bjelle, E.L., Steen-Olsen, K., Wood, R., 2018. Climate change mitigation potential of Norwegian households and the rebound effect. J. Cleaner Prod. 172, 208–217.

Blasch, J., Filippini, M., Kumar, N., 2017. Boundedly rational consumers, energy and investment literacy, and the display of information on household appliances. Resour. Energy Econ. 56, 39–58.

Brännlund, R., Persson, L., 2012. To tax, or not to tax: preferences for climate policy attributes. Clim. Policy 12 (6), 704–721.

Broberg, T., Kažukauskas, A., 2014. Inefficiencies in residential use of energy—a critical overview of literature and energy efficiency policies in EU and Sweden. In: CERE Working Paper, 2014:7. Available at SSRN: <https://ssrn.com/abstract = 2573047; https://doi.org/10.2139/ssrn.2573047>.

Christensen, T.H., Gram-Hanssen, K., de Best-Waldhober, M., Adjei, A., 2014. Energy retrofits of Danish homes: is the energy performance certificate useful? Build. Res. Inf. 42 (4), 489–500.

Climate Council, 2015. Paris COP21: Key Issues for the New Climate Agreement. Tech. Rep. <https://www.climatecouncil.org.au/2015/12/14/paris-cop21-key-issues-for-the-newclimate-agreement/>.

Cornwell, A., Creedy, J., 1996. Carbon taxation, prices and inequality in Australia. Fiscal Stud. 17 (3), 21–38.

Dahl, C.A., 2012. Measuring global gasoline and diesel price and income elasticities. Energy Policy 41, 2–13.

Dieu-Hang, T., Quentin Grafton, R., Martínez-Espiñeira, R., Garcia-Valiñas, M., 2017. Household adoption of energy and water-efficient appliances: an analysis of attitudes, labelling and complementary green behaviours in selected OECD countries. J. Environ. Manage. 197, 140–150.

Ecofys. 2010. EnergySavings 2020—how to triple the Impact of energy saving policies in Europe. In: A Contributing Study to Roadmap 2050: A Practical Guide to a Prosperous, Low-Carbon Europe. <http://www.roadmap2050.eu/attachments/files/EnergySavings2020-FullReport.pdf>.

Espey, M., 1998. Gasoline demand revisited: an international meta-analysis of elasticities. Energy Econ. 20 (3), 273–295.

European Commission, 2013. Employment Effects of Selected Scenarios from the Energy Roadmap 2050 Final Report for the European Commission (DG Energy). <www.camecon.com>.

European Commission, 2017. EU Research & Innovation for and With Cities.

Figus, G., Turner, K., McGregor, P., Katris, A., 2017. Making the case for supporting broad energy efficiency programmes: impacts on household incomes and other economic benefits. Energy Policy 111, 157–165.

Freire-González, J., 2017. Evidence of direct and indirect rebound effect in households in EU-27 countries. Energy Policy 102, 270–276.

Galvin, R., 2012. German federal policy on thermal renovation of existing homes: a policy evaluation. Sustain. Cities Soc. 4 (1), 58–66.

Gardner, G.T., Stern, P.C., 2002. Environmental Problems and Human Behavior. Pearson, Boston.

Gillingham, K., Palmer, K., 2014. Bridging the energy efficiency gap: policy insights from economic theory and empirical evidence. Rev. Environ. Econ. Policy 8 (1), 18–38.
Graham, D.J., Glaister, S., 2002. The demand for automobile fuel: a survey of elasticities. J. Transp. Econ. Policy 36, 1–25.
Gram-Hanssen, K., Bartiaux, F., Jensen, O.M., Cantaert, M., 2007. Do homeowners use energy labels? A comparison between Denmark and Belgium. Energy Policy 35 (5), 2879–2888.
Grigolon, L., Reynaert, M., Verboven, F., 2018. Consumer valuation of fuel costs and the effectiveness of tax policy—evidence from the European car market. Am. Econ. J.: Econ. Policy 10 (3), 193–225.
Grimm, N.B., Faeth, S.H., Golubiewski, N.E., Redman, C.L., Wu, J., Bai, X., et al., 2008. Global change and the ecology of cities. Science 319 (5864), 756–760.
Haines, V., Mitchell, V., 2014. A persona-based approach to domestic energy retrofit. Build. Res. Inf. 42 (4), 462–476.
Halvorsen, B., Frøyen, Y., 2009. Trafikk i kollektivfelt. Kapasitet og avvikling. Elbilens rolle. (Traffic in Public Lanes. Capacity and Handling. The Role of Electric Cars). Report from Asplan Viak.
Halvorsen, B., Larsen, B.M., 2001. The flexibility of household electricity demand over time. Resour. Energy Econ. 23, 1–18.
Hanemann, M., Labandeira, X., Labeaga, J.M., 2013. Energy demand for heating: short run and long run. In: Working Papers 07-2013, Economics for Energy.
Holtsmark, B., Skonhoft, A., 2014. The Norwegian support and subsidy policy of electric cars. Should it be adopted by other countries? Environ. Sci. Policy 42, 160–168.
ILO, EU, IILS, 2013. Green policies in the EU: a review. In: EC-IIS Joint Discussion Paper Series No. 14. pp. 1–33.
Judson, E.P., Maller, C., 2014. Housing renovations and energy efficiency: insights from homeowners' practices. Build. Res. Inf. 42 (4), 501–511.
Kallbekken, S., Aasen, M., 2010. The demand for earmarking: results from a focus group study. Ecol. Econ. 69 (11), 2183–2190.
Kallbekken, S., Kroll, S., Cherry, T.L., 2011. Do you not like Pigou, or do you not understand him? Tax aversion and revenue recycling in the lab. J. Environ. Econ. Manage. 62 (1), 53–64.
Keirstead, J., 2007. Behavioural responses to photovoltaic systems in the UK domestic sector. Energy Policy 35 (8), 4128–4141.
Kotilainen, K., Saari, U.A., 2018. Policy influence on consumers' evolution into prosumers empirical findings from an exploratory survey in Europe. Sustainability 10 (1), 1–22.
Labandeira, X., Labeaga, J.M., 1999. Combining input–output analysis and microsimulation to assess the effects of carbon taxation on Spanish households. Fiscal Stud. 30 (3), 305–320.
Labandeira, X., Labeaga, J.M., Rodríguez, M., 2006. A residential energy demand system for Spain. Energy J. 27, 87–112.
Labandeira, X., Labeaga, J.M., López, X., 2012. Estimation of elasticity price of electricity with incomplete information. Energy Econ. 34, 627–633.
Laes, E., Mayeres, I., Renders, N., Valkering, P., Verbeke, S., 2018. How do policies help to increase the uptake of carbon reduction measures in the EU residential sector? Evidence from recent studies. Renew. Sustain. Energy Rev. 94, 234–250.
Langbroek, J.H.M., Franklin, J.P., Susilo, Y.O., 2017. Electric vehicle users and their travel patterns in Greater Stockholm. Transp. Res. D: Transp. Environ. 52, 98–111.

Lecca, P., McGregor, P.G., Swales, J.K., Turner, K., 2014. The added value from a general equilibrium analysis of increased efficiency in household energy use. Ecol. Econ. 100, 51–62.

Löschel, A., Sturm, B., Vogt, C., 2013. The demand for climate protection-empirical evidence from Germany. Econ. Lett. 118 (3), 415–418.

Maxwell, D., Owen, P., McAndrew, L., Muehmel, K., Neubauer, A., April 2011. Addressing the Rebound Effect, a Report for the European Commission DG Environment.

Meier, H., Rehdanz, K., 2010. Determinants of residential space heating expenditures in Great Britain. Energy Econ. 32 (5), 949–959.

Mills, B., Schleich, J., 2010. What's driving energy efficient appliance label awareness and purchase propensity? Energy Policy 38 (2), 814–825.

Nesbakken, R., 2001. Energy consumption for space heating: a discrete-continuous approach. Scand. J. Econ. 103, 164–184.

Nikodinoska, D., Schröder, C., 2016. On the emissions-inequality and emissions-welfare trade-offs in energy taxation: evidence on the German car fuels tax. Resour. Energy Econ. 44, 206–233.

Ozaki, R., Sevastyanova, K., 2011. Going hybrid: an analysis of consumer purchase motivations. Energy Policy 39 (5), 2217–2227.

Pettifor, H., Wilson, C., Chryssochoidis, G., 2015. The appeal of the green deal: empirical evidence for the influence of energy efficiency policy on renovating homeowners. Energy Policy 79, 161–176.

Prud'homme, R., Koning, M., 2012. Electric vehicles: a tentative economic and environmental evaluation. Transp. Policy 23, 60–69.

Rehdanz, K., 2007. Determinants of residential space heating expenditures in Germany. Energy Econ. 29 (2), 167–182.

Sardianou, E., Genoudi, P., 2013. Which factors affect the willingness of consumers to adopt renewable energies? Renew. Energy 57, 1–4.

Smith, Z., 2000. The petrol tax debate. In: Briefing Note No. 10/2000. The Institute for Fiscal Studies, London.

Sierzchula, W., Bakker, S., Maat, K., Van Wee, B., 2014. The influence of financial incentives and other socio-economic factors on electric vehicle adoption. Energy Policy 68, 183–194.

Solomon, S., Plattner, G., Knutti, R., Friedlingstein, P., 2009. Irreversible climate change due to carbon dioxide emissions. Proc. Natl. Acad. Sci. U.S.A. 106 (6), 1704–1709.

Sorrell, S., Dimitropoulos, J., 2007. UKERC Review of Evidence for the Rebound Effect: Technical Report 5: Energy, Productivity and Economic Growth Studies. UK Energy Research Centre, London.

Sorrell, S., 2018. Energy Sufficiency and Rebound Effects Concept Paper. Retrieved from <https://www.energysufficiency.org/static/media/uploads/site-8/library/papers/sufficiency-rebound-final_formatted_181118.pdf>.

Symons, E., Speck, S., Proops, J., 2000. The effects of pollution and energy taxes across the European income distribution. In: Keele University Working Paper No. 5/2000. Keele, UK.

Tiezzi, S., 2005. The welfare effects and the distributive impact of carbon taxation on Italian households. Energy Policy 33 (12), 1597–1612.

Vaage, K., 2000. Heating technology and energy use: a discrete/continuous choice approach to Norwegian household energy demand. Energy Econ. 22 (6), 649–666.

Vasseur, V., Kemp, R., 2015. The adoption of PV in the Netherlands: a statistical analysis of adoption factors. Renew. Sustain. Energy Rev. 41, 483–494.

Appendix 1

Country	Target subsection	Target	Type of instrument
Austria	H	New Housing Subsidization Scheme	Incentives/Subsidies
Austria	H	Energy tax	Tax
Austria	H	Green Electricity Act	Incentives/Subsidies
Austria	H	Promotion of energy-efficient electronic products, the group for energy-efficient appliances	Education and outreach
Austria	T	Charging for parking cars in limited parking zones in Vienna	Fee/Charge
Austria	T	Toll for alpine roads	Fee/Charge
Austria	T	Vignette for the use of highways	Fee/Charge
Austria	T	Mineral oil tax	Tax
Austria	T	Motor vehicle tax	Tax
Austria	T	Tax on motor vehicle insurance	Tax
Austria	T	Promotion of EVs	Incentives/Subsidies
Austria	T	Vehicle registration tax	Tax
Bulgaria	H	Feed-in tariff for hydro, biomass, or wind power	Incentives/Subsidies
Bulgaria	T	Fuel road charge	Fee/Charge
Bulgaria	T	Fuel excise tax	Tax
Bulgaria	T	Passenger cars excise tax—motor vehicles tax/excise duty	Tax
Bulgaria	T	Road tax	Tax
Bulgaria	T	Promotion of EVs	Tax
Bulgaria	T	Vehicle, vessels, and aircraft tax—motor/transport vehicle tax	Tax
Belgium	H	Subsidies for renewable energy investment—Wallonia	Incentives/Subsidies
Belgium	H	Subsidies for passive house construction and low-energy renovation	Incentives/Subsidies
Belgium	H	Renewable energy support scheme—Wallonia	Incentives/Subsidies
Belgium	H	Tax deductions for investments in energy efficiency and renewable energy—Federal	Incentives/Subsidies

Country	Target subsection	Target	Type of instrument
Belgium	H	Brussels—subsidy schemes for energy efficiency measures	Incentives/Subsidies
Belgium	H	Fiscal deduction to promote energy efficiency	Incentives/Subsidies
Belgium	H	CO_2/RUE policy plan—Flanders	Education and outreach
Belgium	H	Energy consumption labeling of new appliances	Education and outreach
Belgium	T	Promotion of modal shift in transport	Incentives/Subsidies
Belgium	T	Tax deductions on travel to and from home	Incentives/Subsidies
Belgium	T	Supporting alternative mobility—the Bruxell'Air bonus	Incentives/Subsidies
Belgium	T	Choosing an eco-friendly vehicle: the eco-score	Education and outreach
Belgium	H	Fuel excise tax	Tax
Belgium	H	Energy saving and promotion of alternative energy sources	Incentives/Subsidies
Belgium	T	Allowances for environment-friendly vehicles	Incentives/Subsidies
Belgium	T	Promotion of EVs	Incentives/Subsidies
Belgium	T	Excise duty—motor vehicles	Fee/Charge
Czech Republic	H	Green investment scheme	Incentives/Subsidies
Czech Republic	H	Building retrofit subsidies: PANEL programme	Incentives/Subsidies
Czech Republic	H	Electricity tax	Tax
Czech Republic	H	Natural gas tax	Tax
Czech Republic	H	Act on the promotion of the use of renewable energy sources (Act No. 180/2005 Coll.)	Regulatory instruments
Czech Republic	H	Act on the promotion of the use of renewable energy sources (Act No. 180/2005 Coll.)	Incentives/Subsidies
Czech Republic	H	Secondary legislation on the methodology for the purchase of electricity from renewable and CHP	Regulatory instruments
Czech Republic	H	Secondary legislation on the methodology for the purchase of electricity from renewable and CHP	Incentives/Subsidies

Country	Target subsection	Target	Type of instrument
Czech Republic	T	Electronic road-toll fee	Fee/Charge
Czech Republic	T	Highway fee	Fee/Charge
Czech Republic	T	Fuel excise duty	Tax
Czech Republic	T	Road tax	Tax
Czech Republic	T	Promotion of EVs	Tax
Czech Republic	T	Measures to support low-emission cars	Tax
Denmark	H	Promotion of energy saving in buildings	Education and outreach
Denmark	H	Duty on electricity	Tax
Denmark	H/T	Duty on CO_2	Tax
Denmark	T	Environmental duty for passenger cars and vans	Fee/Charge
Denmark	T	Fee on petrol	Fee/Charge
Denmark	T	Scraping charge on passenger cars and vans	Fee/Charge
Denmark	T	Duty on motor vehicle compulsory insurance	Tax
Denmark	T	Duty on petrol	Tax
Denmark	T	Motor vehicle registration duty	Tax
Denmark	T	Motor vehicle weight tax	Tax
Denmark	T	Passenger car fuel consumption tax	Tax
Denmark	T	Promotion of EVs	Incentives/Subsidies
Denmark	T	Energy labeling of new cars	Education and outreach
Estonia	H	Fuel excise tax	Tax
Estonia	T	Motor vehicle excise tax	Tax
Estonia	H/T	Air pollution charge	Tax
Finland	H	Energy grants for residential buildings	Incentives/Subsidies
Finland	T	Motor vehicle tax (Diesel tax)	Tax
Finland	T	Vehicle tax (Sticker tax)	Tax

Country	Target subsection	Target	Type of instrument
Finland	T	Car tax	Tax
Finland	T	Charge on exceeding of GHG emission limits	Tax
Finland	T	Strategic stockpile fee	Tax
Finland	T	Vehicle tax	Tax
Finland	T	Amendment of car tax and annual vehicle tax regimes	Tax
Finland	T	Energy labeling of passenger cars	Education and outreach
Finland	H/T	Excise on fuels and electricity	Tax
France	H	Tax rebate on some environment-friendly house equipment	Incentives/Subsidies
France	H	Renewable energy feed-in tariff: solar PV	Incentives/Subsidies
France	H	Tax on natural gas	Tax
France	T	Bonus for the acquisition of clean vehicles	Incentives/Subsidies
France	T	Promotion of electric and natural gas-powered vehicles	Tax
France	T	Annual tax on motor vehicles with large CO_2 emissions	Tax
France	T	CO_2-related bonus malus system for vehicle registration	Tax
France	T	Promotion of EVs	Incentives/Subsidies
France	T	Mineral oils tax	Tax
France	H/T	Ecological consumption website—ADEME	Education and outreach
Germany	H	KfW housing modernization program	Incentives/Subsidies
Germany	H	KfW build ecologically program	Incentives/Subsidies
Germany	H	Tax on electricity bills	Fee/Charge
Germany	T	Duty on mineral oils	Tax
Germany	T	Promotion of EVs	Incentives/Subsidies
Germany	T	Motor vehicle tax	Tax
Greece	H	Development incentives for renewable energy sources	Incentives/Subsidies
Greece	H	Law 224494	Incentives/Subsidies
Greece	T	Taxon motor vehicle purchases	Tax

Country	Target subsection	Target	Type of instrument
Greece	T	Mineral oil tax	Tax
Greece	T	Promotion of EVs	Tax
Greece	T	Taxon motor vehicle usage	Tax
Hungary	H	Feed-in tariff (Electricity Act)	Incentives/Subsidies
Hungary	H	Efficiency labeling for household appliances	Education and outreach
Hungary	T	Promotion of EVs	Tax
Hungary	T	Vehicle tax	Tax
Ireland	H	New house grants	Incentives/Subsidies
Ireland	H	House of Tomorrow programme	Incentives/Subsidies
Ireland	T	Use of motor taxation to encourage more efficient vehicles	Tax
Ireland	T	Mineral oil tax	Tax
Ireland	T	Motor vehicle tax	Tax
Ireland	T	Promotion of EVs	Tax
Ireland	T	Vehicle registration tax	Tax
Italy	H	Subsidies to promote energy efficiency	Incentives/Subsidies
Italy	H	Feed-in tariff for solar thermodynamic energy	Incentives/Subsidies
Italy	H	Additional regional taxon natural gas consumption	Tax
Italy	H	Additional regional electricity—towns/provinces	Tax
Italy	H	Taxon electrical energy—State	Tax
Italy	H	Tax allowance for efficient fridges	Tax
Italy	H	Demand side management to reduce GHG emissions—ENEL voluntary agreement	Education and outreach
Italy	T	Vehicle certification	Regulatory instruments
Italy	T	Personal income tax allowance for public transport	Incentives/Subsidies
Italy	T	Subsidy program to increase bicycle use	Incentives/Subsidies
Italy	T	Car sharing	Incentives/Subsidies

Country	Target subsection	Target	Type of instrument
Italy	T	Tax for registration of vehicles	Fee/Charge
Italy	T	Excise duty on energy products	Tax
Italy	T	Tax allowance for electric motors	Tax
Italy	T	Promotion of EVs	Tax
Italy	H/T	Excise duty on energy products	Tax
Italy	T	Excise duty on oil/petrol	Fee/Charge
Latvia	H	Fuel excise tax	Tax
Latvia	T	Annual motor vehicle charge	Fee/Charge
Latvia	T	Vehicles excise tax	Tax
Latvia	T	Promotion of EVs	Fee/Charge
Latvia	H/T	Air emissions charge	Fee/Charge
Lithuania	H	EEHP	Incentives/Subsidies
Lithuania	H	Fuel excise tax	Tax
Lithuania	T	Air pollution charge for mobile sources/fuels	Fee/Charge
Lithuania	T	Motor vehicle import duty	Tax
Luxembourg	H	Feed-in tariffs for renewable energy and cogeneration: regalement grand-ducal	Incentives/Subsidies
Luxembourg	H	Feed-in tariffs for renewable energy and cogeneration: law of Oct. 14, 2005	Incentives/Subsidies
Luxembourg	H	Energy performance of residential buildings	Education and outreach
Luxembourg	H	Think climate—financial aid program for energy savings and renewable energy in housing	Incentives/Subsidies
Luxembourg	T	Annual vehicle tax	Tax
Luxembourg	T	Promotion of EVs	Incentives/Subsidies
Luxembourg	T	Promotion of EVs	Tax
Luxembourg	T	Mineral oil tax	Tax

Country	Target subsection	Target	Type of instrument
Malta	H	Electricity charges	Fee/Charge
Malta	T	Registration tax	Tax
Malta	T	Annual motor vehicle license/Promotion of EVs	Fee/Charge
Netherlands	H	Energy tax	Tax
Netherlands	H	Energy tax regime	Incentives/Subsidies
Netherlands	H	MEP: environmental quality of electricity production (Milieukwaliteit van de elektriciteitsproductie)	Incentives/Subsidies
Netherlands	H	Implementation of EU energy performance of buildings directive, energy performance certificate, and energy labeling	Education and outreach
Netherlands	H	SDE +	Incentives/Subsidies
Netherlands	H	Dutch Standard for Smart Meter Requirements (NTA 8130)	Education and outreach
Netherlands	T	EcoDriving (Het nieuwe rijden)	Education and outreach
Netherlands	T	Energy labels on passenger cars	Education and outreach
Netherlands	T	Labeling of vehicle efficiency	Education and outreach
Netherlands	T	Kilometer pricing system for road usage	Tax
Netherlands	T	Tax to promote CHP	Tax
Netherlands	T	Tax for commuters	Tax
Netherlands	T	Tax on passenger cars and motor cycles	Tax
Netherlands	T	Duty on petrol	Tax
Netherlands	T	Promotion of EVs	Incentives/Subsidies
Poland	H	Rationalization of heat consumption in household sector	Incentives/Subsidies
Poland	T	Promotion of EVs	Tax
Poland	H/T	Charge on air pollution	Fee/Charge
Poland	T	Excise tax on energy products	Tax
Portugal	H	Taxation on less efficient light bulbs	Tax
Portugal	H	Regulations on thermal ventilation of buildings	Regulatory instruments

Country	Target subsection	Target	Type of instrument
Portugal	H	Energy efficiency requirements for appliances	Education and outreach
Portugal	T	Regulations for electric mobility in Portugal	Regulatory instruments
Portugal	T	Promotion of EVs	Tax
Portugal	T	Tax on petroleum and energy products	Tax
Portugal	T	Excise tax on motor vehicles	Tax
Romania	T	Fuel road charge	Fee/Charge
Romania	T	Annual vehicle tax	Tax
Romania	T	Vehicle excise tax	Tax
Romania	T	Vehicle import duty	Tax
Romania	T	Promotion of EVs	Incentives/Subsidies
Romania	T	Promotion of EVs	Tax
Romania	H/T	Air emissions charge	Fee/Charge
Romania	T	Fuel excise tax	Tax
Slovak Republic	H	Excise duty on electricity	Tax
Slovak Republic	H	Excise duty on natural gas	Tax
Slovak Republic	H	Eco-labeling scheme	Education and outreach
Slovak Republic	H	Act on energy and amendments	Regulatory instruments
Slovak Republic	T	Road tax	Tax
Slovak Republic	T	Promotion of EVs	Tax
Slovak Republic	T	Tax on permits to enter historical city district with motor vehicle	Tax
Slovenia	H	Energy efficiency tax	Tax
Slovenia	T	Fuel excise tax	Tax
Slovenia	T	Promotion of EVs	Tax
Slovenia	T	Promotion of EVs	Incentives/Subsidies
Slovenia	T	Vehicles excise tax	Tax
Spain	H	Grants for energy efficiency in buildings	Incentives/Subsidies

Country	Target subsection	Target	Type of instrument
Spain	H	Feed-in tariffs for electricity from renewable energy sources	Incentives/Subsidies
Spain	H	Feed-in tariffs for small scale cogeneration/renewable electricity production	Incentives/Subsidies
Spain	H	Tax on electricity	Tax
Spain	H	Plan for the progressive replacement of electricity meters (smart meters)	Regulatory instruments
Spain	H	Building energy certification	Regulatory instruments
Spain	H	Housing labels	Education and outreach
Spain	H	Building energy certification	Education and outreach
Spain	T	Canary islands—Tax on petroleum fuels	Tax
Spain	T	Tax on mineral oils	Tax
Spain	T	Tax on motor vehicle	Tax
Spain	T	Promotion of EVs	Incentives/Subsidies
Spain	T	Car registration tax linked to CO_2 emissions	Tax
Spain	H/T	Andalusia—tax on air pollution	Tax
Sweden	H	Energy tax on electricity	Tax
Sweden	H	Building energy performance certificates	Education and outreach
Sweden	Ho	Building performance standards (Building regulation)	Regulatory instruments
Sweden	H	Energy labeling of domestic appliances and windows	Education and outreach
Sweden	T	Eco car subsidy	Incentives/Subsidies
Sweden	T	Promotion of EVs	Tax
Sweden	T	Energy and CO_2 tax on fuels except petrol	Tax
Sweden	T	Energy and CO_2 tax on petrol	Tax
Sweden	T	Motor vehicle tax	Tax
Sweden	T	Fuel consumption and CO_2 labels for new cars	Education and outreach

Country	Target subsection	Target	Type of instrument
Sweden	T	Mandatory eco-driving for driver's license	Education and outreach
United Kingdom	H	Feed-in tariffs for renewable electricity	Incentives/Subsidies
United Kingdom	H	Pay-as-you-save pilots	incentives/Subsidies
United Kingdom	H	Wales home energy efficiency scheme	Incentives/Subsidies
United Kingdom	H	Reduced VAT for energy saving materials	Tax
United Kingdom	H	Carbon emissions reduction target (Energy efficiency ointment 3)	Regulatory instruments
United Kingdom	H	Low carbon transition plan	Incentives/Subsidies
United Kingdom	H	Decent homes	Regulatory instruments
United Kingdom	H	Energy labeling for new buildings	Education and outreach
United Kingdom	H	Act on CO_2 advice line	Education and outreach
United Kingdom	H	Code for sustainable homes	Education and outreach
United Kingdom	T	The "power shift" program	Incentives/Subsidies
United Kingdom	T	Vehicle excise duty	Tax
United Kingdom	T	Vehicle excise duty: fuel type and CO_2 emission vehicle bands	Tax
United Kingdom	T	Vehicle excise duty	Tax
United Kingdom	T	Promotion of EVs	Incentives/Subsidies
United Kingdom	T	Promotion of EVs	Tax

EEHP, Energy efficiency/housing project.

CHAPTER 4

Energy consumption in the transport and domestic sectors: a household-level comparison between capital cities of Japan, China, and Indonesia

Weiyan Zong[1], Junyi Zhang[1,*], Biying Yu[2], Enjian Yao[3] and Chunfu Shao[3]
[1]Hiroshima University, Higashi Hiroshima, Japan;
[2]Beijing Institute of Technology, Beijing, P.R. China;
[3]Beijing Jiaotong University, Beijing, P.R. China

4.1 Introduction

For promoting a low-carbon life style globally, Asia is playing a more and more important role, as Asia houses more than half of the world population and its economic growth rate is the highest in the world. Even though current urbanization rate of Asian countries is just about 40%, Asia holds 17 megacities (population: 10 million or more) out of 31 in the world (as of 2016). By 2030, there will be 41 megacities around the world, 23 of which will be in Asia (UN, 2016). In Asia the development of megacities has influenced other cities in various aspects significantly, in the sense that other cities tend to mimic the development patterns of megacities. Accordingly, it is important to understand more about megacities in Asia.

In reality, household energy consumption behaviors vary across households and cities. Scientific investigations on such variations are necessary; however, existing efforts are very limited. To date, household energy consumption behaviors have been examined in many existing studies, most of which, however, mainly focused on only a certain specific energy type, such as electricity consumption (Yun and Steemers, 2011: path analysis; Kavousian et al., 2013: factor analysis and regression model), or gas

* Corresponding author.

Transport and Energy Research
DOI: https://doi.org/10.1016/B978-0-12-815965-1.00004-1

consumption (Liu et al., 2018: panel data regression model). There are several exceptions. Abrahamse and Steg (2009) (based on an expert system), Yu et al. (2011a) [multiple discrete-continuous extreme value (MDCEV) model], and Zheng et al. (2014) (aggregate analyses) analyzed the relationships between various types of household energy consumption. Yu et al. (2011b) (Heckman's latent index model) and Hori et al. (2013) (regression model) compared energy consumption across countries at different development stages. Zhang (2004) compared household annual energy consumption in China with that in Japan, Canada, and the United States, based on aggregate analyses and regression models with national statistics in the late 1990s. Yu et al. (2011b) explored the effects of car ownership and self-selection on the household energy consumption in Tokyo, Beijing, Jakarta, and Dhaka. However, as examined by Zhang (2017), various life choices are interdependent, meaning that choices in other life domains would affect household energy consumption behavior. A more comprehensive understanding might be obtained by taking other life choices into consideration instead of only studying energy consumption behavior itself. Such a life-oriented research has been rarely conducted, especially in a comparative way.

This study compares household energy consumption behavior of different energy types (in-home: mainly electricity and gas; out-of-home: mainly gasoline) in three Asian capital cities: Tokyo, Japan; Beijing, China; Jakarta, Indonesia. Japan and China are two main Asian countries contributing to global energy consumption and greenhouse gas emissions, and Indonesia is a typical developing country with a moderate development growth rate. As climatic conditions affect energy consumption, especially, electricity consumption (Nejat et al., 2015), such a comparison between cities in different climatic conditions is of significance and has its global implications. For this study, in 2009, the authors conducted a questionnaire survey in the three capital cities mentioned earlier, with sample sizes of 1194 households in Tokyo, 1024 in Beijing, and 1009 in Jakarta, respectively. In this survey, ownership of various end uses, transport-related attributes, life cycle stages, residential characteristics, commuting modes, and household attributes are investigated together with energy consumption. Because there are one or more types of energy consumption and some households do not consume a certain type of energy, a MDCEV model is built to jointly represent household decisions on different types of energy consumption at different seasons of a year. Such modeling allows for the existence of zero consumption and seasonal comparisons as well as seasonal interdependencies.

In the remainder of the section, Section 4.2 introduces the data used in this study. Section 4.3 describes the MDCEV model. Section 4.4 explains and compares the model estimation results. Finally, the study is concluded in Section 4.5.

4.2 Analysis of household energy consumption in Tokyo, Beijing, and Jakarta

At the time of survey, according to Kutani (2013), (1) the population at city level was 13.0 million in Tokyo, 17.6 in Beijing, and 9.2 in Jakarta; (2) the population density was the highest in Tokyo: 20,901 persons/km^2, followed by Jakarta (13,943) and Beijing (1,440); (3) per capita GRP was 53,202 US$ in Tokyo, 16,738 in Beijing, and 13,909 in Jakarta; (4) road lengths per passenger vehicles were 0.08 km in Tokyo, 0.07 km in Beijing, and 0.03 km in Jakarta; and (5) per 1000 population passenger vehicle stocks were about 250 in Tokyo, 147 in Beijing, and 222 in Jakarta. Even though Beijing had the lowest car ownership level, its growth has been sharp since 2001, and its growth rate was the highest, in comparison with other two cities. As for in-home end uses,[1] regarding the trends of ownership of various durable goods between 2000 and 2009, compared with Tokyo, Beijing was one of the booming markets, which had medium penetration rates of durable goods but had been growing rapidly, while Jakarta showed explosive potentials, even though its penetration rate was still low.

The questionnaire contents include household attributes, household monthly energy consumption at different seasons, ownership, and use of various end uses (vehicles and in-home appliances). All types of energy consumption data were collected in spring and autumn (hereafter, spring/autumn), summer and winter in Tokyo and Beijing, and dry and wet seasons in Jakarta. As explained in the next section, outside goods must be nonzero when applying the MDCEV model. Because there were very few samples whose electricity consumption in spring/autumn or dry season was zero, this study treats electricity consumption in spring/autumn or dry season as the outside goods. Because of this, we deleted samples with zero consumption for outside goods. After filtering out these zero

[1] <http://www.hakuhodo.jp/seikatsusha/research/pdf/2010.pdf> (accessed 23.04.19.).

observations and outliers, the final samples in Tokyo, Beijing, and Jakarta are 1123, 920, and 941 households, respectively.

Dependent variables in the MDCEV models are household monthly consumption of electricity and gas in the domestic sector, and gasoline in the passenger transport sector, at different seasons in the year 2009. All of the energy units were converted to "kW h" for the sake of comparison (Table 4.1). Electricity is mainly used for in-home electric end uses, and gas for cooking, hot water supply and heating, and gasoline for vehicles. Among the three cities, only Jakarta is in a tropical monsoon climate with wet and dry seasons, and its temperature is stably high almost all the year. As a result, energy consumption in Jakarta did not vary significantly between dry and wet seasons. Households in Beijing had a much lower level of electricity consumption than that in other cities. For this, we compared all explanatory variables used in modeling analyses (see Table 4.2: the selection of explanatory variables will be described at the subsection 4.3.2) and found that 20% and 22% of the households in Beijing lived in a public house and a dormitory, respectively, which were much higher than those (2%–7%) in Tokyo and Jakarta. Furthermore, the housing ownership rate in Beijing (32%) was much lower than in those in Tokyo (54%) and Jakarta (81%). Such differences in housing attributes may explain Beijing's lower electricity consumption, to some extent. In Tokyo and Beijing, which have four distinct seasons, the electricity consumption in summer and winter are higher than that in other seasons. This is probably because households need to make more use of energy-consuming equipment for cooling in summer and heating in winter. In Beijing the electricity consumption in summer is the highest among four seasons and much higher than in other seasons, while the gap between spring/autumn and winter is small. Compared with Beijing, there were more apparent seasonal variations in gas consumption in Tokyo with a descending order of winter, spring/autumn, and summer. Gasoline consumption during winter in Tokyo was much higher than that in other seasons. Climate conditions may affect energy consumption. From such a viewpoint, we found that in Tokyo, the overall energy consumption was negatively correlated with temperature (correlation: −0.475) and precipitation (−0.545). In contrast the overall energy consumption in Beijing (Jakarta) showed a positive (negative) but very weak correlation (its absolute values are less than 0.05). However, the correlations between electricity/gas consumption with temperature and precipitation in Beijing were higher: 0.275 and 0.368, respectively.

Table 4.1 Descriptive statistics of household energy consumption, temperature, and precipitation.

City			Tokyo (2009)			Beijing (2009)			Jakarta (2009)		
Consumption	Temperature	Precipitation	Energy consumption kW h/month (use share)	Temperature (°C)	Precipitation (mm)	Energy consumption kW h/month (use share)	Temperature (°C)	Precipitation (mm/month)	Energy consumption kW h/month (use share)	Temperature (°C)	Precipitation (mm/month)
Spring/autumn (dry season: Jakarta)		Electricity	350.44 (100.0%)	10.0−20.1/ 13.5−23.0 (average: 16.9)	98.5−242.0/ 53.0−276.5 (average: 164.0)	144.78 (100.0%)	7.0−22.9/ 6.0−26.0 (average: 14.1)	9.8−37.3/ 9.8−48.8 (average: 25.6)	433.25 (100.0%)	28.0−28.9 (average: 28.5)	67−195 (average: 130)
		Gas	345.94 (95.4%)			241.04 (96.3%)			182.77 (97.4%)		
		Gasoline	527.84 (47.4%)			915.83 (22.9%)			250.09 (73.1%)		
Summer (wet season: Jakarta)		Electricity	447.83 (100.0%)	22.5−26.6 (average: 25.1)	78.5−242.0 (average: 182.2)	200.49 (100.0%)	25.7−27.0 (average: 26.3)	72.1−160.5 (average: 26.3)	430.72 (99.9%)	27.7−29.9 (average: 28.7)	21−572 (average: 198)
		Gas	265.87 (95.5%)			275.76 (96.1%)			181.66 (97.4%)		
		Gasoline	545.00 (47.2%)			952.77 (23.0%)			249.33 (73.0%)		
Winter		Electricity	464.33 (100.0%)	68−9.0 (average: 7.9)	465−142.0 (average: 90.3)	155.24 (100.0%)	− 3.0−1.0 (average: −1.4)	2.3−4.4 (average: 3.1)	None	None	None
		Gas	526.60 (95.5%)			250.68 (96.1%)					
		Gasoline	667.01 (47.5%)			949.20 (22.6%)					
Data source for temperature/precipitation			Tokyo: http://www.data.jma.go.jp/obd/stats/etrn/index.php			Beijing and Jakarta: http://www.data.jma.go.jp/gmd/cpd/monitor/dailyview/index.php (Jakarta: 2010)					

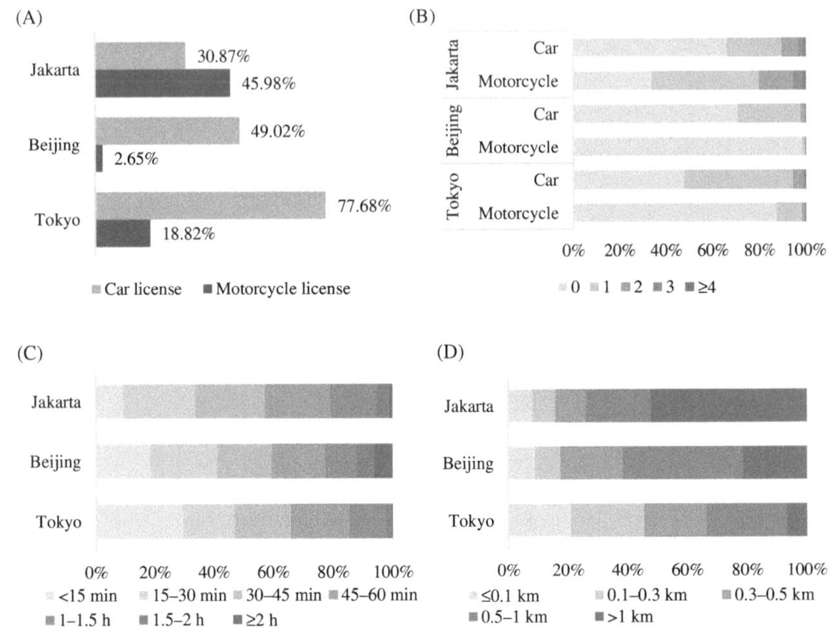

Figure 4.1 Distributions of transport-related attributes in Tokyo, Beijing, and Jakarta: (A) license ownership of motorcycle and car, (B) numbers of cars and motorcycles, (C) commuting time, and (D) distance from residence to nearest transit station.

As for transport-related attributes, this study selected four person-based variables, including motorcycle driving license, car driving license, number of motorcycles, and number of cars, and two location-based indicators, that is, commuting time from home to workplace, and distance from home to the nearest transit station. In this study, as for commuting time, there are totally seven categories: (1) ≤15 minutes; (2) 15–30 minutes; (3) 30–45 minutes; (4) 45–60 minutes; (5) 60–90 minutes; (6) 90–120 minutes; and (7) >120 minutes. Such a categorical scale was adopted considering the ease of respondents' answer and was common for all three cities, for the sake of comparison. Fig. 4.1 shows the distributions of households with different transport-related attributes in three cities. Looking at Fig. 4.1A and B, car was the dominating transport modes in Tokyo and Beijing, while in Jakarta the dominating mode was motorcycle. As shown in Fig. 4.1C, households in Beijing and Jakarta experienced longer commuting time than in Tokyo, while households in Tokyo resided much closer to transit stations than in other cities (see Fig. 4.1D).

4.3 Household energy consumption behavior modeling

4.3.1 Multiple discrete-continuous extreme value model

To depict household energy consumption behavior in both transport and domestic sectors across seasons, seasonal consumption choices of electricity, gas, and gasoline over a year are modeled simultaneously using the MDCEV model with outside goods (Bhat, 2005, 2008). Assume that e_{ni} is the consumption quantity of energy type i ($i = 1, 2, 3, \ldots, I$) by household n. As described earlier, the electricity in spring/autumn or dry season is selected as the outside goods and labeled as the first type of consumption ($i = 1$) in the MDCEV model. Conditioning on the existence of the outside goods, the function of utility (u_n) obtained from all types of energy consumption (including zero consumption) by household n can be formulated as

$$u_n = \frac{1}{\alpha_1}\exp(\varepsilon_1)e_{n1}^{\alpha_1} + \sum_{i=2}^{I}\frac{\gamma_i}{\alpha_i}\exp(\beta x_{ni} + \varepsilon_i)\left\{\left(\frac{e_{ni}}{\gamma_i}+1\right)^{\alpha_i} - 1\right\} \quad (4.1)$$

subject to

$$\sum_{i=1}^{I} e_{ni} = E_n \quad (4.2)$$

where E_n is the energy consumption budget and ε_i ($i = 1, 2, 3, \ldots$) is the error term with a Gumbel distribution.

Based on the previous function, household n maximizes its utility of consuming all types of energy subject to the energy consumption budget. There are two categories of unknown parameters: parameters (β) for explanatory variables (x_{ni}) and satiation parameters of α_i and γ_i. The satiation effect increases with α_i dropping from 1 or with the decreasing of γ_i, meaning that household n is more likely to allocate more budget to energy type i. However, it is challenging to simultaneously estimate α_i and γ_i for a nonoutside goods i ($i = 2, 3, \ldots, I$). Instead, one can just estimate one satiation parameter. In this case, three possible forms of utility functions are provided. If $\gamma_i = 1$, then Eq. (4.1) can be simplified to α profile (Eq. 4.3), where separate α_i values are estimated for all types of energy consumption. If $\alpha_i \rightarrow 0$ ($i = 2, 3, \ldots, I$), α_1 and separate γ_i values for all types of energy consumptions except the outside goods can be estimated (see Eq. 4.4). If α values for all types of energy consumption are set to be equal, the α value and separate γ_i values for all types of energy consumptions except the outside goods can be estimated (see Eq. 4.5).

$$\alpha \text{ profile:} \quad u_n = \frac{1}{\alpha_1}\exp(\varepsilon_1)e_{n1}^{\alpha_1} + \sum_{i=2}^{I}\frac{1}{\alpha_i}\exp(\beta x_{ni} + \varepsilon_i)\left\{(e_{ni}+1)^{\alpha_i} - 1\right\} \tag{4.3}$$

$$\gamma \text{ profile:} \quad u_n = \frac{1}{\alpha_1}\exp(\varepsilon_1)e_{n1}^{\alpha_1} + \sum_{i=2}^{I}\gamma_i\exp(\beta x_{ni} + \varepsilon_i)\left\{\frac{e_{ni}}{\gamma_i} - 1\right\} \tag{4.4}$$

$$\text{Constant } \alpha: \quad u_n = \frac{1}{\alpha}\exp(\varepsilon_1)e_{n1}^{\alpha} + \sum_{i=2}^{I}\frac{\gamma_i}{\alpha}\exp(\beta x_{ni} + \varepsilon_i)\left\{\left(\frac{e_{ni}}{\gamma_i}+1\right)^{\alpha} - 1\right\} \tag{4.5}$$

With the previous information the probability that household n allocates energy consumption to all T_n types at different seasons (T_n is equal to nine in Tokyo and Beijing and six in Jakarta) can be constructed, as shown by Bhat (2005, 2008):

$$P\left(e_{n1}, e_{n2}, e_{n3}, \ldots, e_{nT_n}, 0, 0, \ldots, 0\right) = \left[\prod_{i=1}^{T_n} f_{ni}\right]\left[\sum_{i=1}^{T_n}\frac{1}{f_{ni}}\right]\left[\frac{\prod_{i=1}^{T_n} e^{V_{ni}}}{\left(\sum_{i'=1}^{I} e^{V_{ni'}}\right)^{T_n}}\right](T_n - 1)! \tag{4.6}$$

where $f_{ni} = \left((1 - \alpha_i)/\left(e_{ni} + \gamma_i\right)\right)$.

4.3.2 Selection of explanatory variables

We first introduced six transport-related attributes [motorcycle driving license (1-yes, 0-no), car driving license (1-yes, 0-no), number of motorcycles, number of cars, commuting time (minutes), and distance to a nearest transit station (km)]. And then, we further selected six categories of explanatory variables. First, households at different life cycle stages (Alwin, 2012; Delbosc and Nakanishi, 2017) usually possess different life demands, including energy demand, and thus behave differently. This study specifies five variables to represent life cycle stages: (1) age-group—young household, (2) age-group—elderly household, (3) living only with spouse, (4) living with child(ren), and (5) living with parent(s). Other categories of explanatory variables are in-home end-use attributes, residential attributes, commuting modes, and household attributes. All explanatory variables are summarized in Table 4.2.

Table 4.2 Explanatory variables in household energy consumption models.

Category	Variable	Abbr.	Average value		
			Tokyo	Beijing	Jakarta
Transport-related attributes	Motorcycle driving license (1—yes and 0—no)	MDL	0.19	N/A	0.46
	Car driving license (1—yes and 0—no)	CDL	0.78	0.49	0.31
	Number of motorcycles	NOM	0.15	N/A	0.93
	Number of cars	NOC	0.59	0.31	0.48
	Commuting time (minutes)	CT	35.16	45.20	44.12
	Distance to nearest transit station (km)	DNT	0.51	0.99	1.95
End-use attributes	Number of refrigerators	NOF	1.12	0.99	0.96
	Number of air conditioners	NOA	2.31	1.54	0.87
	Number of electric stoves	NOE	0.50	0.08	0.06
	Number of electric showers	NES	N/A	0.43	N/A
	Number of gas stoves	NOG	N/A	0.07	0.02
	Using gas cooking utensil (1—yes and 0—no)	UGC	N/A	0.86	0.97
Life cycle stage indicators	Age-group—young (1—yes and 0—no)	Y	0.18	0.27	0.39
	Age-group—elderly (1—yes and 0—no)	E	0.37	0.15	0.10
	Living only with spouse (1—yes and 0—no)	LWS	0.24	0.17	0.10
	Living with child(ren) (1—yes and 0—no)	LWC	0.50	0.75	0.57
	Living with parent (1—yes and 0—no)	LWP	0.06	0.17	0.15
Residential attributes	Living in own house (1—yes and 0—no)	LO	0.54	0.32	0.81
	Living in public house (1—yes and 0—no)	LP	0.07	0.20	0.02
	Living in rental house (1—yes and 0—no)	LR	0.35	0.23	0.14
	Living in dormitory (1—yes and 0—no)	LD	0.03	0.22	0.02
	Duration of residence: 0–5 years (1—yes and 0—no)	D5	0.23	0.36	0.43
	Duration of residence: 5–10 years (1—yes and 0—no)	D510	0.18	0.36	0.18
	House area: ≤ 60 m^2 (1—yes and 0—no)	H60	0.49	0.46	0.34
	House area: >100 m^2 (1—yes and 0—no)	H100	0.17	0.15	0.37
Commuting modes	Private transport (1—yes and 0—no)	PRT	0.08	0.26	0.55
	Public transport (1—yes and 0—no)	PUT	0.49	0.32	0.21
Household attributes	Full-time job (1—yes and 0—no)	FT	0.62	0.73	0.75
	Low income (1—yes and 0—no)	LI	0.37	0.31	0.32
	High income (1—yes and 0—no)	HI	0.26	0.34	0.23

N/A, Not available.

4.4 Household-level energy consumption comparisons between three Asian capitals

All three models are estimated by using a maximum likelihood method, programing codes of which are provided by Bhat (2008) in GAUSS software. To select the best modeling profile among Eqs. (4.3)–(4.5), we estimated all the three profiles and then compared their statistical goodness-of-fit. Based on estimation results, the MDCEV model with constant α (see Eq. 4.5) outperforms other profiles and is finally adopted. Estimation results are shown in Tables 4.3–4.6. Adjusted Rho-squared values are 0.2599 in the Tokyo model (Table 4.4), 0.3695 in Beijing (Table 4.5), and 0.1266 in Jakarta (Table 4.6), suggesting that the MDCEV models are good enough regarding to goodness-of-fit.

4.4.1 Satiation effects

All satiation parameters (α and γ_i) are statistically significant at 5% level, indicating that satiation effects surely exist, but the effects vary across energy types and seasons (see Table 4.3). One common observation is that satiation effects of each type of energy do not vary much across seasons; however, the effects vary largely across energy types. In all cities, gasoline consumption possesses the highest gamma values, meaning that a household is most likely to allocate a large proportion of their budgets to gasoline consumption once it decides to use gasoline. The very large satiation effects for gasoline consumption in Tokyo and Beijing coincide with large values of actual gasoline consumption, and the largest satiation effects in Beijing are consistent with the largest gasoline consumption. Thus satiation effects are strongly associated with actual energy consumption. The earlier results support the use of MDCEV model for this study.

4.4.2 Effects of explanatory variables

As shown in Tables 4.4–4.6, all constant terms were estimated to be negative and statistically significant, except one constant. This means that there are omitted factors from the analysis, which show a nonignorable role to reduce energy consumption and need to be further investigated in future.

4.4.2.1 Tokyo

In total, 109 parameters were estimated, out of which total 45 are statistically significant at the 10%, 5%, or 1% level. Among 48 transport-related

Table 4.3 Estimation results of multiple discrete-continuous extreme value models (Part 1).

		Tokyo	Beijing	Jakarta
Satiation parameters				
Alpha				
Electricity	(Spring/autumn, or dry season)	− 2.402**	− 2.130**	− 0.807**
	(Summer, or wet season)			
	(Winter)			N/A
Gas	(Spring/autumn, or dry season)			− 0.807**
	(Summer, or wet season)			
	(Winter)			N/A
Gasoline	(Spring/autumn, or dry season)			− 0.807**
	(Summer, or wet season)			
	(Winter)			N/A
Gamma				
Electricity	(Spring/autumn, or dry season)	0.000	0.000	0.000
	(Summer, or wet season)	3.944**	1.785**	1.942**
	(Winter)	4.298**	1.529**	N/A
Gas	(Spring/autumn, or dry season)	16.677**	6.778**	0.540**
	(Summer, or wet season)	11.997**	8.836**	0.539**
	(Winter)	27.171**	7.129**	N/A
Gasoline	(Spring/autumn, or dry season)	65.201**	137.354**	18.895**
	(Summer, or wet season)	68.695**	142.780**	19.162**
	(Winter)	84.684**	143.776**	N/A
Sample size		1123	920	941
Number of parameters		109	101	94
Initial log-likelihood		− 41212.08	− 29967.07	− 20517.97
Converged log-likelihood		− 30392.31	− 18794.59	− 17868.37
McFadden's Rho-squared		0.2625	0.3728	0.1291
Adjusted Rho-squared		0.2599	0.3695	0.1266

Note: N/A, Not available; **significant at 5% level.

parameters, 14 are significant, most of which are related to gasoline consumption. All significant transport-related parameters are positive, indicating that more car driving licenses, more cars, more motorcycles, and longer commuting time lead to more gasoline and gas consumption. Unfortunately, none of distance-related parameters is statistically significant. End-use attributes are introduced only to electricity, which are all significant: the more the refrigerators, air conditioners, and electric stoves, the more the electricity consumption.

Table 4.4 Estimation results of multiple discrete-continuous extreme value model (Part 2): Tokyo [Unit: × 10 kW h].

Tokyo	Electricity (summer)		Electricity (winter)		Gas (spring)		Gas (summer)		Gas (winter)		Gasoline (spring)		Gasoline (summer)		Gasoline (winter)	
	Parameter	vr (%)	Parameter	vr (%)	Parameter	vr (%)	Parameter	vr (%)	Parameter	vr (%)	Parameter	vr (%)	Parameter	vr (%)	Parameter	vr (%)
Constant terms																
Con	− 4.104***	0.00	− 4.104***	0.00	− 9.035***	0.00	− 9.035***	0.00	− 9.035***	0.00	− 15.000***	0.00	− 15.000***	0.00	− 15.000***	0.00
End-use ownership and usage attributes																
NOF	0.268***	11.56	0.193***	6.79	–	–	–	–	–	–	–	–	–	–	–	–
NOA	0.076***	10.57	–	–	–	–	–	–	–	–	–	–	–	–	–	–
NOE	–	–	0.119***	7.09	–	–	–	–	–	–	–	–	–	–	–	–
Total		22.13		13.88		0.00		0.00		0.00		0.00		0.00		0.00
Transport-related attributes																
MDL	0.058	0.38	0.034	0.14	0.046	0.06	0.057	0.09	0.024	0.02	0.057	0.01	0.072	0.02	0.145	0.08
CDL	0.031	0.12	− 0.073	0.76	0.233*	1.62	0.220*	1.42	0.182	1.01	0.887***	3.30	0.921***	3.56	0.891***	3.30
NOM	0.096	1.20	0.069	0.70	− 0.180	1.02	− 0.144	0.64	− 0.163	0.85	0.228*	0.23	0.237*	0.25	0.197	0.17
NOC	0.002	0.00	0.037	0.45	0.105	0.79	0.047	0.16	0.034	0.08	1.900***	36.66	1.891***	36.34	1.920***	37.10
CT	0.030	1.44	− 0.006	0.07	0.046	0.80	0.101***	3.76	− 0.000	0.00	0.081*	0.35	0.078*	0.32	0.082*	0.36
DNT	− 0.050	0.40	0.002	0.00	− 0.012	0.01	− 0.048	0.09	0.018	0.01	0.046	0.01	− 0.038	0.01	− 0.104	0.06
Total		3.55		2.12		4.29		6.14		1.98		40.56		40.49		41.06
Life cycle stage indicators																
Y	0.109	1.25	0.109	1.41	− 0.002	0.00	− 0.002	0.00	− 0.002	0.00	− 0.134	0.06	− 0.134	0.06	− 0.134	0.06
E	− 0.036	0.22	− 0.036	1.24	− 0.227**	2.09	− 0.227**	2.05	− 0.227**	2.14	− 0.380***	0.82	− 0.380***	0.83	− 0.380***	0.82
LWS	− 0.274**	9.74	− 0.274**	11.01	− 0.178	0.99	− 0.178	0.97	− 0.178	1.01	0.111	0.05	0.111	0.05	0.111	0.05
LWC	− 0.256**	11.82	− 0.256**	13.36	− 0.067	0.19	− 0.067	0.19	− 0.067	0.20	− 0.490***	1.47	− 0.490***	1.47	− 0.490***	1.45
LWP	− 0.235	2.23	− 0.235	2.52	− 0.007	0.00	− 0.007	0.00	− 0.007	0.00	− 0.665***	0.60	− 0.665***	0.61	− 0.665***	0.60
Total		25.26		28.55		3.27		3.21		3.35		3.01		3.02		2.99
Residential attributes																
LO	0.243	10.61	0.243	11.99	0.634*	17.30	0.634*	16.96	0.634*	17.72	1.622***	15.94	1.622***	15.96	1.622***	15.81
LP	0.173	1.36	0.173	1.54	0.730**	5.82	0.730**	5.71	0.730**	5.96	1.914***	5.64	1.914***	5.64	1.914***	5.64
LR	0.450	32.99	0.450	37.29	1.144***	51.29	1.144***	50.29	1.144***	52.53	2.264***	28.26	2.264***	28.29	2.264***	28.02

LD	0.018	0.01	0.018	0.01	0.715*	2.96	0.715*	2.91	0.715*	3.04	2.347***	4.50	2.347***	4.51	2.347***	4.46
D5	0.056	0.41	0.056	0.46	0.252**	1.96	0.252**	1.92	0.252**	2.01	0.074	0.02	0.074	0.02	0.074	0.02
D510	0.113	1.38	0.113	1.55	0.248**	1.59	0.248**	1.56	0.248**	1.63	0.134	0.07	0.134	0.07	0.134	0.07
H60	− 0.008	0.01	− 0.008	0.01	− 0.004	0.00	− 0.004	0.00	− 0.004	0.00	− 0.052	0.02	− 0.052	0.02	− 0.052	0.02
H100	− 0.122	1.54	− 0.122	1.75	− 0.208*	1.08	− 0.208*	1.06	− 0.208*	1.10	− 0.350***	0.43	− 0.350***	0.43	− 0.350***	0.43
Total		48.31		54.60		82.00		80.41		83.99		54.88		54.94		54.47
Commuting modes																
PRT	0.052	0.14	0.052	0.16	− 0.266*	0.90	− 0.266*	0.88	− 0.266*	0.92	0.578***	0.60	0.578***	0.60	0.578***	0.60
PUT	0.038	0.27	0.038	0.30	0.075	0.24	0.075	0.24	0.075	0.25	− 0.184*	0.21	− 0.184*	0.21	− 0.184*	0.21
Total		0.41		0.46		1.14		1.12		1.17		0.81		0.81		0.81
Household attributes																
FT	0.024	0.10	0.024	0.11	− 0.447***	8.17	− 0.447***	8.01	− 0.447***	8.37	− 0.236**	0.32	− 0.236**	0.32	− 0.236**	0.32
LI	0.032	0.17	0.032	0.20	0.151	0.93	0.151	0.91	0.151	0.95	0.016	0.00	0.016	0.00	0.016	0.00
HI	0.021	0.06	0.021	0.07	− 0.078	0.20	− 0.078	0.20	− 0.078	0.20	− 0.301***	0.42	− 0.301***	0.42	− 0.301***	0.42
Total		0.33		0.38		9.30		9.12		9.52		0.74		0.74		0.74

Note: *Significant at 10% level; **significant at 5% level; and ***significant at 1% level; *vr*, variance ratio.

Table 4.5 Estimation results of multiple discrete-continuous extreme value model (Part 3): Beijing [Unit: × 10 kW h].

Beijing	Electricity (summer)		Electricity (winter)		Gas (spring)		Gas (summer)		Gas (winter)		Gasoline (spring)		Gasoline (summer)		Gasoline (winter)	
	Parameter	vr (%)	Parameter	vr (%)	Parameter	vr (%)	Parameter	vr (%)	Parameter	vr (%)	Parameter	vr (%)	Parameter	vr (%)	Parameter	vr (%)
Baseline constant																
Con	− 1.205***	0.00	− 1.205***	0.00	− 5.242***	0.00	− 5.242***	0.00	− 5.242***	0.00	− 13.761***	0.00	− 13.761***	0.00	− 13.761***	0.00
End-use ownership and usage attributes																
NOF	0.479***	13.89	0.256**	4.27	–	–	–	–	–	–	–	–	–	–	–	–
NOA	0.097**	4.98	–	–	–	–	–	–	–	–	–	–	–	–	–	–
NOE	–	–	0.496***	7.71	–	–	–	–	–	–	–	–	–	–	–	–
UES	0.405***	18.64	0.298***	10.85	–	–	–	–	–	–	–	–	–	–	–	–
NOG	–	–	–	–	− 0.124	0.10	− 0.190	0.25	0.078	0.04						
UGC	–	–	–	–	0.871***	9.17	0.518***	3.45	0.857***	8.97	–	–	–	–	–	–
Total		37.51		22.83		9.27		3.69		9.01		0.00		0.00		0.00
Transport-related attributes																
CDL	− 0.022	0.05	− 0.095	0.98	− 0.576***	8.10	− 0.577***	8.64	− 0.601***	8.92	1.014***	7.40	1.073***	8.27	1.053***	8.01
NOC	0.060	0.38	0.206	4.87	0.372**	3.57	0.319*	2.79	0.309*	2.49	2.074***	32.75	2.074***	32.66	2.056***	32.29
CT	− 0.011	0.15	0.031	1.25	0.115***	3.87	0.123***	4.70	0.109***	3.51	0.184***	2.92	0.163***	2.29	0.173***	2.59
DNT	− 0.060	1.18	0.116**	4.73	− 0.041	0.13	− 0.073	0.45	− 0.051	0.21	− 0.003	0.00	0.019	0.01	0.009	0.00
Total		1.76		11.83		15.68		16.58		15.13		43.07		43.22		42.89
Life cycle stage indicators																
Y	− 0.050	0.20	− 0.050	0.21	− 0.006	0.00	− 0.006	0.00	− 0.006	0.00	0.277	0.44	0.277	0.44	0.277	0.44
E	− 0.259*	3.51	− 0.259*	3.78	− 0.439***	2.45	− 0.439***	2.60	− 0.439***	2.47	− 0.154	0.09	− 0.154	0.09	− 0.154	0.09
LWS	0.414**	9.94	0.414**	10.69	1.119***	17.59	1.119***	18.68	1.119***	17.78	1.710***	12.11	1.710***	12.07	1.710***	12.14
LWC	0.361**	9.92	0.361**	10.67	0.881***	14.31	0.881***	15.20	0.881***	14.46	1.426***	11.05	1.426***	11.02	1.426***	11.09
LWP	0.046	0.12	0.046	0.13	0.466***	2.97	0.466***	3.16	0.466***	3.01	− 0.466***	0.88	− 0.466***	0.87	− 0.466***	0.88
Total		23.69		25.49		37.32		39.64		37.72		24.56		24.49		24.64
Residential attributes																
LO	− 0.130	1.48	− 0.130	1.59	0.311	2.04	0.311	2.17	0.311	2.07	0.622	2.41	0.622	2.40	0.622	2.42
LP	− 0.017	0.02	− 0.017	0.02	0.336	1.74	0.336	1.85	0.336	1.76	0.636	1.84	0.636	1.84	0.636	1.85

LR	0.308	6.81	0.308	7.33	0.744***	9.63	0.744***	10.23	0.744***	9.73	1.353***	9.39	1.353***	9.36	1.353***	9.42
LD	− 0.019	0.02	− 0.019	0.03	0.096	0.15	0.096	0.16	0.096	0.15	0.811**	3.21	0.811**	3.20	0.811**	3.22
D5	− 0.359**	11.98	− 0.359**	12.89	− 0.593***	7.92	− 0.593***	8.41	− 0.593***	8.00	− 0.194	0.25	− 0.194	0.25	− 0.194	0.25
D510	− 0.276**	7.10	− 0.276**	7.64	− 0.398***	3.58	− 0.398***	3.80	− 0.398***	3.61	− 0.397	1.05	− 0.397	1.05	− 0.397	1.05
H60	0.241**	5.81	0.241**	6.26	0.437***	4.63	0.437***	4.92	0.437***	4.68	0.193	0.27	0.193	0.27	0.193	0.27
H100	0.102	0.54	0.102	0.58	− 0.420***	2.20	− 0.420***	2.34	− 0.420***	2.22	− 0.106	0.04	− 0.106	0.04	− 0.106	0.04
Total		33.77		36.32		31.89		33.88		32.24		18.46		18.41		18.52
Commuting modes																
PRT	− 0.020	0.03	− 0.020	0.03	− 0.157	0.46	− 0.157	0.49	− 0.157	0.47	1.049***	6.06	1.049***	6.05	1.049***	6.08
PUT	0.065	0.37	0.065	0.40	− 0.211*	0.94	− 0.211*	1.00	− 0.211*	0.95	− 1.111***	7.72	− 1.111***	7.70	− 1.111***	7.74
Total		0.40		0.43		1.41		1.49		1.42		13.78		13.75		13.83
Household attributes																
FT	− 0.130	1.34	− 0.130	1.44	0.220*	0.93	0.220*	0.98	0.220*	0.94	0.125	0.09	0.125	0.09	0.125	0.09
LI	0.134	1.54	0.134	1.65	0.411***	3.50	0.411***	3.72	0.411***	3.54	0.076	0.04	0.076	0.04	0.076	0.04
HI	0.007	0.00	0.007	0.00	− 0.006	0.00	− 0.006	0.00	− 0.006	0.00	− 0.026	0.00	− 0.026	0.00	− 0.026	0.00
Total		2.88		3.10		4.43		4.71		4.48		0.13		0.13		0.13

Note: *Significant at 10% level; **significant at 5% level; and ***significant at 1% level; *vr*, variance ratio.

Table 4.6 Estimation results of multiple discrete-continuous extreme value model (Part 4): Jakarta [Unit: × 10 kW h].

Jakarta	Electricity (wet season)		Gas (dry season)		Gas (wet season)		Gasoline (dry season)		Gasoline (wet season)	
	Parameter	vr (%)	Parameter	vr (%)	Parameter	vr (%)	Parameter	vr (%)	Parameter	vr (%)
Baseline constant										
Con	− 0.906	0.00	− 7.343***	0.00	− 7.343***	0.00	− 6.634***	0.00	− 6.634***	0.00
End-use ownership and usage attributes										
NOF	− 0.138	14.96	–	–	–	–	–	–	–	–
NOA	− 0.038	6.31	–	–	–	–	–	–	–	–
NOE	− 0.051	0.53	–	–	–	–	–	–	–	–
NOG	–	–	0.786***	0.81	0.782***	0.80	–	–	–	–
UGC	–	–	7.706***	76.59	7.713***	76.63	–	–	–	–
Total		22.13		77.39		77.43				
Transport-related attributes										
MDL	− 0.021	0.31	− 0.250*	0.80	− 0.253**	0.82	0.164	0.49	0.145	0.39
CDL	0.051	1.62	0.335**	1.24	0.333**	1.22	0.486***	3.70	0.469***	3.49
NOM	− 0.058	7.57	− 0.014	0.01	− 0.016	0.01	0.206***	2.39	0.200***	2.28
NOC	0.032	1.89	0.129	0.56	0.126	0.53	0.740***	25.88	0.736***	25.87
CT	0.018	1.85	0.151***	2.28	0.151***	2.29	0.169***	4.08	0.170***	4.14
DNT	− 0.013	1.85	− 0.064**	0.84	− 0.064**	0.84	− 0.108***	3.46	− 0.106***	3.37
Total		15.08		5.73		5.72		40.01		39.53
Life cycle stage indicators										
Y	− 0.068	3.24	− 0.350***	1.51	− 0.350***	1.51	− 0.258**	1.17	− 0.258**	1.17
E	− 0.022	0.13	− 0.101	0.05	− 0.101	0.05	− 0.105	0.07	− 0.105	0.07
LWS	–	–	–	–	–	–	–	–	–	–
LWC	− 0.077	4.24	− 0.188*	0.45	− 0.188*	0.45	− 0.044	0.04	− 0.044	0.04

LWP	− 0.041	0.62	− 0.129	0.11	− 0.129	0.11	− 0.184	0.32	− 0.184	0.32
Total		8.23		2.12		2.11		1.59		1.61
Residential attributes										
LO	0.093	3.88	0.128	0.13	0.128	0.13	0.528	3.16	0.528	3.16
LP	0.153	1.29	0.539	0.28	0.539	0.28	0.494	0.34	0.494	0.34
LR	0.053	1.03	− 0.327	0.69	− 0.327	0.69	0.337	1.03	0.337	1.04
LD	− 0.067	0.25	0.133	0.02	0.133	0.02	0.709	0.70	0.709	0.70
D5	− 0.025	0.44	− 0.219*	0.61	− 0.219*	0.61	− 0.205	0.75	− 0.205	0.76
D510	0.021	0.19	0.526***	2.08	0.526***	2.08	0.425***	1.93	0.425***	1.94
H60	0.009	0.05	− 0.167	0.32	− 0.167	0.32	0.044	0.03	0.044	0.03
H100	− 0.020	0.27	0.066	0.05	0.066	0.05	0.009	0.00	0.009	0.00
Total		7.40		4.18		4.18		7.94		7.97
Commuting modes										
PRT	0.163	19.25	0.478***	2.92	0.478***	2.92	1.352***	33.26	1.352***	33.52
PUT	0.032	0.50	0.373**	1.20	0.373**	1.20	0.102	0.13	0.102	0.13
Total		19.75		4.12		4.12		33.39		33.65
Household attributes										
FT	0.195	21.02	0.401***	1.58	0.401***	1.57	0.602***	5.05	0.602***	5.05
LI	0.074	3.49	− 0.654***	4.82	− 0.654***	4.82	− 0.860***	11.85	− 0.860***	11.95
HI	− 0.078	3.22	0.074	0.05	0.074	0.05	0.115	0.18	0.115	0.18
Total		27.73		6.45		6.44		17.08		17.18

Note: *significant at 10% level; **significant at 5% level; and ***significant at 1% level; *vr*, variance ratio.

As for life cycle stages, only 4 out of 15 parameters are significant. Households consuming less gasoline are those elderly persons, living with children and parents. Elderly persons also consume less gas. Those who live only with spouse and live with children further consume less electricity.

Half of the parameters for residential attributes are significant, and it is stated that living in own house, public house, rental house, and dormitory leads to more gas and gasoline consumption. Residential duration also shows a positive influence, but only on gas consumption, the duration shorter than 10 years is associated with less gas use. Households living in a larger house ($>100\,m^2$) consume less gas and gasoline. Furthermore, commuting by private vehicles (public transport) leads to more (less) gasoline consumption, but private vehicles usage results in less gas consumption. Full-time workers consume less gas and gasoline and high-income households consume less gasoline. Finally, transport-related attributes, residential attributes, commuting modes, and household attributes are not influential to electricity consumption.

4.4.2.2 Beijing

In total, 101 parameters were estimated, out of which 56 parameters are significant at the 10%, 5%, or 1% level. Among 32 transport-related parameters, 19 are significant, most of which are related to gasoline and gas consumption. And except the car driving license parameter, all the other significant transport-related parameters are positive. More cars and longer commuting time are associated with more gasoline and gas consumption. Holding a car driving license results in more gasoline consumption but less gas consumption. End-use attributes are introduced only to electricity and gas, which are all significant and positive, except number of gas stoves, meaning that more refrigerators, air conditioners, and showers lead to more electricity consumption, and using gas cooking utensil results in more gas consumption.

As for life cycle stages, 10 out of 15 parameters are significant. Gas consumption utility includes the largest number (four) of significant life cycle stage indicators. Three life cycle stage indicators are influential to gasoline consumption: living only with spouse and living with children lead to more gas and gasoline consumption. Living with parents is associated with a decline in gasoline consumption, but an increase in gas consumption.

Residential attributes mostly affect gas consumption, because of the highest number of significant parameters. Living in rental house and dormitory results in more gasoline consumption. Living in rental house further leads to more gas consumption. But residential duration is negatively related to gas consumption: the duration longer than 10 years is associated with more gas consumption. Housing size shows a mixed effect on gas consumption: living in a larger room ($> 100\ m^2$) results in less gas consumption, but a smaller room is more likely to encourage more gas usage. Those who are living in a larger room are usually those with higher level of income, who may have more out-of-home social activities. Furthermore, commuting by private vehicles (public transport) leads to more (less) gasoline consumption, which is the same as in Tokyo, but the use of private vehicles does not matter to gas consumption. Different from Tokyo, the use of public transport reduces gas usage. The different influences of public transport on gas consumption are interesting. One potential reason may be that travel time by public transport in Beijing is usually longer than in Tokyo, resulting in less time use at home. Household attributes only affect gas consumption, where full-time workers and low-income households consume more gas, but high-income does not matter to gas usage. Finally, none of commuting modes and household attributes is statistically significant in the case of electricity consumption.

4.4.2.3 *Jakarta*

In total, 83 parameters were estimated, out of which 35 are statistically significant at the 10%, 5%, or 1% level. Among 24 transport-related attributes, 18 are significant, most of which are related to gasoline and gas consumption. For gasoline consumption, motorcycle driving license does not matter to its decision, while other transport-related attributes are all influential: the shorter the distance to the nearest transit station, the more the gasoline consumption; car driving license parameter, car ownership, and commuting time are all positively associated with gasoline consumption. Similarly, car driving license and commuting time are positively related to gas consumption. Motorcycle driving license shows an opposite influence on gas consumption. Effects of distance to the nearest transit station on gas consumption are negative.

There are only three significant life cycle stage indicators in Jakarta: living with children leads to less use of gas; younger households consume less gas and gasoline consumption.

Among residential attributes, there are also only three significant parameters: living in a residence for 5–10 years leads to more gas and gasoline consumption, while the duration less than 5 years results in less gas consumption. Commuting by private vehicles leads to more gasoline and gas consumption, while the use of public transport for commuting also results in more gas usage. Full-time workers consume more gas and gasoline, while low income is negatively related to gas and gasoline consumption. But high income is not influential to any energy consumption. Surprisingly, none of explanatory variables is influential to electricity consumption.

4.4.3 Comparison

4.4.3.1 Transport-related attributes

4.4.3.1.1 Gasoline

In the three capital cities, transport-related attributes explain about 40% (ranging between 39.53% and 43.22%) of the total variance of gasoline consumption, indicating it is the most influential factor to gasoline consumption, among six categories of explanatory variables, except in Tokyo, which gasoline consumption is mostly influenced by residential attributes (about 54% of the total variance). The earlier results are understandable, considering that transport accessibility directly affects people's travel-mode choices, including car usage. Looking at the signs and statistical significance of transport-related parameters, it is found that for reducing gasoline consumption, it is necessary to reduce the numbers of car driving licenses and cars as well as the commuting time in Tokyo, Beijing, and Jakarta. For Jakarta, it is further necessary to reduce the number of motorcycles. Unexpectedly, the parameters of distance from residence to the nearest transit station in Jakarta indicate that a reduced distance will further increase the gasoline consumption. As for the reason(s), the data used in this study do not contain any relevant information; however, residences closer to transit stations are usually expensive in these two cities. Such residences are usually owned by richer residents, who are more likely to own a car. Differently, the distance does not matter to the gasoline consumption in Tokyo and Beijing. For this, it may be due to the fact that the distance is already very short and shows very small variations in Tokyo (average distance: 0.51 km, standard deviation: 0.47 km) and Beijing (average: 0.99 km, standard deviation: 0.81 km), which is much shorter than that in Jakarta (average: 1.95 km, standard deviation: 2.00 km).

4.4.3.1.2 Electricity and gas

It is expected that the higher the transport accessibility level, the more frequent the trips made, leading to less time spent at home. As a result, electricity and gas consumption may be reduced. This expectation becomes true for the gas consumption but not for the electricity consumption. Transport-related attributes mostly affect the gas consumption in Beijing (15.13%–16.58% of the total variance, where owning a car diving license reduces gas consumption, but the more the cars owned and the longer the commuting time, the more the gas consumption), followed by that in Jakarta (5.37%–5.72%, where the distance to the nearest transit station is additionally significant: the shorter the distance, the more the gas consumption) and in Tokyo (1.98%–6.14%). On the contrary, in Tokyo, owning a car driving license results in more gas consumption. Thus only the ownership of car driving license in Beijing and Jakarta shows expected effects on gap consumption, and the effects of other transport-related attributes are unexpected. In contrast, in all three cities, transport-related attributes do not affect electricity consumption at all, except electricity consumption in winter in Beijing.

4.4.3.1.3 Summary

Transport-related attributes are mostly associated with gasoline consumption, as expected, where the number of cars is most influential. Positive signs of this transport-related attribute suggest that to reduce gasoline consumption, it is necessary to reduce car ownership. Looking at the influences of commuting modes on energy consumption, the relevant parameters are positive and statistically significant. This indicates that it is also necessary to reduce car usage for realizing an energy-saving society. Increasing the use of public transport also induces less gasoline consumption. Furthermore, promoting job–residence proximity (i.e., shortening the commuting time) is also useful to reduce not only gasoline consumption in three cities but also gas consumption in Beijing and Jakarta. Even though a shorter distance to the nearest transit station is associated with more gasoline consumption in Jakarta, the size of the effects is very small (3.37%–3.46% of the total variance).

4.4.3.2 Most influential factors

4.4.3.2.1 Tokyo

Residential attributes are the most influential factor (variance ratio: 48.31%–83.99%) to all the three types of energy consumption at all seasons, especially in the case of gas, residential attributes explain more than 80% of the total variance.

4.4.3.2.2 Beijing

The most influential factors vary with energy type: gasoline is mostly affected by transport-related attributes (42.89%–43.22%), gas by life cycle stage indicators (37.32%–39.64%), and electricity in summer by end-use ownership and usage attributes and in winter by residential attributes.

4.4.3.2.3 Jakarta

Gas and electricity consumption are mostly influenced by end-use ownership and usage attributes. Differently, gasoline consumption is most affected by transport-related attributes.

4.4.3.2.4 Summary

The abovementioned differences are interesting; however, how to explain such differences is challenging. Tokyo represents a developed city and its findings seem common in developed countries. Larger influences of transport-related attributes on gasoline in the three cities [Beijing and Jakarta (the largest), Tokyo (the second largest)] are intuitive. However, why in Tokyo residential attributes are the most influential to gasoline consumption? Comparing all the eight residential attributes, living in a larger house ($> 100\ m^2$) is associated with less use of gasoline. Living in a larger house indicates that more expenditure needs to be spent on residence and consequently, the trade-off between residence expenditure and gasoline consumption results in less consumption of gasoline. However, the influence of house area being larger than $100\ m^2$ is very small. In other words the influence of the above trade-off is minor. In contrast, living in own house, a public house, a rental house, and a dormitory all lead to more gasoline consumption, among which living in a rental house shows the most remarkable influence. With the previous results, it seems that residential behavior is positively related to use of vehicle, that is, gasoline consumption, but if the spending on residence is larger than a certain threshold, an inverse relationship occurs, even though its influence on gasoline consumption is very limited.

4.4.3.3 Life cycle stage factors

This category of explanatory variables shows a larger influence on the electricity consumption in Tokyo and all three types of energy consumption in Beijing; however, its influence is minor on energy consumption in Jakarta. In Tokyo and Beijing, living only with spouse and living with children are much more influential than the other life cycle stage variables. In Beijing, its gas consumption is more influenced by life cycle stage variables than other two types of energy consumption. Thus effects of life cycle stage variables vary across cities. Most surprisingly, living only with spouse and living with children lead to more energy consumption in Beijing, but less consumption of electricity in Tokyo. The Beijing case may reflect the China's traditional cultures that households usually care more about parents' and children's lives.

4.4.3.4 Other factors

In Tokyo, low income does not affect any energy consumption, but high income is associated with less gasoline consumption. However, in Beijing and Jakarta, high income is not influential at all. Low-income households in Beijing use more gas; however, low-income households in Jakarta use less gas and gasoline. Full-time workers consume less gas and gasoline in Tokyo, more gas in Beijing and Jakarta, and more gasoline in Jakarta. Thus Tokyo and Jakarta show a completely different picture about the influence of full-time employment. As for end-use ownership and usage attributes, their influences on electricity consumption are not small, explaining 13.88% – 22.13% of the total variance in Tokyo, 22.83% – 37.51% in Beijing, and 22.13% in Jakarta.

4.4.3.5 Seasonal effects

This study estimated energy consumption at all seasons simultaneously. Because clarifying the effects of transport-related attributes is one of the main research purposes, we set all transport-related indicators differing over seasons and assume the influences of other explanatory variables being invariant over seasons. Even though transport-related attributes are assumed season specific, our results show that the seasonal variations are small in all three cities. Because of this, influences of the other five categories of explanatory variables do not change largely over seasons, as a whole. But we surely observe some seasonal variations, which is summarized as follows:

- *Tokyo*: End-use ownership and usage attributes explains 22.13% of the total variance of electricity consumption in summer, but just 13.88% in winter. Commuting time explains 3.76% of the total variance of gas in summer, but its influences in other seasons are almost ignorable.

- *Beijing*: The variance explained by distance to the nearest transit station accounts for 4.73% of the total variance of electricity consumption in winter: the longer the distance, the more the consumption.
- *Jakarta*: Probably due to the small seasonal variations of all the three types of energy consumption, all the six categories of explanatory variables show almost similar influences over seasons.

4.5 Conclusion

This study has compared household energy consumption and its association with transport-related attributes and other factors in the three Asian capital cities, by using data collected by the authors in 2009. The current Tokyo and Jakarta have not changed dramatically since 2009, in terms of transportation development. In Beijing, there are several policy changes: (1) since 2012, Beijing has implemented a lottery-based vehicle license plate allocation policy, but the vehicle ownership still increased from 4.02 million in 2009 to 5.91 million in 2017; (2) bike sharing has become more and more popular, where the famous Beijing-based bicycle sharing company *ofo* was founded in 2012; and (3) the length of subways increased from 228 km in 2009 to 608 km in 2017. These changes may result in different influences of transport-related attributes on household energy consumption; however, considering that the transportation stocks in 2009 were already huge, the new developments mentioned above may not be sufficient to bring structural changes in the relationship between transport-related attributes and energy consumption. In this sense the findings from this study still have their significant implications to the present policymaking.

The major findings are summarized as follows:

First, the MDCEV model is effective to represent household energy consumption in the Asian context and seasonal influences of transport-related attributes on energy consumption have been confirmed.

Second, transport-related attributes are most influential to gasoline consumption, which explains 39.53%–43.22% of the total variance, within which the person-based transport-related attributes are mostly influential to gasoline consumption. It is concluded that, for reducing gasoline consumption, it is necessary to reduce car usage and promote job–residence proximity. The location-based transport-related attributes are against our expectation, influences of which on energy consumption are very small. Better transport accessibility is also beneficial to reduce gas consumption, but not electricity consumption. Comparing three cities, transport-related attributes mostly affect the gas consumption in Beijing (15.13%–16.58% of the total variance), followed by that in Jakarta (5.37%–5.72%) and in Tokyo (1.98%–6.14%). In all three cities, transport-related attributes do not affect electricity consumption at all, except electricity consumption in winter in Beijing. Detailed transport-related attributes show different influences on different types of energy consumption in the three cities. Influences of transport-related attributes show some seasonal variations, which are not that large.

Third, the most influential factors to energy consumption vary with energy types in Beijing and Jakarta. In contrast, residential attributes are the most influential factors to all three types of energy consumption in Tokyo.

However, there are some limitations in our research. Due to data availability, only limited transport-related attributes are used in this study. More transport-related attributes should be investigated. Even though there are some new transportation developments in the three capital cities, it is unclear whether or not such new developments may alter people's energy consumption habit. For this, panel survey should be conducted to trace people's behavioral changes. Furthermore, as for the influence of transport-related attributes on household energy consumption, there are probably some mediators such as time use and activity participation, which may need careful investigation. Finally, it is also necessary to compare more cities, especially in developing cities.

References

Abrahamse, W., Steg, L., 2009. How do socio-demographic and psychological factors relate to households' direct and indirect energy use and savings? J. Econ. Psychol. 30 (5), 711–720.

Alwin, D.F., 2012. Integrating varieties of life course concepts. J. Gerontol. Ser. B 67 (2), 206–220.

Bhat, C.R., 2005. A multiple discrete–continuous extreme value model: formulation and application to discretionary time-use decisions. Transp. Res. B: Methodol. 39 (8), 679–707.

Bhat, C.R., 2008. The multiple discrete-continuous extreme value (MDCEV) model: role of utility function parameters, identification considerations, and model extensions. Transp. Res. B: Methodol. 42 (3), 274–303.

Delbosc, A., Nakanishi, H., 2017. A life course perspective on the travel of Australian millennials. Transp. Res. A: Policy Pract. 104, 319–336.

Hori, S., Kondo, K., Nogata, D., Ben, H., 2013. The determinants of household energy-saving behavior: survey and comparison in five major Asian cities. Energy Policy 52, 354–362.

Kavousian, A., Rajagopal, R., Fischer, M., 2013. Determinants of residential electricity consumption: using smart meter data to examine the effect of climate, building characteristics, appliance stock, and occupants' behavior. Energy 55, 184–194.
Kutani, I. (Ed.), 2013. Chapter 2, Study on energy efficiency improvement in the transport sector through transport improvement and smart community development in the urban area. In: ERIA Research Project Report 2012-29. pp. 5–14. Available from: <http://www.eria.org/RPR_FY2012_No.29_Chapter_2.pdf> (accessed 23.04.19.).
Liu, G., Dong, X., Jiang, Q., Dong, C., Li, J., 2018. Natural gas consumption of urban households in China and corresponding influencing factors. Energy Policy 122, 17–26.
Nejat, P., Jomehzadeh, F., Taheri, M.M., Gohari, M., Majid, M.Z.A., 2015. A global review of energy consumption, CO_2 emissions and policy in the residential sector (with an overview of the top ten CO_2 emitting countries). Renew. Sustain. Energy Rev. 43, 843–862.
UN, 2016. The World's Cities in 2016. Available from: <http://www.un.org/en/development/desa/population/publications/pdf/urbanization/the_worlds_cities_in_2016_data_booklet.pdf> (accessed 23.04.19.).
Yu, B., Zhang, J., Fujiwara, A., 2011a. Representing in-home and out-of-home energy consumption behavior in Beijing. Energy Policy 39 (7), 4168–4177.
Yu, B., Zhang, J., Fujiwara, A., 2011b. Comparative analysis on household energy consumption behavior in Asia megacities by considering the effects of car ownership and self-selection. J. East. Asia Soc. Transp. Stud. 9, 724–739.
Yun, G.Y., Steemers, K., 2011. Behavioural, physical and socio-economic factors in household cooling energy consumption. Appl. Energy 88 (6), 2191–2200.
Zhang, Q., 2004. Residential energy consumption in China and its comparison with Japan, Canada, and USA. Energy Build. 36 (12), 1217–1225.
Zhang, J., 2017. Life-Oriented Behavioral Research for Urban Policy. Springer, Tokyo.
Zheng, X., Wei, C., Qin, P., Guo, J., Yu, Y., Song, F., et al., 2014. Characteristics of residential energy consumption in China: findings from a household survey. Energy Policy 75, 126–135.

CHAPTER 5

Travel choices, preferences and energy implications in the United States

Wei-Shiuen Ng*
International Transport Forum, Organisation for Economic Co-operation and Development (OECD), Paris, France

5.1 Introduction

The transportation sector is one of the fastest growing contributors to global carbon dioxide (CO_2) emissions and currently constitutes 23% of global carbon emissions, with an annual increase of 2.5% between 2010 and 2015 (IEA, 2017). Although technological advancements have the capacity to create low-carbon transportation systems, travel behavior and demand have to be modified simultaneously in order to achieve lower emissions. For urban passenger transportation, technology can contribute to up to 70% of CO_2 emission reduction, while policies that trigger changes in behavior constitute the rest 30% (ITF, 2017). However, the interconnection between policies that change technology and behavior can be complex, leading to varying levels of mitigation impact. Even if there are existing technologies that will help transition to energy efficient and low-carbon transportation systems, the choice to purchase an electric vehicle or to use public transportation services is still the key to a successful transition. Broad transportation reforms are required to influence travel behavior, either in the form of modal share or travel distance through the implementation of land use, urban design, planning, and pricing policies.

Global transportation activities have increased over the past few decades due to the growing use of motor vehicles, increasing urban population, economic development, rising incomes, expanding number of households, technology advancement, declining prices, and changes in land use and urban planning (Southworth, 2001; Greene, 1996; Pickrell, 1999; Heanue, 1998). The transportation sector in the United States

* Corresponding author.

Transport and Energy Research
DOI: https://doi.org/10.1016/B978-0-12-815965-1.00005-3

(US), together with American travel behavior, has changed substantially since the end of World War II. Private car use for urban travel, in particular, has increased considerably, leading to an increase from 66.9% to 87.9% of work trips share from 1960 to 2000 (Pucher and Renne, 2003). This change in behavior could be explained by the decrease in transportation costs over time as technology advances and subsequent changes in urban land-use patterns, which encouraged urban sprawls across the country (Pickrell, 1999), leading to dispersed residential areas and employment centers in suburban areas. Meanwhile, the shares of public transit and walking have declined, falling from 12.6% to 4.7% and 10.3% to 4.7%, respectively during the same 40 years from 1960 to 2000 (Pucher and Renne, 2003). Highway vehicle miles traveled (VMT) had an annualized growth rate of 2.3% between 1980 and 1999, which is higher than increases in population, employment rate, and income level over the same period of time (Southworth, 2001).

In addition to costly local air, land, and noise pollution (Quinet, 1997; Tietenberg and Lewis, 2009), the transportation sector has become a significant contributor to global climate change. It is a rapidly growing source of CO_2, mainly due to its dependence on fossil fuel consumption. More than 90% of the fuel use in the US transportation sector came from petroleum products in 2017 (U.S. EIA, 2018). In the same year, transportation consumed 29% of the total US energy consumption (U.S. EIA, 2018) and emitted approximately 29% of the total US CO_2 emissions. In fact, only transportation CO_2 emissions increased in 2017 compared to other energy end-use sectors, that is., industrial, residential, and commercial sectors (U.S. EIA, 2018). Passenger vehicles, including cars and light trucks, contributed to more than half of the CO_2 emissions in the transportation sector (BTS, 2009). Passenger transportation, especially road transportation, is still a major source of energy use and emissions.

The environmental impact of transportation, when measured in emissions of any pollutant, is directly related to total distance traveled, mode choice, the energy intensity of the transportation mode chosen, and emissions factor. Advanced vehicle technology and alternative transportation fuel can only affect the latter two variables, energy intensity and emissions factor, but not how total transportation activity is formed, the distance traveled, or how certain transportation modes are chosen over others. In other words, technology can only help solve part of the problem.

Studies have shown that transportation activity could be influenced by external factors, such as congestion or land-use planning (Cervero and Landis, 1995; Randall, 2000), while mode choice is often determined by discrete individual preferences that could be shaped through pricing strategies (Pickrell, 1999; Gómez-Ibáñez, 1999) or governmental regulations. Pricing strategies with the potential to shift behavior can be applied according to vehicle type, for example, vehicle taxation or insurance, fuel type, such as fuel taxation and fee, the use of transportation infrastructure, such as road, toll and parking pricings, or even distance traveled through VMT taxation. Other policies and measures include the provision of transit incentives, employer-based transit and shuttles, the creation of ridesharing programs, encouraging walking and bicycling, and developing flexible work schedules that include telecommuting. These measures can be categorized as transportation demand management (TDM) measures. Several innovative TDM measures have also been developed over the past few years, focusing on the use of technology to influence travel behavior through the creation of mobile phone applications that can provide more information on each mode choice and other travel decisions. Such measures have also created economic incentives to reward less energy intensive transportation mode choices, such as transit, carpooling, walking, or bicycling.

Travel demand and behavior thus play an increasingly important role in determining future emissions, as travel patterns, which can be affected by various factors, are capable of influencing total distance traveled and determining transportation mode choice. Reductions in CO_2 emissions will occur when travel demand shifts to alternative transportation modes that are less energy intensive and when distance traveled declines. Since energy consumption, CO_2 emissions, transportation systems, and land-use patterns are intrinsically related (Hankey and Marshall, 2010), transportation plays an important role in reducing total energy use.

5.1.1 Transportation demand in the form of mode choice and travel distance

This section focuses on urban passenger transportation demand and presents a range of significant variables that will influence demand. Fuel price is often the most direct variable for transportation demand by shifting mode choice or reducing travel distance. It is not surprising that some of the transportation demand elasticity studies that are reviewed have different results. Different methodologies, models, data, timeline, and assumptions can indeed present varying outcomes. For example, most studies on price elasticity of gasoline demand have relied on consumer behavior data from the 1970s and 1980s and may no longer represent current demand elasticities (Hughes et al., 2008). Changes in behavior over time will naturally lead to different demand elasticity estimates. Studies on gasoline price elasticities of transportation demand using disaggregate data from the 1970s to 1980s have found household responses to gasoline price changes to be within the range of −0.43 to −0.67 in the short term (Archibald and Gillingham, 1980; Greene and Hu, 1986; Greening et al., 1995; Dahl and Sterner, 1991; Walls et al., 1993). Long-run estimates could be as high as −1 for the US (Sterner et al., 1992; Dahl, 1995).

More recent studies, on the other hand, have found elasticities to be slightly more inelastic, between −0.2 and −0.3 in the short term and between −0.6 and −0.8 in the long term (Graham and Glaister, 2002). In addition to changes in behavior over the past few decades, these differences found in demand elasticity studies could also be attributed to improvements in data analysis and demand modeling. The level of data, either disaggregate or aggregate, micro or macro-level, the type of data, which could be cross section, time series or pooled, and vehicle technology and fuel efficiency all play an important role in estimating the price elasticity of gasoline and VMT demand (Graham and Glaister, 2002). Increasing fuel efficiency will have a larger impact on long-term price elasticity, in particular, as gasoline efficiency of the vehicle fleet and driving conditions requires time for adaptation and change (Baltagi and Griffin, 1983).

Most studies of gasoline demand elasticity are conducted on a national or international level. However, this may not be true for other variables, especially variables that are associated with public transportation demand. Urban cities have unique characteristics that can shape specific demand elasticities, which could be inapplicable to other regions lacking similar land-use or sociodemographical characteristics. Public transportation demand relies heavily on existing transit supply, transit mode share, population centrality, and vehicle ownership (Bento et al., 2005). Therefore cities, such as New York, Boston, San Francisco, or Chicago, where there are good transit networks and high urban density, are expected to have different transit demand elasticity estimates when compared to San Jose, Los Angeles, New Orleans, or San Diego. Income, race, and education have also been found to have significant impact on commute mode choice (Bento et al., 2005; Sarmiento, 2000).

5.2 Land use, public transportation, and pricing policies

Land use, public transportation and pricing policies are three well-recognized strategies that have been shown to reduce greenhouse gas (GHG) emissions within the transportation sector (Cambridge Systematics, Inc., 2009; Randall, 2000; Pickrell, 1999; Weissman and Corbett, 1992; Kain, 1999).

5.2.1 Transportation and land-use planning

Urban land-use patterns in the US have changed drastically over the past few decades partly because transportation costs have decreased over time due to technological advancements. According to transportation demand theory, the distance between household location and workplace will therefore increase as a result of lower transportation costs (Pickrell, 1999), leading to urban sprawl and people moving further away from employment centers. Transportation and land-use planning's impact on each other are undeniably interrelated (Cervero and Landis, 1995; Randall, 2000), and key transportation policies would therefore have to consider land-use development and activities in order to effectively lower the environmental impact of transportation. According to microeconomics theory, when transportation costs are high, travel demand will decrease, leading to shorter commuting distances, less congestion, and even changes in housing location patterns.

Transportation activity and land-use planning strategies that can change the density of urban cities, diversity of activities in neighborhoods or communities, and distance traveled will be able to reduce VMT and CO_2 emissions subsequently. In addition, any land-use planning policies or measures that can increase public transportation supply and create incentives to maintain and increase public transportation demand can achieve the goal of carbon reductions. Examples of such policies are transit-oriented development and transit improvement through scheduling and increasing frequency, as well as providing better transit coverage.

The dynamic relationship between transportation and land use is linked by the concept of accessibility, which can be defined as the measurement of spatial interaction between activities or land uses (Handy, 1993) or the number of opportunities available within a certain distance or travel time (Hanson, 2004). Land-use patterns have the potential to affect both the type and range of activities and opportunities in a city, while transportation systems can determine the level of accessibility.

The supply of accessibility and even mobility, which is the movement between activities (Hanson, 2004), is heavily dependent upon land-use planning, including the type of investment available. Investments, either private or public, are required for transportation infrastructure and urban development projects, which form the basic supply of accessibility and mobility. Decisions involving urban transportation investment and planning should not be taken lightly, as transportation infrastructure not only determines mobility but it also structures space and controls trade within cities and affects employment and housing locations, while consuming significant resources and causing environmental problems (Short and Kopp, 2005).

The social and economic benefits of transportation infrastructure and planning are considerable. Public investments in transportation generate long-term economic benefits that include increases in output and productivity, reductions in production costs, and increases in income, property values, employment, and real wages, as well as reductions in noncommercial travel time, improved access, and quality of life (Bhatta and Drennan, 2003). Transportation infrastructure projects have found to result in positive and statistically significant economic outputs, measured in gross state product or manufacture output (Munnell and Cook, 1990; Eisner, 1991; Coughlin et al., 1991; Garcia-Mila et al., 1996). Economic benefit can also be reflected by increased productivity, measured in output per unit of labor. Transportation investment, such as highway capital, is positively correlated with productivity (Aschauer, 1989; Fernald, 1999), and at the same time, it has a negative relationship with reductions in costs of production (Nadiri and Mamuneas, 1994; Morrison and Schwartz, 1996; Holleyman, 1996). Although the impact of transportation infrastructure can vary according to the type of industry, economic benefits, in terms of employment and income, have been found to increase when the level of public transportation capital rises (Mofidi and Stone, 1990; Duffy-Deno and Eberts, 1991; Luce, 1994).

Transportation serves a critical role in ensuring the daily operations of urban life. Since goods and services in cities are located separately, transportation is required for moving goods and people to allow social and economic activities to take place. The majority of trips are not undertaken for their own consumption value or pleasure driving, and since people do not usually travel unless they are trying to accomplish nontravel activities, urban transportation is a derived demand (Hanson, 2004; Small, 1992). This observation creates complexity to the understanding of travel demand, which can be influenced by various types of activities and in turn affects labor supply, firms' location and technology decisions, urban development, and even trip chaining processes (Small, 1992).

Personal mobility is fundamentally determined by time and monetary cost budgets that are often fixed constraints (Zahavi, 1981; Schafer and Victor, 2000; Golob et al., 1981). Time spent in motorized modes has been shown to increase with income and mobility, and at the same time, travel money budget rises with motorization (Zahavi, 1981; Zahavi and Talvitie, 1980; Gunn, 1981; Hupkes, 1982). As a result, increasing income affects mode choice and urban mobility directly. This could explain the transformation of cities and urban form over the past few decades when the average income level has increased, which coincides with the growth of automobile use and ownership.

5.2.2 Public transportation development

The role of public transportation has to be strengthened in US cities where there is high urban density, in order to provide good transportation alternatives and help reduce energy use and CO_2 emissions. Implementing public transportation in areas with high urban population density and user demand is a way to reduce congestion, and by having the capacity to hold more passengers than a private automobile, it can help reduce emissions. Given the current level of vehicle efficiencies, this assumption will hold true if public transportation systems have sufficient demand and can attract enough passengers to offset the energy consumed by their buses or trains. In most cases, public transportation could provide transportation services at a lower cost to the user than driving private vehicles. Although rail transit is significantly more costly than buses to implement, rail transit will be a more efficient transportation mode compared to driving if there is high ridership available.

Improving public transportation systems implies the development of faster travel modes with greater convenience through scheduling and adjusting frequency. Speed can be further enhanced by the use of dedicated roadways and enabling smoother transfers between vehicles and different transportation modes. As the speed of vehicles increases, public transportation modes will be able to serve more passengers and maintain ridership demand. Increase in public transportation efficiency and ridership can also be achieved by investment that will ultimately lead to economic benefits through high economic rates of return in dense cities

(Cambridge Systematics Inc. and Apogee Research, 1996). The benefits of public transportation will therefore not be constrained within improving traffic conditions but could also provide greater advantages to the economic growth of the region.

The basic relationship between public transportation service level changes and impact on ridership remain stable over time. Positive results of service elasticities may reflect service quality and also successful service restructuring or external factors such as a booming economy (National Academies of Sciences, Engineering, and Medicine, 2004). The degree of system-wide ridership response to changes in service is greater in small cities and suburbs, which also implies that greater elasticities are found when the initial public transportation service levels tend to be lower than average. In other words, when efficient public transportation services are supplied, demand for it will increase.

5.2.3 Pricing instruments

Transportation pricing policies are tools that can successfully manage travel demand, reduce congestion and travel time, as well as decrease vehicular emissions (Greene and Schafer, 2003). Studies have shown that pricing is one of the most effective measures in lowering travel demand, in terms of distance traveled, and thus reducing CO_2 emissions (Parry and Small, 2005; Cambridge Systematics, Inc., 2009; Barbour and Deakin, 2012). Pricing policies can regulate demand and influence behavior through changes in travel distance, number of trips, and transportation modes, while appropriate parking pricing can reduce vehicle usage and extensive cruising, especially when unregulated parking options are unavailable. Such changes in travel demand and behavior are reflected in the price elasticity of travel demand calculated for different types of pricing policy implementation. Since there are many types of transportation pricing, some are certainly more efficient and equitable than others (Levinson, 2010). Despite the numerous types of pricing strategies, such as road pricing, VMT fee, time varied congestion charging, parking fees, insurance fees, and pricing schemes based on emission levels, and their known impact on travel demand and activity patterns (Cambridge Systematics, Inc., 2009), few cities have implemented adequate pricing policies.

The importance of pricing in transportation was noted by Pigou (1912), who emphasized the need for government intervention in order to internalize externalities through taxes or subsidies. Vickrey (1954) further developed this concept, recognized the validity of equity in pricing, and critiqued the lack of adequate pricing in urban transportation, which he believed could affect the development of communities and property values. Empirical studies on how transportation pricing could affect welfare across different levels of income groups began to appear in the 1970s in the professional literature, as arguments about whether to deregulate taxis, airfares, and other public services began to appear. In the economics literature, there were numerous studies examining the optimal pricing of urban roads, highway investment, travel costs, highway speed, value of time, and the capacity provided during peak and off-peak periods by the late 1970s (Keeler and Small, 1977; Smeed, 1968; Forsyth, 1977). Small's study (1983) on the impact of congestion toll on income distribution and its associated welfare effects in the San Francisco Bay Area was probably one of the first empirical studies that focused on how congestion pricing can influence on welfare and equity.

Currently, there are only a few cities in the world that have congestion pricing in operation. Singapore, being the first city to do so, started adopting congestion pricing policy with an Area Licensing Scheme in 1975, as an effort to curb congestion and reduce travel delays, which were common trends occurring in its neighboring Southeast Asian capital cities (Santos and Fraser, 2006). This scheme had successfully reduced congestion by lowering peak hour traffic by 65% (Wilson, 1988), while average travel speed had increased by almost 90% (Phang and Toh, 1997). The paper based, manual Area Licensing Scheme was replaced by Electronic Road Pricing in 1998, which reduced traffic volume in the restricted zone by an additional 15% (Menon, 2000) with its varying charging times on central roads, expressways, and outer roads (Santos and Fraser, 2006).

Many other cities have seriously considered congestion pricing and have produced thorough analysis but few have been able to put such a policy in place. An example is the transportation pricing study conducted for the state of California in 1996 (Deakin and Harvey, 1996). This study focused on four major metropolitan areas in California, the San Francisco Bay Area, Los Angeles, Sacramento, and San Diego. Although results of this study have shown that transportation pricing measures are highly efficient in reducing congestion, air pollution, and energy use, there has not been any general implementation of congestion pricing in the state. There has, however, been congestion-based pricing on a series of high occupancy toll lanes, in Orange County, San Diego, and the San Francisco

Bay Area. In addition, there are tolls on eight toll bridges in the San Francisco Bay Area (Caltrans, 2012) and five other toll projects either in review or construction (Perez and Lockwood, 2006). Variable pricing has been implemented on the San Francisco–Oakland Bay Bridge, which has a peak-off–peak-toll differential (MTC, 2010) and on State Route 91 Express Lanes in Los Angeles, which boasts a variable toll schedule (Sullivan, 2003) that can also be found in San Diego, on the South Bay Expressway.

London's congestion charging scheme started in 2003, as a way to reduce the significant congestion problem in the city. It is an area licensing system with a fixed fee for all vehicles entering the congestion zone. This is one of the differences between the London scheme and Singapore's, as the charges in Singapore are imposed per trip, while in London a payment will last for the entire day, allowing vehicles to enter and exit the zone as many times as they wish (Santos et al., 2010). After the implementation of this scheme, traffic volume entered the zone when congestion pricing was in operation reduced by 18% (TfL, 2005). Travel time decreased by 30% during the first year of the implementation (TfL, 2003), while bus riders increased by 18% (TfL, 2007).

A more recent example of a successful implemented congestion pricing policy can be seen in Stockholm, which first imposed congestion pricing as a trial to reduce congestion and emissions, and later reintroduced it permanently in 2007 due to popular demand (Eliasson, 2008). Like London, it is a cordon toll system, with a congestion zone surrounding Stockholm City, using 18 entry and exit points (Santos et al., 2010). Congestion charges differ according to the time of day. Vehicle distance driven in the congestion zone decreased by 16%, while traffic volume also decreased by 5% outside the city center, and traffic emissions reduced by 10%–15% between 2005 and 2006, the year of the launch of the congestion pricing scheme (Eliasson, 2008). The decrease in vehicle distance and traffic volume in the congestion zone and beyond could be due to the 24% of work trips by car that switched to transit across the cordon (Franklin, 2007), which would affect both traffic entering and surrounding the cordon.

Parking pricing is another effective yet underused policy. Parking pricing policies can apply to commuter, noncommuter, and residential parking and can address a variety of financial, social, economic, and environmental objectives. In particular, parking pricing policies can generate revenue for operators, serve as tools to support commercial success and residential quality of life, and at the same time, help manage travel demand, reduce congestion and travel time (Shoup 2005), as well as decrease vehicular emissions (Greene and Schafer, 2003).

A number of studies have observed the short-term impact of parking pricing. Three separate studies in San Francisco, Toronto, and Dublin have shown that for every 10% increase in parking price, there would be an average of 3% decrease in demand for parking spaces (Kulash, 1974; Gillen, 1977; Kelly and Clinch, 2009). This elasticity estimate is the most commonly found, but there are studies showing larger decreases in parking demand as price increases. For example, Dueker et al. (1998) found that in urban Portland, the price elasticity of demand for commuter parking was −0.58 for single occupancy vehicles and −0.43 for carpools. Therefore the overall impact may depend on the type of parking studied, that is, on-street, off-street parking garages for commuter or shopping, how parking demand is defined, that is, the number of vehicles parked (Kulash, 1974) or the number of auto trips (Gillen, 1977), time of day studied, or whether there is alternative free or lower priced parking available, as is sometimes the case. Most studies of current parking pricing have focused on its impact on parking space demand (Kulash, 1974; Gillen, 1977; Kelly and Clinch, 2009) and fewer empirical studies of parking pricing changes have considered mode choice impacts. Surveys tracking parking pricing changes in Los Angeles city center and suburbs have shown that when employers stopped paying for parking, the number of solo drivers decreased substantially, between 19% and 81%, depending on the location. Likewise, the use of private vehicle as a commuting mode had decreased by 15%–38% after the removal of parking subsidies (Willson and Shoup, 1990).

Additional studies have considered parking pricing policies' impact on congestion. Free parking reduces the financial incentives to drive less (Jansson, 2010) and increases congestion both from increased traffic flow and the search for parking, also known as cruising. A review of 16 studies on 11 cities conducted between 1927 and 2001 has found that about 30% of the vehicles in the central business districts were cruising for parking (Shoup, 2007). Cruising, that is, driving to find parking, should be reduced when possible, as it could lead to substantial increases in distance traveled, fuel use, and emissions (Shoup, 2005).

The implementation of pricing policies can be sensitive and difficult to practice, but incentives could be created to encourage production and consumption of high fuel efficiency vehicles. Transit vehicles can also be equipped with advanced fuel technology, such as fuel cells, and both the private and transit vehicles can be priced accordingly by how much emissions they produce. Similarly, vehicles that use alternative fuels often need different forms of infrastructure from conventional gasoline vehicles. Thus land-use planning policies can place hydrogen or natural gas refueling stations in more convenient locations to enable better access and incentives for alternative fuel vehicle users.

5.3 Travel choices and preferences in California

The impact of different transportation policies on travel behavior will depend on existing travel patterns and mode choice, in addition to transportation alternatives and services. Commuting choices are not only dependent upon the availability of feasible transportation modes but also on the socioeconomic characteristics of the employees and their employment and life circumstances, both of these can affect travel choices.

The reasons for using private vehicles as a primary mode choice have been studied extensively, and they are usually due to travel time, convenience, flexibility, physical effort, the desire for personal space, travel costs, the need for control (Tertoolen et al., 1998; Gardner and Abraham, 2007; Van Vugt et al., 1995), as well as factors that are based on the experience of driving and perceived uncertainty, safety, excitement, and enjoyment (Ory and Mokhtarian, 2005; Anable and Gatersleben, 2005; Ellaway et al., 2003). The motivation for private vehicle and public transportation is often determined by the same set of factors.

Transportation behavior studies have identified various types of commuters or drivers, some with a higher level of intention to switch to another mode than others. For example, Jensen (1999) categorized transportation users into six mobility types, including three car driving specific groups, namely, passionate drivers, everyday drivers, and leisure time drivers, as well as three types of bicycling or public transportation users, including users of the heart, users of convenience, and users of necessity. Using attitudinal analysis and applying the theory of planned behavior, Anable (2005) derived six psychographic groups, each representing a distinct combination of preferences, worldviews, and attitudes. Similarly to Jensen's (1999) study, these six groups consisted of both car-owning and noncar-owning segments. The four car-owning segments include malcontented motorists, complacent car addicts, die hard drivers, and aspiring environmentalists, while the noncar-owning segments are car-less crusaders and reluctant riders (Anable, 2005). The main difference between malcontented motorists and complacent car addicts is that the former group perceives the use of public transportation to be constraining but at the same time feeling unhappy with car use, while the latter group thinks that the use of nondriving modes is feasible, yet they are content with their existing driving behavior. For noncar-owning segments, car-less crusaders feel positive with their noncar mode choices, but reluctant riders are only choosing public transportation because of health or financial reasons, and they would choose to drive in the future if possible (Anable, 2005). Her study concluded that segmenting users by sociodemographic factors alone is inadequate in constructing transportation behavior, which is far more complex than profiling personal characteristics.

Gotz (2003) also defined mobility types based on attitudinal and lifestyle. He identified five mobility styles that include (1) the traditional domestic, (2) reckless car fans, (3) the status oriented automobilists, (4) the traditional nature lovers, and (5) the ecologically resolute. Another study focused on the groups of bicyclists based on their attitudes has identified a typology that included fair weather, lifestyle, practical, and idealist bicyclists (Davies et al., 1997). These studies have all presented a wide range of users with varying demand for each specific mode and attitudes that reflect different perceived image, status, and constraints that go beyond sociodemographic factors. Most of these studies only included leisure trips (Anable, 2005), and none of them examined the frequency of driving or frequency of other transportation mode choice. Current studies do not recognize the fact that different user groups may not choose the same mode for every trip, and some groups are more likely to use a combination of different modes on different days of the week than others.

Using data collected from a total of 17 focus groups, conducted in 2013, with 164 participants who live and work in the San Francisco Bay Area in California, this section provides a more detailed overview of commuting mode choices, including their driving frequency, based on different preferences. Despite the relatively integrated transportation and land-use planning, wide range of public transportation options and stringent pricing policies compared to other parts of the country, driving still has a high transportation mode share of 75% in the San Francisco Bay Area (MTC, 2016), which is very similar to the national 76% (U.S. Census, 2016).

It is well settled that socioeconomic factors are important determinants of mode choice, which are also affected by the availability of modal options and residential location (Ben-Akiva and Lerman, 1985;

Cervero, 2002). Despite the relatively high level of public transportation services in the San Francisco Bay Area compared to the other US metropolitan areas, many residents nonetheless live in areas where taking public transportation to work would require several time consuming transfers for whom the total travel time is not competitive with driving. Others live close enough to work that walking and biking are options along with public transportation (bus and train) and driving. The decision to either drive or use other forms of transportation is more complex for employees with good options than for those with limited alternatives. In addition, the ability to schedule flexible working hours also influences mode choice, as employees can arrive later at their workplace or work from home part of the week and drive on other days (Ng, 2017). Major life changes, such as death in family, divorce, moving further away from work, and expansion of family, have contributed to mode shift in the past, mostly shifting from public transportation or walking to driving.

There are several reasons why certain transportation modes are chosen. It is also common to use a combination of different modes to travel, and these differences may vary with the frequency of driving trips. Some focus group participants rarely choose any other mode but drive alone, while others rarely choose other mode but walk, bike, or use transit. However, most of the regular users of transit or nonmotorized modes drive occasionally too, with driving frequency ranging from once or twice a week to a few times a year. The type of drivers can be categorized by the frequency of driving trips, leading to four different categories, namely, (1) regular drivers, who drive to work everyday and have specific reasons for doing so; (2) regular but flexible drivers, who drive to work every day but are willing to use other transportation modes; (3) occasional drivers, who drive to work less than once a week; and (4) nondrivers, who do not drive to work at all. Reasons for each of these four driving behavior, which will ultimately affect travel patterns, are described in the following sections.

5.3.1 Regular drivers

Driving alone was the most popular transportation mode choice amongst all focus group participants, and it was also the only mode that at least one participant used in every focus group. However, this does not imply that all of them are regular drivers, that is, driving to work every day of the workweek. Participants' reasons for driving include the superior comfort of the automobile, concerns about safety, and for some, the need to transport or be available on short notice for dependents. Another group of participants cited low transit accessibility from their homes to the workplace. In addition, some participants reported that cost factors made driving preferable to using transit. They cited the availability of free parking close to work and transit fares that brought transit out of pocket cost to the same or higher cost as that of driving alone. In discussing convenience, focus group participants included the walk to and from the bus stop or transit station as an inconvenience. Other inconveniences cited by participants included having to coordinate travel times with transit schedules. However, the most important convenience factor favoring driving was travel time. For example, using AC Transit (Alameda-Contra Costa Transit District) buses would take twice as long to commute as driving would for some participants.

In discussing the greater comfort of driving the ability to be in a private space and not commute by public transportation, which is subject to crowded spaces and other commuters, is desirable to some participants. Several participants mentioned comfort as the key reason why they drive to work. Driving is also perceived to be a safer transportation mode than other modes, especially for female participants. For transit and walking, safety has to do with the environment around the transit stations and bus stops, as well as the fear of crime. The choice to drive everyday to work also depends on the availability of other modes for an individual. Some of the participants live in communities with limited travel options to work. If transit services were unavailable and biking or walking is infeasible, driving alone becomes the only option that commuters deem feasible.

5.3.2 Regular but flexible drivers

Approximately 28% of the participants reside 3 miles or less from their workplace, which is equivalent to a 20-minute bicycle ride. Another 20% of the respondents live within a 30-minute commute by public transportation. Thus a significant percentage of the participants are able to travel to work by transportation modes other than driving alone. Many of the drivers classified in the "regular but flexible" category live in residential locations where public transportation services are available or where walking and biking are feasible commuting options. Most participants (64%) do not need to schedule their arrival and departure times around someone else's schedule, that is, their travel schedule is independent. Some of the participants in this category expressed the preference to bicycle to work as a form of exercise or want to reduce their own transportation emissions by using public transportation. Most are in one-car households (which constitute 56% of the survey sample) and need to share their vehicle with other family members. The weather condition, lack of dependents, schedules, and feeling guilty about driving were all listed as reasons for not driving on a daily basis. Even though many in this group would like to drive less, they do not all see the existing alternatives as adequate.

"Nobody wants to always drive but the alternatives are painful." (Regular but Flexible Driver, Focus Group)

5.3.3 Occasional drivers

The most common reason for not driving regularly is due to the cost of driving being higher than using other transportation modes. Driving alone could be more expensive than transit for the trip from the participants' residence to their workplace in approximately 97% of the cases if the cost of parking was included. Without taking the cost of parking into consideration, 6% of the participants' cost of driving will be less than transit, given their residential locations.

Weather conditions also can lead to a decision to drive or not. Cold, rainy weather is a deterrent to transit use. In addition, some participants are more reluctant to use public transportation when it gets dark earlier in the winter and would drive more than in the summer. Occasional drivers also drive to work for special events, which include personal and professional appointments, social engagements, or work events. In addition, some participants will also drive when they are running late in the mornings, working late in the evenings, or carrying heavy or bulky things to work.

The physical condition of the commuter is another factor that directly affects transportation mode choice. While some physical limitations are permanent, others can be unpredictable and can vary on a daily basis. When some participants "do not feel like" using their primary modes of transportation, for example, biking or walking, they will then choose to drive. Some participants simply drive occasionally because they prefer to use other transportation modes and would drive as minimally as possible. This is mostly due to the negative attributes associated with driving, such as waiting in traffic or looking for parking.

5.3.4 Nondrivers

There were two focus group participants who did not have a valid driver's license and did not know how to drive. There were also participants who live close enough to walk or bike to work and have never driven to work. Another common reason participants gave for being nondrivers is that they much prefer to use other transportation modes and simply do not enjoy driving to work and finding parking once they arrive at work.

> *"I find parking to be stressful. I do not like to drive all the way to work and I would not do it even if it were very cheap." (Non-Driver, Focus Group)*

5.4 The complexity of travel behavior and policy implications

Travel behavior is complex even for individuals who have identified driving as their primary mode choice, as driving frequency could vary. The majority of focus group participants (approximately 95%) in the Californian case study would have driven alone to work at some point in time, some more than others. Regardless of what they have reported their primary mode choice to be, employees who usually use transit, bicycle, or walk also drive to work at a frequency rate between 2 and 3 days a week to a few times a year. Carpoolers and motorcycle riders tend to be the most consistent with their behavior. Participants who are able to use a combination of modes clearly have the options to do so, and their transportation behavior reflects complex decision-making processes based on various factors that vary on a daily basis. Such transportation behavior is harder to predict but is also most likely to be influenced by external factors, including changes in professional or personal schedules. They also do not have very strong preferences on one particular mode choice or are more flexible. Hence, it is possible for them to use transit, bicycle, or walk to work on most days, but drive on other days when they prefer to do so. On the other hand, respondents who have a strong preference on their chosen mode are less likely to change and have already formed habitual behavior over time. This applies to parking location choice too. Respondents with preferred parking locations are less likely to find alternative parking spaces unless there are significant changes in parking pricing.

The impact of land use, public transportation, and pricing policies on mode choice will therefore vary, and they may not be very effective in influencing current regular drivers, who have chosen to drive alone not because they were unaware of the existing policies but because of other factors that are usually unrelated to cost or alternative services. These are noncost-related factors that are just as or more important than cost and these factors, such as personal and professional schedules, travel time, and the reliability or perceived reliability of public transportation services, will still be significant even when transit cost is fully subsided.

It is assumed that some policies will be more effective in reducing solo driving trips when coupled with other transportation policies. For example, transit incentives provided by the employer will trigger higher transit rides if there are policies supporting transit services in the region. Hence, transportation policies are most effective when there are complementary measures, either in the form of pricing policies or well-developed transit networks, available in its neighboring areas too. Therefore although transit incentives were found to be attractive, they may only be attractive for current public transportation users, and improvements in services would have to occur in order to attract more users. This implies that transportation policies that are targeted at increasing transit subsidies may not be effective when existing transit services are unsatisfactory.

It is common for travelers to have tried at least one different transportation mode other than their primary mode choice. Based on their experience with various mode choices, most of them expressed the notion that their current mode or current combination of modes was their most preferred choice, as found in the focus group discussion session discussed in the previous section. They are already familiar with their options and have already chosen one that they perceive to be the best fit for them, given their schedules, lifestyle, residential location, and annual household income. Any future changes in transportation and parking policies have to address their travel needs to encourage substantial shifts in mode choice. In addition, since they are already aware of their travel options, additional information on existing transportation alternatives may not create significant shifts. Switching from one type of driver to another requires a shift in attitude, in addition to the implementation of innovative transportation policies and measures. Nevertheless, drivers in the "regular but flexible" category tend to be more likely to change their travel behavior and to drive less than "regular drivers."

Having different sets of transportation policies or programs that are targeted at specific groups of drivers will be more effective than having one generic program. For example, since "regular but flexible" drivers are fairly independent and do not have schedules that depend on someone else's schedule, programs that support the improvement of public transportation services in the form of convenience (e.g., less connections) or shorter travel time would be more attractive than programs that focus on increasing the frequency of services. Similarly, travelers who drive regularly because of their families' schedules would be more attracted to programs that offer dynamic transportation services.

The increase in shared mobility is an example of more dynamic transportation services that are demand responsive. In the US and many other countries, ride-hailing services have grown exponentially since 2009 to more than 250 million users globally within its first 5 years, while car sharing has more than 2 million members in North America and almost 5 million globally (Clewlow and Mishra, 2017). Although such shared mobility services have proven to have the potential to disrupt future travel behavior (ITF, 2015), their actual impact on current energy use is still unclear, as approximately half of ride-hailing trips have been shown to be previously made by walking, biking, transit, or avoided altogether in the US cities (Clewlow and Mishra, 2017). In fact, 49%–61% of ride-hailing trips would have not been made at all, or by walking, biking, or public transit and ride-hailing users have reported a decrease in public transportation use (Clewlow and Mishra, 2017). Hence, ride-hailing services are likely to lead to an increase in distance traveled, traffic volume, congestion, energy use, and emissions. These trends could be avoided if ride hailing is coupled with the concept of carpooling, vehicle electrification and automation, as well as the shifting of the energy source away from fossil fuels (Fulton et al., 2017; Sperling, 2018). A combination of sharing, electrification, and automation can reduce global energy use from urban passenger transportation by more than 70%, by significantly reducing the number of vehicles needed to meet transportation demand (Fulton et al., 2017).

5.5 Vehicle choice and energy implications

Travel behavior and energy use are linked not only by mode choice and distance traveled but also directly through vehicle choice, which will determine the fuel type and subsequent energy intensity. This section will present an overview of vehicle technologies with different levels of energy intensity and fuel economy, as well as associated behavior and attitudes toward the purchase and ownership of noninternal combustion engine vehicles. Attitudes related to the characteristics of vehicle type or driving in general reflect the unique preferences of each individual. Individuals' preferences on various choices, such as comfort, convenience, and flexibility, are common variables that determine vehicle choice (Johansson et al., 2006). Behavior and attitudes toward alternative fuels and electrification, as well as recharging infrastructure and range limitations are still significant barriers to a more rapid rate of electric vehicle adoption.

5.5.1 Vehicle electrification trends

The electrification of personal mobility, in the form of cars or motorcycles, has accelerated dramatically over the past decade, and it has the potential to alter the emerging personal mobility landscape. Current private motor vehicle ownership and usage trends in developing and developed regions have shown little evidence of decreasing. In fact, the global growth rate of private vehicle ownership, which usually reflects economic growth, has increased substantially over the past decade. Global new passenger vehicle sales hit 70.85 million in 2017, which is an increase of more than 50% from 2005 (OICA, 2017). More than 1 billion motorized vehicles are already being driven in the world today, and existing projections have shown that there could be more than 2 billion passenger vehicles by 2050 (ITF, 2017), while the total number of all types of motor vehicles is expected to already pass 2 billion by 2030 (Sperling and Gordon, 2009). More vehicles are being produced today than at any other time in history, with China, the US, Japan, Germany, South Korea, and India taking the lead.

Although electric vehicles only constitute 0.2% of the global vehicle stock currently (IEA, 2017), the market for electric vehicles has started to gain momentum as reflected by the sales of new electric cars in 2017, which surpassed 1 million for the first time (IEA, 2018). Advanced markets of electric cars in terms of electric cars sales share, include Norway, Iceland and Sweden, while China and the US remain to be the two largest electric car markets in the world (IEA, 2018). Sales of electric two wheelers have also increased, reaching 30 million in 2017 (IEA, 2018).

The main benefits of electric personal mobility are its low local air pollutants and decarbonizing potential, which is dependent upon the original source of the electricity, as well as its positive impact on energy security. However, electric vehicles do have a strong competitor, namely, the internal combustion engine vehicle, which has a much superior range thus far and is able to provide good performance with its high energy density oil-derived liquid fuels (Pollet et al., 2012). The costs of internal combustion engine vehicles also start at a much lower level than electric vehicles.

5.5.2 Vehicle technologies and consumer attitudes

There are currently different levels of vehicle electrification, some of which have been available in the market longer than others. They include hybrid electric vehicle (HEV), plug-in HEV (PHEV), battery electric vehicle (BEV), and hydrogen fuel cell electric vehicle (FCEV). The adoption rate of electric vehicles varies according to the type of technology, but there are similar barriers across all electric vehicle technologies, one of them being consumer behavior and attitude.

Generally, a hybrid vehicle powertrain can combine any two power sources, where one component is usually used for storage and the other is for the conversion of a fuel into usable energy (Pollet et al., 2012). HEV combines the best features of two power sources, while overcoming their individual disadvantages (Pollet et al., 2012). The hybrid technology is the only major electric vehicle success between the late 1990s and mid-2000s (Dijk et al., 2013), which saw the mass commercialization of low-carbon vehicles via alternative powertrain technology. Between 1997 and 2011, Toyota sold more than 3 million Prius worldwide (Dijk et al., 2013), and more than 12 million HEVs have been sold worldwide since then. In fact, HEVs have already been accepted as a mainstream vehicle option in the US and Japan.

A PHEV is very similar to an HEV. However, a PHEV can be conveniently connected to the grid and has a larger battery pack than a HEV. Its battery pack can be fully charged by plugging the vehicle into a standard electrical outlet. The batter pack is also the primary source of power for short distances, unlike HEVs, which still depend on petroleum fuels to generate electricity (Amjad et al., 2010). For longer distances, once the battery has been depleted, PHEVs would then switch to a hybrid mode. Energy can also be recaptured from braking (i.e., regenerative braking), the engine is turned off when the vehicle is stationary and the internal combustion engine can run at a more constant and efficient speed (Amjad et al., 2010).

PHEVs were introduced to limited production in 2004 and to mass production in 2011 (Al-Alawi and Bradley, 2013). Despite its recent growth in the market, there are still several design and technological challenges to overcome before achieving better energy efficiency and performance. The design factors include reduction in weight, volume, and cost, which technological limitations are related to the energy storage system (i.e., batteries), motor drives, and associated power electronics (Amjad et al., 2010).

The development of the BEV has accelerated since the mid-2000s and there is now a new momentum from 2005 onwards due to the growing concerns of climate change and energy security (Dijk et al., 2013). BEVs have an electric motor powered by a battery instead of an internal combustion engine and use batteries for their energy source instead of liquid fuels (Pollet et al., 2012). BEVs are usually plugged to a charging station when they are not in use (Tie and Tan, 2013; Peng et al., 2012; Amjad et al., 2010). They are highly efficient, do not produce tailpipe emissions, and can be charged overnight on electricity produced by any type of power station (Amjad et al., 2010; Shareef et al., 2016; Castro et al., 2017). However, electricity storage is still relatively costly, and the charging of batteries could be time consuming resulting in the limited range of such vehicles and the need for charging infrastructure to be installed in order for significant market penetration (Andwari et al., 2017). On the other hand, the most challenging barrier to overcome is not the technology itself but the social acceptance of BEVs. The relatively higher capital cost than the average internal combustion engine vehicle is an initial barrier for consumers, and since BEVs have a lower range than conventional vehicles and charging could take a long time, consumers face "range anxiety" as they fear they will be unable to complete their journey (Andwari et al., 2017). Inadequate charging infrastructure hence further raises the market barrier.

Factors that could determine the uptake of future BEVs include the technological readiness and cost effectiveness of BEV components over time and the satisfaction of consumers' needs and range requirements. While significant progress has been made in developing batteries for electric vehicles, major challenges remain the same. These include cost reduction, safety improvement, lifespan expansion, shortening of charging time and providing better charging facilities, and reducing the size and weight of the battery pack (Pollet et al., 2012).

The FCEV uses fuel cells to generate electricity from hydrogen and oxygen, the electricity is then used to drive the vehicle or stored in batteries or ultracapacitors. Since fuel cells generate electricity from chemical reactions, there is no fuel combustion, and water is the only by-product. Although the FCEV uses a different energy source and has higher energy density (hence, range is less of a problem) than BEVs, it shares many of the same components as BEVs, such as electric motors and power controllers or inverters. While BEVs use energy stored in the battery, FCEVs use fuel cells as they are superior to batteries in many ways. The major advantages are that fuel cells are lighter and smaller and can produce electricity as long as the fuel is supplied. These two technologies are expected to coexist in the future due to their similarities (Pollet et al., 2012). The BEV is suitable for short range and small vehicles, while the FCEV is suitable for medium–large and long-range vehicles (Pollet et al., 2012). Unlike other electric vehicles, their range and refueling processes are comparable to conventional vehicles, though a bulkier hydrogen tank than an equivalent gasoline tank would be required to achieve same range.

Although FCEVs possess many advantages, they also have certain limitations that are related to the fuel cell stack itself, mostly in terms of cost and durability, and its fuel, including hydrogen production, transportation, and storage (Pollet et al., 2012). Fuel cells are also costly and currently produced in very small numbers, but mass production should reduce their cost by an order of magnitude (IEA, 2018). Refueling a hydrogen tank may take significantly less time than fully charging a battery, but electricity is already a widely used energy and recharging infrastructure for BEVs that will be based on the existing power grid is likely to be faster with lower risk than developing new hydrogen production, transmission, and refueling infrastructure (Offer et al., 2010).

BEVs and FCEVs emit zero tailpipe emissions, including CO_2 and air pollutants, while PHEVs have zero emissions when operating on electricity (Sperling, 2018). However, the current high cost of such vehicles, range anxiety, and lack of adequate fueling stations have prevented a greater market share of electric vehicles in general. Consumers' behavior and choices remain as a major market barrier, along with technological challenges.

5.6 Conclusion

Transportation mode choice is a reflection of a complex decision-making process and preferences, particularly for commuters who have more than one transportation alternative. Since mode choice can directly affect energy use and transportation emissions, shifting mode choice to less carbon intensive alternatives will ultimately reduce carbon emissions. Understanding the frequency of driving is critical when analyzing the various factors that lead to different commute mode choices. This is mainly because most commuters in the US drive, albeit at different levels of frequency, even when their primary mode choice is a nondriving alternative. Reducing the frequency of driving can be just as important as shifting mode choice, and it is also a less drastic change in transportation behavior, for example, driving for three days a week instead of five days a week, which is more acceptable in the short term.

Understanding transportation behavior and preferences can help one develop effective transportation policies, especially those that will trigger mode shifts. Mode choice is often constrained by the availability of alternatives and in areas which are well served by a wide range of transportation services, travel preferences may differ according to different travel needs and demand. Although travel cost is important to most individuals when considering their commute choices, it is not always the sole determinant. Hence, moderate changes in transportation pricing, either in the form of parking pricing or transit subsidy, may not achieve long-term mode shifts for commuters who are regular drivers without feasible travel alternatives. Similarly, for those who prefer to use public transportation because of noncost-related reasons, shifting to driving will not occur under circumstances when driving becomes less costly. Personal preferences and perceptions based on past experiences and current needs, commute constraints caused by dependents for example and residential location, which affects travel alternatives, are all influential to mode choice. The impact of transportation policies on transportation demand will thus be region specific and not always reflect the relatively high levels of elasticity estimate found in the previous studies.

By identifying different types of travel choices and preferences, transportation planners will be better equipped to better design mode choice models and to forecast changes in future travel behavior, energy use, and CO_2 emissions under the context of different transportation policies and therefore to design more effective transportation policies. This is especially critical due to current disruptions and innovations in the field of transportation, where new services and technologies, such as shared mobility, vehicle electrification and automation, could have the potential to significantly reduce the transportation sector's dependence on fossil fuels.

References

Al-Alawi, B.M., Bradley, T.H., 2013. Review of hybrid, plug-in hybrid, and electric vehicle market modeling studies. Renew. Sustain. Energy Rev. 21, 190–203.

Amjad, S., Neelakrishnan, S., Rudramoorthy, R., 2010. Review of design considerations and technological challenges for successful development and deployment of plug-in hybrid electric vehicles. Renew. Sustain. Energy Rev. 14, 1104–1110.

Anable, J., 2005. 'Complacent car addicts' or 'aspiring environmentalists'? Identifying travel behaviour segments using attitude theory. Transp. Policy 12, 65–78.

Anable, J., Gatersleben, B., 2005. All work and no play? The role of instrumental and affective factors in work and leisure journeys by different travel modes. Transp. Res., A 39 (2–3), 163–181.

An dwari, A.M., Pesiridis, A., Rajoo, S., Martinez-Botas, R., Esfahanian, V., 2017. A review of Battery Electric Vehicle technology and readiness levels. Renew. Sustain. Energy Rev. 78, 414–430.

Archibald, R., Gillingham, R., 1980. An analysis of the short-run consumer demand for gasoline using household survey data. Rev. Econ. Stat. 62, 622–628.

Aschauer, D.A., 1989. Is public expenditure productive? J. Monetary Econ. 23, 177–200.

Baltagi, B.H., Griffin, J.M., 1983. Gasoline demand in the OECD: an application of pooling and testing procedures. Eur. Econ. Rev. 22 (2), 117–137.

Barbour, E., Deakin, E.A., 2012. Smart growth planning for climate protection: evaluating California's Senate Bill 375. J. Am. Plann. Assoc. 78 (1), 70–86.

Ben-Akiva, M., Lerman, S.R., 1985. Discrete Choice Analysis. Theory and Application to Travel Demand. The MIT Press, Cambridge, MA.

Bento, A.M., Cropper, M.L., Mobarak, A.M., Vinha, K., 2005. The effects of urban spatial structure on travel demand in the United States. Rev. Econ. Stat. 87 (3), 466–478.

Bhatta, S.D., Drennan, M.P., 2003. The economic benefits of public investment in transportation: a review of recent literature. J. Plann. Educ. Res. 22, 288–296.

BTS, 2009. Transportation, energy and the environment. National Transportation Statistics 2009. U.S. Department of Transportation, Bureau of Transportation Statistics.

Caltrans, 2012. California Toll Schedules. California Department of Transportation.

Cambridge Systematics, Inc, 2009. Moving Cooler: An Analysis of Transportation Strategies for Reducing Greenhouse Gas Emissions. Urban Land Institute, Washington DC.

Cambridge Systematics Inc. and Apogee Research, 1996. Measuring and valuing transit benefits and disbenefits: summary. In: Transit Cooperative Research Program Report 20, Transportation Research Board. National Academy Press, Washington, DC.

Castro, T.S., de Souza, T.M., Silveira, J.L., 2017. Feasibility of electric vehicle: electricity by grid × photovoltaic energy. Renew. Sustain. Energy Rev. 69, 1077–1084.

Cervero, R., 2002. Built environments and mode choice: toward a normative framework. Transp. Res., D 7, 265–284.

Cervero, R., Landis, J., 1995. The transportation-land use connection still matters. Access 7, 2–10.

Clewlow, R., Mishra, G.S., 2017. Disruptive Transportation: The Adoption, Utilization, and Impacts of Ride-Hailing in the United States. Institute of Transportation Studies, University of California, Davis, CA, Research Report UCD-ITS-RR-17-07.

Coughlin, C.C., Terza, J.V., Aromdee, V., 1991. State characteristics and the location of foreign direct investment within the United States. Rev. Econ. Stat. 73 (4), 675–683.

Dahl, C.A., 1995. Demand for transportation fuels: a survey of demand elasticities and their components. J. Energy Lit. 1 (2), 3–27.

Dahl, C., Sterner, T., 1991. Analyzing gasoline demand elasticities: a survey. Energy Econ. 13 (3), 203–210.

Davies, D.G., Halliday, M.E., Mayes, M., Pockock, R.C., 1997. TRL Report 266 Attitudes to Cycling—A Qualitative Study and Conceptual Framework. Transport Research Laboratory.

Deakin, E., Harvey, G., 1996. Transportation Pricing Strategies for California: An Assessment of Congestion, Emissions, Energy, and Equity Impacts. California Environmental Protection Agency. Air Resources Board Research Division, Final Report.

Dijk, M., Orsato, R., Kemp, R., 2013. The emergence of an electric mobility trajectory. Energy Policy 52, 135–145.

Dueker, K.J., Strathman, J.G., Bianco, M.J., 1998. Strategies to Attract Auto Users to Public Transportation. Transportation Research Board, Washington, DC, TCRP Report 40.

Duffy-Deno, K.T., Eberts, R.W., 1991. Public infrastructure and regional economic development: a simultaneous equations approach. J. Urban Econ. 30, 329–343.

Eisner, R., 1991. Infrastructure and regional economic performance: comment. N. Engl. Econ. Rev. (September/October), 47–58.

Eliasson, J., 2008. Lessons from the Stockholm congestion charging trial. Transp. Policy 15 (6), 395–404.

Ellaway, A., Macintyre, S., Hiscock, R., Kearns, A., 2003. In the driving seat: psychosocial benefits from private motor vehicle transport compared to public transport. Transp. Res., F 6 (3), 217–231.

Fernald, J.G., 1999. Roads to prosperity? Assessing the link between public capital and productivity. Am. Econ. Rev. 89 (June), 619–638.

Forsyth, P.J., 1977. The pricing of urban transportation: some implications of recent theory. In: Hensher, D.A. (Ed.), Urban Transport Economics. Cambridge University Press, Cambridge.

Franklin, J.P., 2007. Decomposing the distributional effects of roadway tolls. In: Paper 07-2991. 86th Annual Meeting of the Transportation Research Board, Washington, DC.

Fulton, L., Mason, J., Meroux, D., 2017. Three Revolutions in Urban Transportation. Sustainable Transportation Energy Pathways of the Institute of Transportation Studies, UC Davis and the Institute for Transportation and Development Policy.

Garcia-Mila, T., McGuire, T., Porter, R.H., 1996. The effect of public capital in state-level production functions reconsidered. Rev. Econ. Stat. 78 (February), 177–180.

Gardner, B., Abraham, C., 2007. What drives car use? A grounded theory analysis of commuters' reasons for driving. Transp. Res., F 10 (3), 187–200.

Gillen, D.W., 1977. Estimation and specification of the effects of parking costs on urban transport mode choice. J. Urban Econ. 4, 186–199.

Golob, T.F., Beckmann, M.J., Zahavi, Y., 1981. A utility-theory travel demand model incorporating travel budgets. Transp. Res., B 15 (6), 375–389.

Gómez-Ibáñez, J.A., 1999. Pricing. In: Gomez-Ibanez, J., Tye, W.B., Winston, C. (Eds.), Essays in Transportation Economics and Policy: A Handbook in Honor of John R. Meyer. Brookings Institution Press, pp. 99–136, Chapter 4.

Gotz, K. A research concept for mobility styles. In: Paper Presented at the 10th International Association of Travel Behaviour Research Tri-Annual Conference, Lucerne, 10–15 August 2003.

Graham, D.J., Glaister, S., 2002. The demand for automobile fuel: a survey of elasticities. J. Transp. Econ. Policy 36 (1), 1–25.

Greene, D., 1996. Transportation and Energy. Eno Foundation, Washington, DC.

Greene, D., Hu, P., 1986. A Functional Form Analysis of the Short-Run Demand for Travel and Gasoline by One-Vehicle Households. Transportation Research Board,

National Research Council, Washington, DC, pp. 10–15, Transportation Research Record 1092.
Greene, D.L., Schafer, A., 2003. Reducing Greenhouse Gas Emissions From U.S. Transportation. Pew Center on Global Climate Change.
Greening, L., Jeng, H., Formby, J., Cheng, D., 1995. Use of region, life-cycle and role variables in the short-run estimation of the demand for gasoline and miles travelled. Appl. Econ. 27, 643–656.
Gunn, H.F., 1981. Travel budgets—a review of evidence and modeling implications. Transp. Res. 15A, 7–23.
Handy, S., 1993. Regional Versus Local Accessibility: Implications for Nonwork Travel. TRB, National Research Council, Washington, DC, pp. 58–66, Transportation Research Record 1400.
Hankey, S., Marshall, J.D., 2010. Impacts of urban form on future US passenger-vehicle greenhouse gas emissions. Energy Policy 38, 4880–4887.
Hanson, S., 2004. The context of urban travel: concepts and recent trends. In: Hanson, S., Giuliano, G. (Eds.), The Geography of Urban Transportation, third ed. The Guilford Press, pp. 3–29, Chapter 1.
Heanue, K., 1998. Harmonizing transportation and community values. ITE J. 68 (11), 32–35.
Holleyman, C., 1996. Industry studies of the relationship between highway infrastructure investment and productivity. Logist. Transp. Rev. 32 (1), 93–117.
Hughes, J.E., Knittel, C.R., Sperling, D., 2008. Evidence of a shift in the short-run price elasticity of gasoline demand. Energy Journal, International Association for Energy Econ. 29 (1), 113–134.
Hupkes, G., 1982. The law of constant travel time and trip-rates. Futures 14 (1), 38–46.
IEA, 2017. Tracking clean energy progress 2017. Energy Technology Perspectives 2017. International Energy Agency. OECD/IEA.
IEA, 2018. Global Electric Vehicle (EV) Outlook 2018. International Energy Agency. OECD/IEA.
ITF, 2015. Urban Mobility System Upgrade: How Shared Self-Driving Cars Could Change City Traffic. OECD Publishing, Paris.
ITF, 2017. ITF Transport Outlook 2017. OECD Publishing, Paris.
Jansson, J.O., 2010. Road pricing and parking policy. Res. Transp. Econ. 29, 346–353.
Jensen, M., 1999. Passion and heart in transport—a sociological analysis on transport behaviour. Transp. Policy 6, 19–33.
Johansson, M.V., Heldt, T., Johansson, P., 2006. The effects of attitudes and personality traits on mode choice. Transp. Res., A 40, 507–525.
Kain, J.F., 1999. The urban transportation problem: a reexamination and update. In: Gomez-Ibanez, J., Tye, W.B., Winston, C. (Eds.), Essays in Transportation Economics and Policy: A Handbook in Honor of John R. Meyer. Brookings Institution Press, pp. 359–401, Chapter 11.
Keeler, T.E., Small, K.A., 1977. Optimal peak-load pricing, investment, and service levels on urban expressways. J. Pol. Econ. 85, 1–25.
Kelly, J.A., Clinch, J.P., 2009. Temporal variance of revealed preference on-street parking price elasticity. Transp. Policy 16, 193–199.
Kulash, D., 1974. Parking Taxes as Roadway Prices: A Case Study of the San Francisco Experience. The Urban Institute, Washington, DC.
Levinson, D., 2010. Equity effects of road pricing: a review. Transp. Rev. 30 (1), 33–57.
Luce Jr., T.F., 1994. Local taxes, public services, and the intrametropolitan location of firms and households. Public Finance Q. 22 (2), 139–167.
Menon, G., 2000. ERP in Singapore: a perspective one year on. The ERP experience in Singapore. Traffic Eng. Control 41 (2), 40–45.

Mofidi, A., Stone, J., 1990. Do state and local taxes affect economic growth? Rev. Econ. Stat. 72, 689–691.
Morrison, C.J., Schwartz, A.E., 1996. Public infrastructure, private input demand, and economic performance in New England manufacturing. J. Bus. Econ. Stat. 14, 91–101.
MTC, 2010. Commission Approves Toll Hike Package for Earthquake Safety. Metropolitan Transportation Commission, Transportation News. <http://www.mtc.ca.gov/news/info/toll_increase.htm> (accessed 01.09.18).
MTC, 2016. Workers by means of transportation to work (1980–2000). In: MTC/ABAG Bay Area Census. Metropolitan Transportation Commission and the Association of Bay Area Governments.
Munnell, A.H., Cook, L.M., 1990. How does public infrastructure affect regional economic performance? N. Engl. Econ. Rev. (September/October), 11–32.
Nadiri, M.I., Mamuneas, T.P., 1994. The effects of public infrastructure and R&D capital on the cost structure and performance of U.S. manufacturing industries. Rev. Econ. Stat. 76 (February), 22–37.
National Academies of Sciences, Engineering, and Medicine, 2004. Traveler Response to Transportation System Changes Handbook, Third Edition: Chapter 9, Transit Scheduling and Frequency. Washington, DC: The National Academies Press. https://doi.org/10.17226/23433.
Ng, W., 2017. Work schedule flexibility and parking preferences. J. Transp. Land Use 10 (1), 57–75.
Offer, G.J., Howey, D., Contestabile, M., Clague, R., Brandon, N.P., 2010. Comparative analysis of battery electric, hydrogen fuel cell and hybrid vehicles in a future sustainable road transport system. Energy Policy 38 (1), 24–29.
OICA, 2017. Sales Statistics. International Organization of Motor Vehicle Manufacturers.
Ory, D.T., Mokhtarian, P.L., 2005. When is getting there half the fun? Modeling the liking for travel. Transp. Res., A 39A (2–3), 97–124.
Parry, I.W.H., Small, K.A., 2005. Does Britain or the United States have the right gasoline tax? Am. Econ. Rev. 95 (4), 1276–1289.
Peng, M., Liu, L., Jiang, C., 2012. A review on the economic dispatch and risk management of the large-scale plug-in electric vehicles (PHEVs) penetrated power systems. Renew. Sustain. Energy Rev. 16, 1508–1515.
Perez, B., Lockwood, S., 2006. Current Toll Road Activity in the U.S.: A Survey and Analysis. Office of Transportation Studies, FHWA, Washington DC.
Phang, S., Toh, R.S., 1997. From manual to electronic road congestion pricing: the Singapore experience and experiment. J. Transp. Geogr. 10 (1), 29–38.
Pickrell, D.H., 1999. Transportation and land use. In: Gomez-Ibanez, J., Tye, W.B., Winston, C. (Eds.), Essays in Transportation Economics and Policy: A Handbook in Honor of John R. Meyer. Brookings Institution Press, pp. 403–435, Chapter 12.
Pigou, A.C., 1912. Wealth and Welfare. MacMillan and Co., Limited, London.
Pollet, B., Staffell, I., Shang, J.L., 2012. Current status of hybrid, battery and fuel cell electric vehicles: From electrochemistry to market prospects. Electrochim. Acta 84, 235–249.
Pucher, J., Renne, J.L., 2003. Socioeconomics of urban travel: evidence from the 2001 NHTS. Transp. Q. 57 (3), 49–77.
Quinet, E., 1997. Full social cost of transportation in Europe. In: Greene, D., Jones, D., Delucchi, M. (Eds.), The Full Costs and Benefits of Transportation. Contributions to Theory, Method and Measurement. Springer-Verlag, pp. 69–112.
Randall, C., 2000. The influence of urban form on travel: an interpretive review. J. Plann. Lit. 15 (1), 3–23.

Santos, G., Fraser, G., 2006. Road pricing: lessons from London. Econ. Policy 21 (46), 263–310.

Santos, G., Behrendt, H., Teytelboym, A., 2010. Part II: Policy instruments for sustainable road transport. Res. Transp. Econ. 28, 46–91.

Sarmiento, S., 2000. Household, gender and travel. In: Proceedings from the Second National Conference on Women's Travel Issues. Federal Highway Administration.

Schafer, A., Victor, D.G., 2000. The future mobility of the world population. Transp. Res., A 34, 171–205.

Shareef, H., Islam, M., Mohamed, A., 2016. A review of the stage-of-the-art charging technologies, placement methodologies, and impacts of electric vehicles. Renew. Sustain. Energy Rev. 64, 403–420.

Short, J., Kopp, A., 2005. Transport infrastructure: investment and planning. Policy and research aspects. Transp. Policy 12, 360–367.

Shoup, D., 2005. The High Cost of Free Parking. American Planning Association Planners Press.

Shoup, D., 2007. Cruising for parking. Access 30, 16–22.

Small, K.A., 1983. The incidence of congestion tolls on urban highways. J. Urban Econ. 13, 90–111.

Small, K.A., 1992. Travel demand. Urban Transportation Economics. Harwood Academic Publishers, pp. 5–46, Chapter 2.

Smeed, R., 1968. Road Pricing, the Economic and Technical Possibilities: Report of a Panel Set Up by the Ministry of Transport. H.M.S.O, London.

Southworth, F., 2001. On the potential impacts of land use change policies on automobile vehicle miles of travel. Energy Policy 29, 1271–1283.

Sperling, D., 2018. Three Revolutions: Steering Automated, Shared, and Electric Vehicles to a Better Future. Island Press.

Sperling, D., Gordon, D., 2009. Two Billion Cars: Driving Toward Sustainability. Oxford University Press.

Sterner, T., Dahl, C., Franzen, M., 1992. Gasoline tax policy: carbon emissions and the global environment. J. Transp. Econ. Policy 26, 109–119.

Sullivan, E.C., 2003. Implementing value pricing for U.S. Roadways. EJTIR 3 (4), 401–413.

Tie, S.F., Tan, C.W., 2013. A review of energy sources and energy management system in electric vehicles. Renew. Sustain. Energy Rev. 20, 82–102.

Tietenberg, T., Lewis, L., 2009. Regional and global air pollutants: acid rain and atmospheric modification, Environmental and Natural Resource Economics, eighth ed. Pearson Education, Inc, pp. 413–437, Chapter 17.

Tertoolen, G., van Kreveld, D., Verstraten, B., 1998. Psychological resistance against attempts to reduce private car use. Transp. Res., A 32 (3), 171–181.

TfL, 2003. Impacts Monitoring. First Annual Report. Transport for London. <http://www.tfl.gov.uk/assets/downloads/Impacts-monitoring-report1.pdf> (accessed 23.03.19).

TfL, 2005. Central London Congestion Charging: Impacts Monitoring. Third Annual Report. Transport for London. <http://www.tfl.gov.uk/assets/downloads/ThirdAnnualReportFinal.pdf> (accessed 23.03.19).

TfL, 2007. Central London Congestion Charging: Impacts Monitoring. Fifth Annual Report. Transport for London. <http://www.tfl.gov.uk/assets/downloads/fifth-annual-impacts-monitoring-report-2007-07-07.pdf> (accessed 23.03.19).

U.S. Census, 2016. 2016 American Community Survey (ACS). United States Census Bureau.

U.S. EIA, 2018. U.S. Energy-Related Carbon Dioxide Emissions 2017. U.S. Energy Information Administration. U.S. Department of Energy.

Van Vugt, M., Meertens, R., Van Lange, P., 1995. Car versus public transportation? The role of social value orientations in a real-life social dilemma. J. Appl. Soc. Psychol. 25, 358–378.

Vickrey, W.S., November, 1954. The economizing of curb parking space. Traffic Eng. 62–67.

Walls, M., Krupnick, A., Hood, C., 1993. Estimating the demand for vehicle-miles-traveled using household survey data: results from the 1990 Nationwide Personal Transportation Survey. In: Resources for the Future Discussion Paper ENR 93-25, Washington DC.

Weissman, S., Corbett, J., 1992. Land Use Strategies for More Livable Places. The Local Government Commission, Sacramento, CA.

Wilson, A.G., 1988. Land-use/transport interaction models: past and future. J. Transp. Econ. Policy 32 (1), 3–26.

Willson, R.W., Shoup, D.C., 1990. Parking subsidies and travel choices: assessing the evidence. Transportation 17, 141–157.

Zahavi, Y., 1981. The UMOT-Urban Interactions. US Department of Transportation, Washington, DC, DOT-RSPA-DPB 10/7.

Zahavi, Y., Talvitie, A., 1980. Regularities in travel time and money expenditures. Transp. Res. Rec. 750, 13–19.

CHAPTER 6

Children's travel behavior and implication to transport energy consumption of household: a case study of three Australian cities

Hitomi Nakanishi*
University of Canberra, Canberra, ACT, Australia

6.1 Introduction

Increasing numbers of children are driven to school in developed countries. Factors that affect parental decisions on travel to school are distance and travel time to/from school, concerns about traffic safety, and children's ability to reach school without any catastrophe (Faulkner et al., 2010; Lang et al., 2011). In addition, the convenience to parents in escorting children to school whilst managing their work and household duties is a significant factor (Dowling, 2000). While an increasing amount of studies examine children's travel to school and the factors that are associated with parental travel mode choice using both quantitative and qualitative methods (Buliung et al., 2012), few studies focus on the implications for energy consumption and parents' perceptions. Passenger cars are major contributors to energy consumption and greenhouse gas emissions (González Palencia et al., 2012). It is important to understand how trends of children driven to school are associated with energy consumption and what is the barrier to reducing car use for travel to school. Existing quantitative studies have enhanced the understanding of the relationship between land use and transport energy consumption for school commuting and found that school locations, the built environment, and access to public transport are significant factors that could promote more active transport of children (Saunders et al., 2008; Marique et al., 2013).

* Corresponding author.

Transport and Energy Research
DOI: https://doi.org/10.1016/B978-0-12-815965-1.00006-5

However, these studies did not consider chain trips of parents who escort children, that is, from home to school and from school to work. Also these studies are mainly from European cities where urban areas are relatively of high density, and access to public transport is easier compared to US and Australian cities. In addition, quantitative studies have limitations in understanding the narratives of families, which are underlying determinants of choice and reflections on individuals' lifestyles (Nakanishi and Black, 2016). Organizing travel time for both children and parents is becoming more challenging in society where acceleration is embedded in our general pace of life, and more people are feeling rushed (Rosa, 2003). Under this circumstance, qualitative interviewing is a suitable method of understanding how parents organize the temporal rhythms of children's everyday life under various constraints (Southerton, 2006).

This chapter focuses on children's travel to school and their parents' perceptions of their decisions on travel mode choice, environmental issues, and the cost of travel, which is currently overlooked in literature. The study was conducted in Australia that is a country of 23 million people and provides low-density suburban living to residents. In Australia, it is common that children are driven to local schools, even when they are within walking/cycling distance. In Australia in 1971, 57.7% of children of ages 5–9 years and 44.2% of children of ages 10–14 years walked to school, and 22.8% of children of ages 5–9 years and 12.2% of children of ages 10–14 years were driven to school. By 1991 the share of walking had decreased to 35.3% for ages 5–9 years and 33.1% for ages 10–14 years, and those being driven had increased to 53.9% for ages 5–9 years and 32.7% for ages 10–14 years (van der Ploeg et al., 2008). Recent data from the Australian Bureau of Statistics shows that 46.6% of Australian children of ages 10–18 years were driven to school in 2013. This proportion is higher compared with other countries (Carver et al., 2013). The fact that many children are driven to school is causing congestion around schools, increasing costs to parents through higher car ownership, loss of independent mobility for children leading to inactivity, lack of social connection and various health problems (van der Ploeg et al., 2008), and even an accident in school yard. The kilometers of passenger cars traveled (PCKT) in Australia have keep increasing in Australia (see Fig. 6.1). Simultaneously transport cost is contributing significantly to the financial stress of households in middle and outer suburbs of Australian cities in particular for those with children (Vidyattama et al., 2013). It is important to understand the factors that are affecting the increased number of

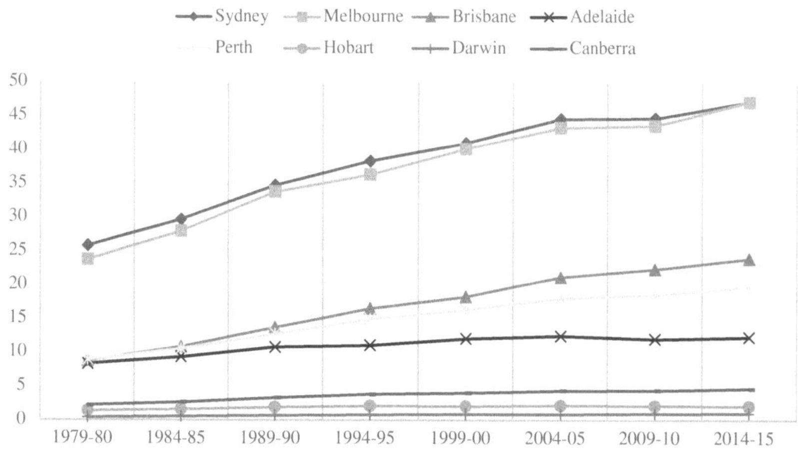

Figure 6.1 Total passenger cars kilometers traveled in capital cities in Australia. *Bureau of Infrastructure, Transport and Regional Economics (BITRE), 2017, Yearbook 2017: Australian Infrastructure Statistics, Statistical Report, BITRE, Canberra ACT.*

children driven to school to identify effective approaches to reduce car use and energy consumption of families. Qualitative interviews with 36 parents from Sydney, Melbourne, and Canberra provide insights into their parental decisions and dilemma on children's transport, their daily routine, main concerns associated with children's travel, and perceptions on environmental issues and cost. A typology approach was taken to interpret the findings of travel choice to enhance the understanding of the relationship with energy use.

This chapter is structured as follows. The next section reviews the literature on increasing car travel by children and the energy consumption associated with this. It is followed by a description of methodology, city context, and participants. Finally, the findings, implications for policy, and areas of future research are discussed.

6.2 Literature review

6.2.1 Increasing car travel of children

The trend in many nations toward increased number of children driven to school and declining active transport and independent mobility for children has been the focus of a large body of research. This research has identified multiple and interrelated factors influencing changing patterns of children's travel to school (Tranter and Sharpe, 2012;

McCarthy et al., 2017). Those factors can be categorized into parental concerns about safety; the complexity of daily activity schedules; social pressures on parents; barriers to using alternative transport modes; distances to school/childcare; increasing reliance on cars for urban transport alongside declines in public transport services; and changes in the built environment, including the closure of local shops, schools, and services.

The most commonly identified factor influencing children's travel is parental concern about safety. The two main concerns relate to traffic and stranger danger (children getting abducted or molested) (Whitzman et al., 2010). A common response by parents to these concerns is an individualistic response—to drive or accompany their children, or restrict their mobility, rather than let them travel independently. Research in New Zealand found that parents mainly cited concerns about traffic safety and stranger danger as likely reasons that their children had less freedom to explore neighborhood environments than they had experienced as children (Witten et al., 2013). Traffic volume and speed, the quality of footpaths and other infrastructure, and the prospect of negotiating road intersections make parents reluctant to let their children walk or cycle unsupervised (Schlossberg et al., 2005; Pont et al., 2013; Stone et al., 2014). Parental perceptions and anxieties about "stranger danger" and personal injury to children are also common in Australia (Rudner, 2012; Moghtaderi et al., 2013) and elsewhere, but some argue that those concerns exaggerate the reality of the dangers children are likely to face (McMillan, 2005; Baslington, 2009).

The second factor identified is time, and this is associated with parents' perception on traffic safety, such as "because it is too dangerous for child (ren) to go to school independently, I (we) need to escort them to school." In this case, parents need to manage time for drop-offs and pick-ups of children no matter what travel mode they take. And in many cases, parents view the car as the appropriate transport mode to enable them to juggle their other responsibilities and constraints on their time (Sheller and Urry, 2000; Schlossberg et al., 2006; McDonald, 2008; Wen et al., 2008; Faulkner et al., 2010; McLaren and Parusel, 2014). Distance to school could be one factor in the decision to drive children to school. However, in a survey of parents living within 2 mi of their children's schools in California, United States found that largest proportion of students (46%) whose school was within walking distance were driven to school (McDonald and Aalborg, 2009). This is because the time cost of school travel depends not only on the child's travel time, but also on the parent's travel time because walking with children takes almost twice the time for parents (McDonald and Aalborg, 2009). The critical implication of children being driven to school is that it means that in many cases, their parents, who could take other modes of transport to work choose to drive to work, that is, drop off children at school and parents continue their journey to work.

6.2.2 Household energy consumption associated with children's travel

Travel habits are formed through the repetitive behavior of travelers who seek to make their travels easier and less risky (Gärling and Axhausen, 2003). A child is a significant factor influencing travelers to reconstruct their travel behavior. Dependence on the car is accelerated at this life stage (Ryley, 2006; Nakanishi and Black, 2016). Once children reach school age, many parents choose car transport to avoid traffic risk. For parents, driving to school becomes a habitual behavior. Consequently, children's independent mobility is increasingly being replaced by adult-dependent and car-dependent mobility. Families who live in outer suburbs are more car dependent and less likely to walk or cycle (Burke and Bonham, 2010) in Australian cities. As a result, household transport energy consumption is affected. Although limited, some studies have focused on the impact of children being driven to school on households' transport energy consumption.

Saunders et al. (2008) estimated transport energy consumption using defined behavior rules, allowing only children and parents who were more than 400 m from the school with no tram/bus stops within 200 m, to use car transport. The study area is the Oststadt district of Karlsruhe, Germany. It is estimated that 17.33 L/week of petroleum is consumed by those children (and accompanying parents) who go to school by car. In this study, number of students who go to school by car is kept to minimum, because cycling and walking are very safe, and many residents are located within 200 m of a tram stop. While this study indicates the extra energy consumption associated with children's car travel to school, it has limitations in quantifying the increasing number of car journeys to school. Singleton (2014) explored the CO_2 emissions associated with school commuting in the United Kingdom. The result shows estimated average daily CO_2 emissions of 1184 g in Sevenoaks; 1115 g in Canterbury; 1110 g in Tonbridge and Malling; 1086 g in East Lindsey; and 969 g in South Holland. Higher average emissions appeared to be in lower density and more rural areas. A study by Van Ristell et al. (2013) simulating the impact of mode share and associated CO_2 emissions if children in the United Kingdom traveled to the nearest school as opposed to their current school shows a 10% drop in car modal share, a 1.1% fall in the total car miles traveled, and −290.66 t petrol use (total pupil population). A study from the United States found that increased walking to school contributes to an 8% decline in car use and a 1%–12% decrease in emissions and that the saving in transportation costs is significant (Marshall et al., 2010).

Studies have shown that the type and size of car owned is associated with family size and age of children (Vincent et al., 2004; McLaren and Parusel, 2014; Nakanishi and Black, 2016; McCarthy et al., 2017). It can therefore be argued that families have additional pressure to purchase a larger vehicle to cope with young children, sports activities on the weekend, fitting prams and strollers, etc. These larger vehicles are likely to consume more fuel per kilometer. Bhat and Sen (2006) show American families with children are more likely to use SUVs and minivans that are less energy efficient. In European context, Marique et al. (2013) discussed that energy consumption for school commuting is strongly dependent upon the school level. Their findings suggest that the concentration of tertiary educational institutions in urban centers leads to higher energy consumption and decentralization of nursery and primary schools throughout the region allows for very low local energy consumption. From a consumer perspective, Grønhøj (2006) found that environmental consequences of car use are not a central concern of families in deciding travel mode in his study of Danish families.

Although methods, hypotheses, and estimated results vary among studies, the overall conclusion is that children being driven to school significantly increases households' fuel consumption and CO_2 emissions. However, it is still unknown how parents are perceiving their decisions and implications to environment and potential dilemma of parents driving their children to school.

6.3 Method

6.3.1 Perceived mobility necessity

This study was designed based on "perceived mobility necessity (PMN)," which is an extended version of the theory of planned behavior (Haustein and Hunecke, 2007). PMN focuses on people's perceptions of mobility-related consequences of their personal living circumstances, which is suitable for the purpose of this study. PMN is related to actual travel demand. Parents and children have higher PMN, and their mobility is affected by various living circumstances, such as availability of public transport and car, road conditions, and parking spaces. Parents tend to have significant resistance to change their travel behavior (Murtagh et al., 2012). In this study, interview questions were designed to understand various living circumstances of parents and children that are closely related with their mobility.

6.3.2 Study area and context of energy use

This study was conducted in three Australian cities: Sydney, Melbourne, and Canberra. Sydney is the largest city in Australia with a population of 4.7 million, followed by Melbourne with 4.3 million. Canberra is considerably smaller with a population of 390,000. Sydney and Melbourne are both characterized by a dense central business district (CBD) area surrounded by sprawling suburban neighborhoods. Canberra, the national capital, is a polycentric structure comprising self-contained towns with a small CBD area. It is rapidly expanding with the development of new neighborhoods in north and west. Canberra has a stable population of young families because of the employment opportunities in national government. Currently, a light rail corridor is under construction in the northern side. In all three cities, the car is the main transport mode.

In Australia the total PCKT in capital cities have been increasing constantly. Sydney and Melbourne have the highest PCKT, whilst Canberra has relatively low PCKT as shown in Fig. 6.1.

Along with the increase in PCKT, energy use of "cars," which includes household cars, consumes the most on a much larger scale than other vehicles, as shown in Fig. 6.2 that presents the energy use in national level.[1]

In terms of methods of travel to work in the three cities, the car (as driver) is dominant, although Sydney has more than 20% of passengers using public transport. Canberra has the highest percentage of "bicycle or walked only" travelers, although use of car is still high and use of public transport is low (Tables 6.1 and 6.2).

6.3.3 Data collection and analysis

In this study, qualitative primary data is used to capture the daily travel behavior of children and parents and develop in-depth insights into social attitudes and perceived barriers to walking to school that are sometimes overlooked

[1] Data for each capital city is not available.

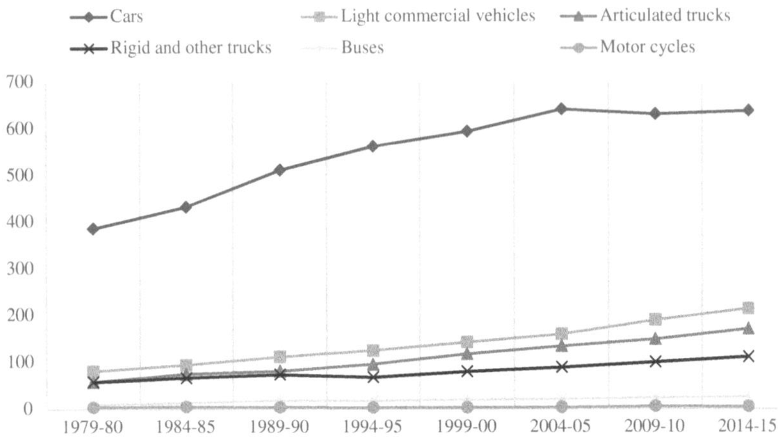

Figure 6.2 Road transport energy use, by vehicle type. *Bureau of Infrastructure, Transport and Regional Economics (BITRE), 2017, Yearbook 2017: Australian Infrastructure Statistics, Statistical Report, BITRE, Canberra ACT.*

Table 6.1 Method of travel to work (2016 census).

	Car, as driver (%)	Train, bus, tram, or ferry (%)	Bicycle or walked only (%)
Sydney	65.5	20.9	5.9
Melbourne	74.4	13.4	5.4
Canberra	74.9	7.1	8.4

Note: Single method of travel only, excludes "worked from home," "not stated," "did not go to work," and all multiresponse categories.
Source: Australian Bureau of Statistics. <http://www.abs.gov.au/ausstats/abs@.nsf/mediareleasesbyReleaseDate/7DD5DC715B608612CA2581BF001F8404?OpenDocument> (retrieved 02.08.18).

through quantitative analysis (Lang et al., 2011; Müggenburg et al., 2015). Purposive sampling technique was used in recruiting interviewees from a mix of locations, socioeconomic status, family structures, and children's ages. Thirty-six participants who have children at the local primary school or/and secondary school[2] in inner and outer suburbs participated in the semistructured interviews from Sydney (9), Melbourne (8), and Canberra (19) (see Appendix B for list of participants). Interviews were conducted during the period of February to December 2014 in Canberra and in June 2015 in Sydney and Melbourne. The questions were designed based on PMN as discussed earlier. They are related to the participants and their partners' family

[2] Children of participants were above 6 or 7 years old and lived in walking/cycling distance of their schools.

Table 6.2 Characteristics of participants.

Characteristics of participants		Number
Gender of interviewee	Male	10 (27.8%)
	Female	26 (72.2%)
Family and work structure	Both parents working or studying full-time	15 (41.7%)
	One parent working or studying full-time, one part-time	12 (33.3%)
	One parent working or studying full-time, one parent at home	3 (8.3%)
	Single parent working full-time	2 (5.6%)
	Single parent working part-time/casual time	3 (8.3%)
	One parent working casual time, one parent at home	1 (2.8%)
Number of children in household	One	10 (27.8%)
	Two	17 (47.2%)
	Three	5 (13.9%)
	Four	4 (11.1%)
Car ownership	One	17 (47.2%)
	Two	19 (52.8%)
Residential district	Inner City	21 (58.3%)
	Outer City	15 (41.7%)
Average distance to school	Sydney	983.3 m
	Melbourne	1412.5 m
	Canberra	768.4 m

structure, the location of their home, employment status and work location, age of their children, car ownership and transport use, weekday and weekend schedules, and their priorities in decision-making (see Appendix A).

Most interviews were conducted in the participants' lunch break, reflecting the fact that they had commitments related to their children before and after work. Each interview lasted between 45 and 60 minutes. All participants were generous in offering further help if the researcher wished to ask follow-up questions by phone or email later. The collected data was rich and saturated enough to be analyzed.[3] Some participants had children who went to schools outside of walking/cycling distance. The cases of these children were

[3] In a qualitative study, the number of participants varies and is often small scale (e.g., Faulkner, Richichi et al., 2010; Dowling, 2000). Data saturation, where the collection of new data does not lead to new ideas or greater clarity on a phenomenon, occurred after 30 interviews. After 30, it was determined that the interviews were not revealing or leading to alternative reflections or suggestions. At this stage, the data was ready to be coded [Creswell (2013). Research design: qualitative, quantitative, and mixed methods approaches. Sage Publications].

excluded from the data. In total, 49 children's cases were used for the study.[4] The average age of children in the data was 9.7 years old.

Interview recordings were transcribed verbatim. A narrative inquiries approach was taken for data analysis in order to take into account the trade-offs and the interrelationship of factors that families consider which underlie travel behavioral choice (Connelly and Clandinin, 1990; Nakanishi and Black, 2016). Each transcript was read several times to develop a full understanding of the context of each individual and the range of nuances in the responses. The interview raw data was translated into meaningful explanatory codes recurring across interviews (Schwanen, 2008). The coding was done by using qualitative analysis software NVivo 11 Pro and further refined several times through repeated inspection. The coding scheme was then used to compare and contrast individual narratives that formed the basis of the findings.

6.4 Findings

6.4.1 Travel behavioral typologies of children and parents

A typology approach, which has been recently used in transport studies (Jones et al., 2014; Delbosc and Nakanishi, 2017), was applied to synthesize the findings of travel behavior of the children and parents. Four typologies emerged from the interview data as shown in Fig. 6.3: (1) children are driven to school, parents drive to work after school drop-offs, (2) children are driven to school, parents return to home or park at a station, then use public transport to work or work from home, (3) both children and parents walk/cycle/use public transport to school and work (parents often go to school with children and then go to work), and (4) children walk/cycle to school either independently or escorted by parents, parents drive to work. Except for the two cases in which children go to school independently, children of participants were escorted to school by car, walking, or cycling.

The number of each type by city is shown in Table 6.3, together with average age and average distance to school.

Type A: Children are driven to school, parents drive to work after school drop-offs

[4] A total of 65.7% of students were enrolled in government schools in 2018 (Australian Bureau of Statistics, 2017a,b https://www.abs.gov.au/ausstats/abs@.nsf/mf/4221.0). Therefore it can be said that children who go to public schools in their catchment area maybe the largest among school children.

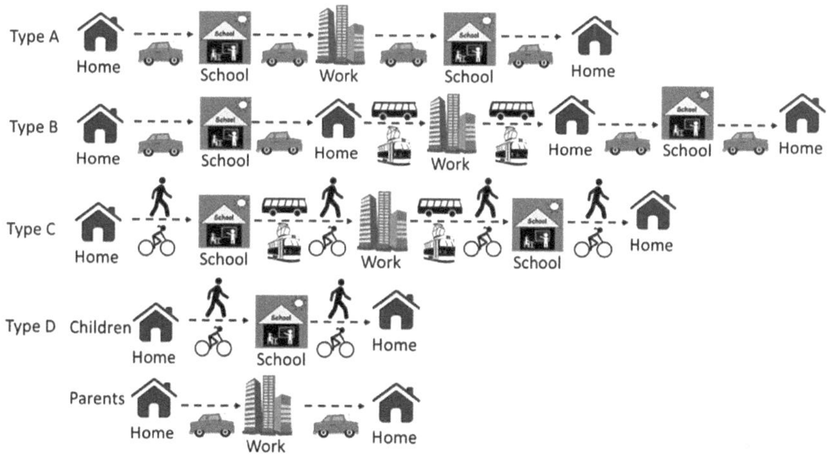

Figure 6.3 Travel behavioral typology of children and parents.

Table 6.3 The features of children in each type of travel behavior by city.

	Sydney	Melbourne	Canberra	Total of three cities (%)	Average age of children	Average distance to school (m)
Type A	3	4	8	15 (41.7%)	9.9	1142.9
Type B	2	2	0	4 (11.1%)	6.8	1375.0
Type C	2	0	6	8 (22.2%)	10.5	842.9
Type D	2	2	5	9 (25.0%)	9.8	583.3

This mode has the largest number among participants. The typical routine of this type is one parent drives children to school (sometimes to more than one school when children go to different schools), drops them off, the parent drives to work. After school, one or other parent picks up the children by car and takes them home or to extracurricular activities. If parents share drop-offs and pick-ups, they usually run two cars. In this case the necessity to drop off and pick up children results in parents always driving to work even if they could use alternative transport. For this type, the main reason they drive children to school is to save time. Comments from parents are

... the only reason I use the car to take him to school is because we tend to try to be just right on time ... Actually if we walk to school it takes about 10, 15 minutes, so the distance is quite walkable distance, but still I take him to school by car.

Canberra, male, both parents working or studying full-time, one child.

She [wife] uses the car to get to and from work because she has to drop the kids at school.

Sydney, male, one parent working full-time, one parent part-time, two children.

It (school) is literally around the corner. It's only about maybe an eight minute walk. But yeah, it's a couple of minutes in the car.

Melbourne, female, One parent working full-time, one parent part-time, three children.

. . . she gets dropped, I drive to before school care every day, despite the fact it's completely walking distance . . . because I know just head off. I then drive into work, which takes me about an hour and a half depending on traffic.

Sydney, female. single parent working part-time, two children.

Type B: Children are driven to school, parents return to home or park at a station, then use public transport to work or work from home

This group is the smallest in the sample and occurs only in Sydney and Melbourne where parking fees are expensive in the CBD area, and traffic congestion is a big issue. Parents drive their children to school (average time taken is about 5 minutes) in the morning, return home or park at the nearest station, and go to work. In the afternoon, they either go home and drive to school or drive from the station to pick up children. Time pressure is also a main factor for this type. Rather than walking/cycling to school, they save time by driving children to school. But because of the parking issue and traffic congestion in the CBD area, parents leave their cars at home or nearest station and take public transport rather than driving to work from the school.

Kids are driven to school . . . because it's just a five minute drive. And then I'll drive to the train station.

Sydney, female, both parents working or studying full-time, two children.

Type C: Both children and parents walk/cycle/use public transport to school and work

This type is seen most in Canberra. The majority of families that are categorized into this type live in the inner area. They value active transport highly and structure their life to ensure minimum car use.

We just walk and pick them up. Typical workday, one of us walk the kids to school, or riding, and the other one collecting them.

Canberra, female, both parents working or studying full-time, two children.

My wife will, on Mondays and Tuesdays, she will start work at sometime between quarter past 9 and 9.30, which allows her to walk with my son to school, and then allow her enough time to ride in to her work and start work.

Canberra, male, both parents working or studying full-time, two children.

Type D: Children walk/cycle to school either independently or escorted by parents, parents drive to work

Families of this type manage children's transport in one of two ways. Some allow children to walk or cycle to school without their parents' supervision while their parents make their own way to work by car. Alternatively, parents walk/cycle to school with their children, return home, then drive to work. Half the families in this type live in the outer areas and their own commuting distance is a consideration whilst they prioritize their children's active transport to school.

> *Now we're trying to get them to be a little bit independent.*
>
> **Canberra, male, one parent working or studying full-time, one part-time, two children.**

> *We walk to school . . . I come home and I get in the car and go to work . . .*
>
> **Melbourne, female, one parent working part-time, one child.**

6.4.2 Use of car and fuel consumption associated with children being driven to school

The main factor leading parents to drive children to school is the time pressure—children need to be at school by a certain time, and parents need to arrive at work on time. The average distance to school of Types A and B are a little more than 1 km that is still a walking/cycling distance. However, if parents need to accompany their children, it takes more time. Some parents are comfortable with the level of safety in their environment, both in terms of traffic danger and stranger danger, which could be associated with the age of children. However, the average age of children of Type A is similar to that of Type D. Nonetheless, the distance to school is shortest for Type D, which is consistent with literature discussing how parents often restrict how far children can venture from home unaccompanied (Valentine, 1997; Black et al., 2001).

While the main concern of parents is time, what could be underestimated is the fuel consumption associated with children's travel. As can be seen in Fig. 6.3, households who use cars to transport children (Types A and B) have doubled the car use of households where only parents use the car for commuting. If children are not driven to school, fuel consumption associated with school travel is zero. This means that using a car for children's travel to school is causing households to consume more fuel on a daily basis. For example, a household of participant 5 (distance to school: 1000 m) that belongs to Type A is driving total of 2 km to and from school a day, which makes up 10 km of car travel to school per week. As Australian students go to school for 40 weeks a year, this totals 400 km per annum. This household consumes an average of 0.2 L of fuel every day[5]—between home and school, which is 1 L a week and 40 L a year. Depending on the type of vehicle, the fuel consumption could be greater. The CO_2 emissions, estimated for this household from travel to school for an average-sized petrol car, are 0.07 t a year.[6] In addition to the fuel consumption and CO_2 emissions from travel to school, Type A households use extra fuel for the commuting parent to travel from school to work, often during peak hour traffic. The total consumption could be a substantial amount.

Table 6.4 provides an indication of how 1 km journey to school (one way), that is, 400 km a year, triggers additional consumption of fuel and CO_2 emissions by different vehicle type. In Australia, many families prefer SUVs that consume more petrol than small or hybrid cars. CO_2 emissions of hybrid luxury cars are low, but initial cost of these cars is often not within the budget of families with average income.

In the following section the perceptions of parents on environmental impacts and energy consumption associated with children's travel to school are examined through interview data.

6.4.3 Perceptions on environmental/energy issues and cost

Type C participants who do not drive to school and work commented that they considered the environmental issues and petrol cost in making decisions of their travel mode, such as

> *In terms of our priorities, it's mostly environmental, it's also about health, obviously. Our kids are very physically active. I know a lot of kids are, get in the car, drive to school . . .*
>
> **Type C, Canberra.**

> *. . . depending on where you live . . . I'm saving on the parking fees, so I'm avoiding parking fees, but given I drive probably close to 40 kilometres a day or would have, the bus fare's probably comparable to petrol costs, and then there's not the wear and tear on the car as well. So I guess that was just another sort of factor that I considered.*
>
> **Type C, Canberra.**

[5] Average rate of fuel consumption per passenger vehicle was 10.6 L/100 km in Australia for the 12 months ending June 30, 2016: <http://www.abs.gov.au/ausstats/abs@.nsf/mf/9208.0/> (accessed 29.07.18).

[6] Calculated on the Carbon Footprint Calculator [<https://www.carbonfootprint.com/calculator.aspx> (accessed 30.07.18)]. The calculation is for an average-sized petrol car.

Table 6.4 Comparison of fuel consumption and CO_2 emissions by vehicle type.

Vehicle type	Popular model	Fuel type	Fuel consumption (L/100 km)	Fuel consumption of 400 km travel (L)	CO_2 emissions of 400 km travel (MT)
Small car	Suzuki Swift	91 or 95 RON	4.6–5.3	18.4–21.2	0.06
Small SUV	Mazda CX-3	Diesel, 91 or 95 RON	3.6–6.1	14.4–24.4	0.13
Medium car	Toyota Camry Hybrid	91 or 95 RON	4.2–8.3	16.8–33.2	0.13
Medium SUV	Mitsubishi Outlander	Diesel or 91 RON	1.9–5.5	7.6–22.0	0.22
Luxury	Audi Q7 e-tron	Diesel or 98 RON	1.8–2.9[a]	7.2–11.6	0.06

Complied by the author referring to carsguide (2017). Cars are all automatic transmission which is the most popular choice in Australia.
[a]Popular luxury cars are predominantly hybrid.

For Type B participants, time matters most.

> *I mean I can ride (bike) to work as well, so I'm trying to do that a bit now . . . and then I started getting stressed about it because I couldn't find the time, and then I thought no, I have to find something that fits in with what I'm already doing, so I'm trying to, but some weeks I just haven't got the energy even though it (school) is just down the road . . . Because I'm a teacher I have to carry my books and I have to actually plan it in, I can't just say oh it's a nice day I might ride today.*
>
> **Type B, Canberra.**

> *Every day I think of time and every day I think of that I want everyone to be happy . . . Because if you're rushing, you are not very fun.*
>
> **Type B, Sydney.**

Participants who drive children to school are aware that their journey to school and work involves time in traffic.

> *I would probably spend . . . Uh . . . Maybe 10 or 15 percent of the journey in traffic.*
>
> **Type A, Melbourne.**

However, time pressure is still a main concern. Participants who have access to public transport said that they considered using it but it is not possible if they try to save time.

> *she (wife) can't take public transport because she has to drive because of the time frame.*
>
> **Type A, Sydney.**

Other participants described that waiting time for public transport is a barrier for a short distance travel.

> *. . . You know, half an hour waiting for a bus when they could have walked home in fifteen minutes. But by the end of the day with full bags, the kids aren't really interested in walking for fifteen minutes . . .*
>
> **Type A, Melbourne.**

6.4.4 Travel to extracurricular activities

All participants stated that their children have extracurricular activities. The majority of them stated that the mode of travel to these activities is by car, regardless of their typology. The main reason is the distance to these activities.

> *. . . swimming is by car. Karate's at his school so it's walking. Um, ballet is car so we drive to that one as well.*
>
> **Type D, Melbourne.**

Because my daughter plays the cello. And the teacher is in the other side of Canberra. So we use a car for that.

Type C, Canberra.

For some participants, the need to take their children to extracurricular activities and time pressure associated with this is affecting their travel mode choice to/from school. For example,

when we've got nights like when we've got activities like on a Wednesday for instance, I have to drive to pick him up because I need to get them, get home straight away (from school) and get them ready before we go.

Type A, Melbourne.

This participant stated that her children have different activities during week and use car for these activities:

that (basketball) will be two nights of training and that will be like on a Tuesday and a Thursday and then he'll play on a Friday night. And then during the winter, he plays football as well. That will be another night of training and then playing on a Sunday. And then girl, she does, it's called Stage School Australia. So it's like a performing arts program. So that's on a Wednesday night and then she has basketball training on a Thursday night and swimming as well for half an hour and then she plays basketball on a Saturday morning.

Type A, Melbourne.

From the interviews, it is evident that the locations of extracurricular activities are beyond walking distance and parents chauffeur children by car, which causes extra consumption of petrol. Distances to these activities vary, but it is reasonable to conclude that the energy consumption associated with this travel is significant.

6.5 Discussion and conclusion

This study examined Australian children being driven to school and its association with energy use. Qualitative interviews with parents from Sydney, Melbourne, and Canberra provided further understanding of parental decisions on children's travel mode to school, focusing on the constraints and perceptions on environmental issues. This study categorized the travel and mobility patterns into four types and compared the circumstances of children and parents of each type.

Concerns about traffic safety around children's independent travel are leading parents to escort children to school. Because most parents work, time pressure is a significant concern as they escort children to school and try to arrive at work on time. Many parents choose to drive their children to save time. This finding is consistent with the literature. In addition, many parents drive children to after-school activities. Parents are aware of adding to traffic congestion and understand that their children could walk to school, but priority on time-saving outweighs other considerations. Expensive parking fee works as a barrier for some parents. However, petrol cost for relatively short distance travel between home and school may be accepted by parents for the sake of saving time. Mackett (2002) described this as an increase in the pace and complexity of modern life. The complexity of everyday schedules is interrelated with barriers to using alternative transport modes, because both factors are grounded in the management of time. For working parents, "daily travel must fit in children's activities alongside parents' work, so walking, cycling or public transport are almost never time-feasible" (Strazdins and Loughrey, 2008). The finding from this study verifies this statement in the low-density Australian-built environment. Implications for policy are twofold: (1) transport and land use integrated policy considering school catchment area and (2) energy policy.

6.5.1 Implication to transport and land use integrated policy considering school catchment area and traffic safety

Even if the school is within walking distance, parents often don't feel comfortable allowing their children to travel unaccompanied because of the concerns for traffic safety. The distance that they allow their children to venture from home independently is much shorter than the distance to school. In this study, it is found that only children in Type D are allowed to go to school independently with the average distance to school being less than 600 m. Participants commented that the location of schools is not always convenient, and they are often some distance away: "It's a smaller number of larger high schools, and that of course means that there's a lot more travel pressure for the students each day." (Type A, Melbourne). Public transport doesn't seem to support children to get around. "... public transport in Melbourne isn't great. Not where we live. It's all right in certain areas where there's trains and things like that. It's fine for getting into the city but it's quite hard to get around." (Type A, Melbourne). To address this issue, a more integrated approach is necessary. The definition of school catchment areas/local enrollment areas is not clear,[7] and this needs to be addressed to ensure that students who attend the local school are within walkable distance. Infrastructure such as walking paths and cycling networks to school that are not attached to high-speed roads should be provided to enhance traffic safety. This may boost the confidence of children and parents. Once this integrated approach is in place, a soft measure such as the "walking school bus" system could be considered. This scheme is particularly effective in Japan, where 98% of children walk/cycle to public schools independently (Mori et al., 2012) (see Appendix C). In addition, implementing lower speed limits (30 km/h or lower) on residential streets around schools (not just in front of schools, but in a wider area, for example, within 200 m radius of school site) would be effective in securing safe travel to school.

Children who live on the edge of a school boundary may still feel that their school is too far to walk/cycle. School buses could be provided even during morning peak hours only, which may be effective in reducing pressure, congestion, and energy use under limited resources.

6.5.2 Implication to energy policy

Even with the integrated land use and transport measures, there would be still many parents who drive children to school, due to time pressure and also "parental traffic safeguarding" (McLaren and Parusel, 2011), which has thus far not been investigated in depth. There is also the effect of habitual behavior—once it becomes a habit, it is hard to change a behavior in everyday life. A scheme such as car sharing may not be preferred by families who have complex schedules, but there could be some incentives that could enable them to reduce petrol consumption. For example, Australian families tend to prefer SUVs or people movers that allow multiple activities and have big boot areas (Drive, 2018). These cars are largely made to run on unleaded petrol. It is suggested that governments collaborate with industry so that SUVs and people movers that families prefer, to provide energy options other than petrol (e.g., hybrid, electronic). Models with new technology tend to be expensive for families, but the reduced cost of energy in the car's lifetime could be an incentive. Governments

[7] For example, in the case of Sydney, the New South Wales (NSW) Government provides the Public School Finder by which individuals can search for their local school by entering their home address. The author used this by randomly inputting addresses of inner area homes and found that distance to school varies considerably (from around 1 to 30 km). The NSW Government's School Finder can be found here: <https://education.nsw.gov.au/school-finder> (accessed 02.08.18).

could also consider subsidizing families who choose low-carbon cars that use cleaner energy. For example, the Japanese government has a scheme to provide reduced tax rates to owners of low-emission cars. There could also be a system where governments could provide a bonus or a tax reduction to families with low car usage and those who run only one car.

For children, there could be a scheme such as "active travel footprint"—children who walk or cycle to school record their walked/cycled distance and according to the distance, then they could be awarded at school and receive a prize. The cost could be covered by local governments, and this could, in turn, contribute to supporting the personal development of children.

In addition, housing affordability is a major concern for Australian families. The median house price is about $1,150,000 in Sydney, $914,518 in Melbourne, and $727,914 in Canberra (Domain, 2018). For young families, it is not easy to live close to the city center where schools, offices, and necessary services are within walking/cycling distance. Governments could consider some support for those families who wish to live in the inner area, with the trade-off of not owning a car (or having limited car use). Suitable housing options for families in the inner area also need to be provided.

The limitations of this study and future research direction need to be noted. Because of the nature of qualitative research, this study does not identify definitive trends among Australian families, nor statistical analysis. Rather, this study aimed to provide insights into the everyday travel of children and parents and perceptions on their travel mode decisions. In future research, a questionnaire survey on the detailed schedule of travel and travel mode choice would be useful in understanding the causality of each factor affecting their decisions. Furthermore, investigating parents' priorities on convenience, environment, and travel cost (including the types of cars they own) would provide more insights into policy implications. A conjoint analysis would be a useful method. There is little research on the influence of children's future travel mode choices after being driven to school every day when young. Children are described as "transport's canaries," and it is unlikely that they will be motivated to develop active travel after years of car travel to school (Freeman and Quigg, 2009). However, there are an increasing number of studies finding that the millennial generation is less car dependent and multimodal, and Australian millennials show a similar tendency (Delbosc and Nakanishi, 2017). It would be useful to examine how children who were driven to school choose their travel mode when they get older. In addition, relationship of travel mode choice and impact on children's health is outside of the scope of this study. Extensive research has already been done in this area.

Nonetheless, this chapter provides the advanced understanding of children being driven to school and the associated energy use. There are various ways to examine travel behavior, but this study applied a qualitative approach. Qualitative research has the advantage in exploring travel behavior as a routine of everyday life through narratives. This chapter demonstrated how travel behavior and travelers' views could be analyzed through narratives. It is suggested that more qualitative studies in travel behavior research would be applied, in a society where peoples' life is becoming more complicated.

Acknowledgments

This research was approved by the Committee for Ethics in Human Research of University of Canberra, Australia (Project number 13–186). The author wishes to thank Paul Tranter, Anne Daly, and Pauline Busch for their valuable comments on the research.

Appendix A: Interview script

Q1. Please let us know about your current life circumstances (family structure, age of children, residential location, car ownership, etc.).

Q2. Please let us know about your/your partner's work circumstances (occupation, working hours, work location, etc.).

Q3. Please let us know your household work and parenting circumstances (sharing of household work and child care, if there is any help from family or nanny, etc.)

Q4. Please let us know your/your partner's schedule on a typical weekday (excluding holidays) and travel diary (including children's travel and possible chauffeuring of children, mode of transport, distance covered).

Q5. Please let us know your/your partner's schedule on a typical weekend (excluding holidays) and travel diary (including children's travel and possible chauffeuring of children, mode of transport, distance covered).

Q6. Please let us know your priorities in making decision on how, where to live and work and time use.

Q7. Did transport considerations affect your choice about where to live and how to work?

Q8. Are there any concerns about your schedule in weekday/weekend, especially in organizing both your schedule and child(ren)? (e.g., transport-related concerns such as taking child(ren) to school, going to work, picking up child(ren) from school, time—budget concerns such as organizing your work, household duties, and looking after children). How do you address these?
Q9. Do you have any suggestions for transport providers/government that might help make things easier for families? (e.g., improving the school bus service, better public transport system, and flexible work options/scheme)
Q10. Is there anything else we need to know regarding your life, work, and parenting circumstances?
Q11. Is there anybody else we should talk to?

Appendix B: List of interview participants

Participant number	Gender	City[a]	Age of children who go to local school	Number of car at household	Home location (inner/outer area)	Typology
P1	Female	S	15	2	Inner	D
P2	Female	S	10	2	Inner	D
P3	Female	S	7, 10	2	Inner	A
P4	Female	S	7, 19	1	Outer	A
P5	Male	S	6	1	Inner	A
P6	Male	S	6	1	Inner	C
P7	Male	S	6, 8	2	Inner	B
P8	Female	S	7	1	Inner	B
P9	Female	S	9	1	Inner	C
P10	Female	M	16, 18	2	Inner	A
P11	Female	M	8, 11	1	Outer	A
P12	Female	M	7	2	Inner	D
P13	Female	M	8	1	Outer	D
P14	Male	M	8	2	Outer	A
P15	Male	M	7	1	Outer	B
P16	Female	M	7, 9	2	Outer	A
P17	Female	M	6	1	Inner	B
P18	Female	C	7, 10	1	Inner	C
P19	Male	C	8	2	Inner	A
P20	Female	C	7	2	Outer	A
P21	Female	C	8	1	Inner	D

(Continued)

Participant number	Gender	City[a]	Age of children who go to local school	Number of car at household	Home location (inner/outer area)	Typology
P22	Female	C	13	1	Inner	C
P23	Female	C	6, 16, 17	2	Outer	A
P24	Female	C	9	2	Outer	A
P25	Female	C	7, 9	2	Outer	D
P26	Female	C	6	2	Outer	A
P27	Male	C	10, 13	1	Inner	C
P28	Female	C	10, 12	2	Outer	A
P29	Female	C	10	2	Inner	C
P30	Female	C	12	2	Inner	D
P31	Female	C	9	1	Outer	D
P32	Female	C	10	2	Outer	A
P33	Female	C	10	1	Outer	C
P34	Male	C	9	1	Inner	C
P35	Male	C	7	2	Inner	A
P36	Male	C	10, 11	2	Outer	D

[a] *S*, Sydney; *M*, Melbourne; *C*, Canberra.

Appendix C: Photo of Japanese children on walking school bus

Photo taken by the author in June 2018.

References

Australian Bureau of Statistics, 2017a. Method of travel to work 2016 census. <http://www.abs.gov.au/ausstats/abs@.nsf/mediareleasesbyReleaseDate/7DD5DC715B608612CA2581BF001F8404?OpenDocument> (retrieved 02.08.18).

Australian Bureau of Statistics, 2017b. Survey of motor vehicle use, Australia, 12 months ended 30 June 2016. <http://www.abs.gov.au/ausstats/abs@.nsf/mf/9208.0/> (retrieved 29.07.18).

Baslington, H., 2009. Children's perceptions of and attitudes towards, transport modes: why a vehicle for change is long overdue. Child. Geogr. 7 (3), 305–322.

Bhat, C.R., Sen, S., 2006. Household vehicle type holdings and usage: an application of the multiple discrete-continuous extreme value (MDCEV) model. Transp. Res. B: Methodol. 40 (1), 35–53.

Black, C., Collins, A., Snell, M., 2001. Encouraging walking: the case of journey-to-school trips in compact urban areas. Urban Stud. 38 (7), 1121–1141.

Buliung, R., Sultana, S., Faulkner, G., 2012. Guest editorial: special section on child and youth mobility – current research and nascent themes. J. Transp. Geogr. 20 (1), 31–33.

Burke, M.I., Bonham, J., 2010. Rethinking oil depletion: what role can cycling really play in dispersed cities? Aust. Plann. 47 (4), 272–283.

Carver, A., Watson, B., Shaw, B., Hillman, M., 2013. A comparison study of children's independent mobility in England and Australia. Child. Geogr. 11 (4), 461–475.

Connelly, F.M., Clandinin, D.J., 1990. Stories of experience and narrative inquiry. Educ. Res. 19 (5), 2–14.

Creswell, J.W., 2013. Research Design: Qualitative, Quantitative, and Mixed Methods Approaches. Sage Publications.

Delbosc, A., Nakanishi, H., 2017. A life course perspective on the travel of Australian millennials. Transp. Res. A: Policy Pract. 104, 319–336.

Domain, 2018. House price report March 2018. <https://www.domain.com.au/product/house-price-report-march-2018/> (retrieved 02.08.18).

Dowling, R., 2000. Cultures of mothering and car use in suburban Sydney: a preliminary investigation. Geoforum 31 (3), 345–353.

Drive, 2018. Family car research & buying guide. <https://www.drive.com.au/what-car-should-i-buy/family-car> (retrieved 02.08.18).

Faulkner, G.E., Richichi, V., Buliung, R.N., Fusco, C., Moola, F., 2010. What's "quickest and easiest?": parental decision making about school trip mode. Int. J. Behav. Nutr. Phys. Act. 7 (1), 62.

Freeman, C., Quigg, R., 2009. Commuting lives: children's mobility and energy use. J. Environ. Plann. Manage. 52 (3), 393–412.

Gärling, T., Axhausen, K.W., 2003. Introduction: habitual travel choice. Transportation 30 (1), 1–11.

González Palencia, J.C., Furubayashi, T., Nakata, T., 2012. Energy use and CO_2 emissions reduction potential in passenger car fleet using zero emission vehicles and lightweight materials. Energy 48 (1), 548–565.

Grønhøj, A., 2006. Communication about consumption: a family process perspective on 'green' consumer practices. J. Consum. Behav. 5 (6), 491–503.

Haustein, S., Hunecke, M., 2007. Reduced use of environmentally friendly modes of transportation caused by perceived mobility necessities: an extension of the theory of planned behavior 1. J. Appl. Soc. Psychol. 37 (8), 1856–1883.

Jones, H., Chatterjee, K., Gray, S., 2014. A biographical approach to studying individual change and continuity in walking and cycling over the life course. J. Transp. Health 1 (3), 182–189.

Lang, D., Collins, D., Kearns, R., 2011. Understanding modal choice for the trip to school. J. Transp. Geogr. 19 (4), 509–514.
Mackett, R.L., 2002. Increasing car dependency of children: should we be worried? Proceedings of the Institution of Civil Engineers-Municipal Engineer. Vol. 151. No. 1. Published for the Institution of Civil Engineers by Thomas Telford Services, London, c1992-, 2002.
Marique, A.-F., Dujardin, S., Teller, J., Reiter, S., 2013. School commuting: the relationship between energy consumption and urban form. J. Transp. Geogr. 26, 1–11.
Marshall, J.D., Wilson, R.D., Meyer, K.L., Rajangam, S.K., McDonald, N.C., Wilson, E.J., 2010. Vehicle emissions during children's school commuting: impacts of education policy. Environ. Sci. Technol. 44 (5), 1537–1543.
McCarthy, L., Delbosc, A., Currie, G., Molloy, A., 2017. Factors influencing travel mode choice among families with young children (aged 0–4): a review of the literature. Transp. Rev. 37, 1–15.
McDonald, N.C., 2008. Household interactions and children's school travel: the effect of parental work patterns on walking and biking to school. J. Transp. Geogr. 16 (5), 324–331.
McDonald, N.C., Aalborg, A.E., 2009. Why parents drive children to school: implications for safe routes to school programs. J. Am. Plann. Assoc. 75 (3), 331–342.
McLaren, A.T., Parusel, S., 2011. Parental traffic safeguarding at school sites: Unequal risks and responsibilities. Can. J. Soc. 36 (2), 161–184.
McLaren, A.T., Parusel, S., 2014. Watching like a hawk': gendered parenting in automobilized urban spaces. Gender Place Cult. 22, 1–19 (ahead-of-print).
McMillan, T.E., 2005. Urban form and a child's trip to school: the current literature and a framework for future research. J. Plann. Lit. 19 (4), 440–456.
Moghtaderi, F., Burke, M., Tranter, P., Armit, C., 2013. Understanding Australian parents' attitudes about their children's travel behaviour: results from the CATCH and iMATCH projects. In: State of Australian Cities Conference.
Mori, N., Armada, F., Willcox, D.C., 2012. Walking to school in Japan and childhood obesity prevention: new lessons from an old policy. Am. J. Public Health 102 (11), 2068–2073.
Müggenburg, H., Busch-Geertsema, A., Lanzendorf, M., 2015. Mobility biographies: a review of achievements and challenges of the mobility biographies approach and a framework for further research. J. Transp. Geogr. 46, 151–163.
Murtagh, N., Gatersleben, B., Uzzell, D., 2012. Self-identity threat and resistance to change: evidence from regular travel behaviour. J. Environ. Psychol. 32 (4), 318–326.
Nakanishi, H., Black, J.A., 2016. Travel habit creation of the elderly and the transition to sustainable transport: exploratory research based on a retrospective survey. Int. J. Sustain. Transp. 10 (7), 604–616.
Pont, K., Wadley, D., Ziviani, J., Khan, A., 2013. The influence of urban form and family decision making on children's travel to school. J. Urban Des. 18 (3), 363–382.
Rosa, H., 2003. Social acceleration: ethical and political consequences of a desynchronized high–speed society. Constellations 10 (1), 3–33.
Rudner, J., 2012. Public knowing of risk and children's independent mobility. Prog. Plann. 78 (1), 1–53.
Ryley, T., 2006. Use of non-motorised modes and life stage in Edinburgh. J. Transp. Geogr. 14 (5), 367–375.
Saunders, M.J., Kuhnimhof, T., Chlond, B., da Silva, A.N.R., 2008. Incorporating transport energy into urban planning. Transp. Res. A: Policy Pract. 42 (6), 874–882.
Schlossberg, M., Phillips, P.P., Johnson, B., Parker, B., 2005. How do they get there? A spatial analysis of a 'sprawl school' in Oregon. Plann. Pract. Res. 20 (2), 147–162.

Schlossberg, M., Greene, J., Phillips, P.P., Johnson, B., Parker, B., 2006. School trips: effects of urban form and distance on travel mode. J. Am. Plann. Assoc. 72 (3), 337–346.
Schwanen, T., 2008. Managing uncertain arrival times through sociomaterial associations. Environ. Plann. B: Plann. Des. 35 (6), 997.
Sheller, M., Urry, J., 2000. The city and the car. Int. J. Urban Reg. Res. 24 (4), 737–757.
Singleton, A., 2014. A GIS approach to modelling CO_2 emissions associated with the pupil-school commute. Int. J. Geogr. Inf. Sci. 28 (2), 256–273.
Southerton, D., 2006. Analysing the temporal organization of daily life:: social constraints, practices and their allocation. Sociology 40 (3), 435–454.
Stone, M., Larsen, K., Faulkner, G.E., Buliung, R.N., Arbour-Nicitopoulos, K.P., Lay, J., 2014. Predictors of driving among families living within 2 km from school: exploring the role of the built environment. Transp. Policy 33, 8–16.
Strazdins, L., Loughrey, B., 2008. Too busy: why time is a health and environmental problem. N S W public health bull. 18 (12), 219–221.
Tranter, P., Sharpe, S., 2012. Disney-Pixar to the rescue: harnessing positive affect for enhancing children's active mobility. J. Transp. Geogr. 20 (1), 34–40.
Valentine, G., 1997. "Oh Yes I Can." "Oh No You Can't": children and parents' understandings of kids' competence to negotiate public space safely. Antipode 29 (1), 65–89.
van der Ploeg, H.P., Merom, D., Corpuz, G., Bauman, A.E., 2008. Trends in Australian children traveling to school 1971–2003: burning petrol or carbohydrates? Prev. Med. 46 (1), 60–62.
Van Ristell, J., Quddus, M., Enoch, M., Wang, C., Hardy, P., 2013. Quantifying the transport-related impacts of parental school choice in England. Transportation 40 (1), 69–90.
Vidyattama, Y., Tanton, R., Nepal, B., 2013. The effect of transport costs on housing-related financial stress in Australia. Urban Stud. 50 (9), 1779–1795.
Vincent, C., Ball, S.J., Pietikainen, S., 2004. Metropolitan mothers: mothers, mothering and paid work. Women's Studies International Forum. Elsevier.
Wen, L.M., Fry, D., Rissel, C., Dirkis, H., Balafas, A., Merom, D., 2008. Factors associated with children being driven to school: implications for walk to school programs. Health Educ. Res. 23 (2), 325–334.
Whitzman, C., Romero, V., Duncan, M., Curtis, C., Tranter, P., Burke, M., 2010. Links between children's independant mobility, active transport, physical activity, and obesity. In: Waters, E., Swinburn, B., Uauy, R., Seidell, J. (Eds.), Preventing Childhood Obesity: Evidence, Policy, and Practice. Wiley-Blackwell, Oxford, pp. 105–112.
Witten, K., Kearns, R., Carroll, P., Asiasiga, L., Tava'e, N., 2013. New Zealand parents' understandings of the intergenerational decline in children's independent outdoor play and active travel. Child. Geogr. 11 (2), 215–229.

CHAPTER 7

Residential location and commuting mode choices: intrahousehold interaction modeling and its implications for energy policy

Junyi Zhang*
Hiroshima University, Higashi Hiroshima, Japan

7.1 Introduction

People consume energy at home and when using a vehicle for travel. Household members living in the suburbs may spend more time outside their home on weekdays, if they work at central urban areas, resulting in less energy consumption at home. Household members living closer to transit stations may use public transport systems more frequently, leading to less energy consumption by car. In this context, Yu et al. (2011) showed that in-home and out-of-home household energy consumption was not independent of each other, and Yu et al. (2012) further confirmed that residential location choices were interacted with ownership and usage of various in-home energy-consuming end uses and vehicles. Thus residential locations and travel modes are closely related to household energy consumption. Accordingly, a better understanding about such household decisions has important implications for both transport and energy policy making. For households with two or more members, intrahousehold interactions cannot be ignored. To this end, Yu et al. (2013) built a household time use and energy consumption behavior model with intrahousehold interactions, together with various interactivity interactions. Intrahousehold interactions may be observed in various ways; however, convincing scientific insights are limited. Therefore this study attempts to present additional ways of representing intrahousehold interactions.

* Corresponding author.

Transport and Energy Research
DOI: https://doi.org/10.1016/B978-0-12-815965-1.00007-7

Group choice survey and modeling methods, especially in the context of household decisions, have been attracted more and more attentions in transportation research. Two relevant special issues were published (Bhat and Pendyala, 2005; Timmermans and Zhang, 2009). The papers included in these two special issues mainly focused on modeling approaches, such as utility-based modeling, latent class modeling, rule-based modeling, and microsimulation approaches. The topics covered household task and time allocation, car ownership, mode choice, activity generation/participation, and scheduling behavior. Since the last special issue, about 10 years have already elapsed; however, no significant progress has been observed with respect to the development of behaviorally oriented group decision models in the literature, with two exceptions: a linear-in-means model with a simultaneous spatial autoregressive Tobit process by Kim and Parent (2016) and a generalized parallel constrained choice model by Weiss and Habib (2018).

In terms of residential location and travel choices, there are only limited studies dealing with intrahousehold interactions from a perspective of survey methods, with the exception of a study by Zhang and Fujiwara (2009), who collected both independent and joint choices related to residential location and commuting mode in the context of Japan. The data collected by Zhang and Fujiwara will be used in this study. Because of the lack of direct surveys about intrahousehold interactions, choice models used in relevant existing studies are all based on normative behavior assumptions. As a result, how to aggregate individual preferences into group choices has not been adequately examined. To make a joint decision, group members need to make various negotiations. Before making a joint decision, each member may already have his/her own decision (i.e., independent choice). If this is the case, group members may have to negotiate with each other with reference to their own independent decision outcomes before the joint decision.

No choice models, however, can be found, which accommodate decision outcome–based negotiations in either the transport or energy literature. Existing choice models with intrahousehold interactions in the context of transport and energy behavior deal with *alternative-level* joint decisions, including Zhang et al. (2009), Zhang and Fujiwara (2009), Kim and Parent (2016), and Weiss and Habib (2018). However, as pointed out by Weiss and Habib (2018), the multilinear utility-based intrahousehold interaction model by Zhang et al. (2009) and Zhang and Fujiwara (2009) is actually a non-RUM (random utility maximization) model. In the multilinear utility model, each member's relative influence in group decision and intermember interactions are simultaneously reflected. The study by Weiss and Habib (2018) has improved our understanding about multilinear utility; the authors suggest to avoid the use of the multilinear utility if the RUM principle is assumed but stated that the multilinear utility has been successfully applied to capture various time use behaviors by specifying time use utility to be positive (see Zhang et al., 2002, 2005). Considering that utility is a relative concept, it is not unrealistic to specify the utility to be positive (or other range of real values) for representing discrete choice behaviors, if keeping the positivity could resolve the problems of the multilinear utility within the RUM principle. This needs more theoretical studies.

Furthermore, group decisions may occur at the attribute level. For example, in the context of household decisions, husband and wife may care about different attributes of choice alternatives, and as a result, they may negotiate with each other by focusing on each alternative attribute, rather than on the whole utility. In other words, some alternative attributes may be within the scope of group decision, while other attributes may not be relevant to group decision at all. If this is the case, all existing *alternative-level* intrahousehold interaction models may be problematic, because they may produce seriously biased model estimations. Unfortunately, no study has been done in the literature to examine the attribute-level group decision.

To fill the preceding research gaps, this study develops two additional types of group choice models at the household level, that is, attribute-based and decision outcome-based intrahousehold interaction models. To this end, this study uses a stated preference (SP) survey that quantitatively captures how to aggregate individual preferences into group choices by focusing on household residential location and commuting mode choices. In the SP survey, husband and wife in a household are targeted, where husband and wife were first asked to make independent choices and then to make a joint decision, under several hypothetical choice scenarios. Details of the preceding SP will be given in the next section.

7.2 A stated preference survey

7.2.1 Differences from existing research

With regard to SP applications in the context of residential choice, Timmermans et al. (1992) and Borgers and Timmermans (1993) conducted a pioneering study based on a hierarchical information integration (HII) theory, which is one type of SP approaches. The HII theory adopts a two-stage survey method. At the first stage, attributes influencing residential choices are classified into several categories. The couple is asked to separately evaluate influences of the categories and their original attributes on residential choices, respectively. After that, influence scores of original attributes under each category are calculated based on statistical methods. At the second stage, the couple is asked to jointly choose one of the alternatives in a choice set based on the preceding influence score of each category. To integrate the two members' preferences, an additive-type household utility was adopted. The advantage of the HII theory is its ease of implementing the SP survey at each stage. However, the couple's evaluations of attributes at the first stage may change depending on the choice alternatives shown in the second stage. As argued by Zhang et al. (2004), the utility of an alternative in a choice set may change with the existence of other alternatives in the choice set. In other words, evaluations of the attributes of the alternative may be altered by the shown alternatives. To overcome the shortcoming of the HII-based SP survey, the SP survey for this study targeted households made up of only couples, where husband and wife were asked first to answer SP questions separately and then to make joint decisions. Details are explained as follows.

7.2.2 Target city: Hiroshima, Japan

The SP survey was conducted in the Hiroshima metropolitan area in 2005, when the population in the core city, Hiroshima City, was 1.15 million (ranked at the 10th place in Japan). In 1994 Hiroshima City opened a transit line, called the ASTRAMLINE, which connects the northwestern part (subcenter) of Hiroshima City to the city center. The ASTRAMLINE was expected to attract a modal shift from cars, thereby mitigating traffic congestion. After the opening of the ASTRAMLINE, ridership increased gradually up to the year 2001; however, since 2001 and until the time of the survey, its ridership declined because of the influences of factors such as the opening of a new road and a direct bus line from the northwestern part of the city to the city center. It seemed difficult to further expect increases in the number of users. Long-term measures to increase the population along the ASTRAMLINE were required from a planning perspective. The preceding situation was the motivation behind the preceding SP survey, which investigated the possibility of promoting residential relocation along the ASTRAMLINE from a long-term perspective.

7.2.3 Survey design

The survey consists of three parts. The first part collects attributes of households and their members (e.g., gender, age, employment status, working place, household income, ownership of driving licenses, and passenger cars), residential attributes [current residential location, monthly rent of apartments (only respondents living in a rental house were targeted), total living spaces, residential duration; distances to neighboring supermarkets, stations, and city center], travel behavior (daily travel mode, travel time and cost, and daily shopping frequency), and intention to move in the future. The second and third parts are the SP questions that investigate respondents' stated choices under several hypothetical choice situations.

The second part investigates individual members' independent choices (i.e., the couples were asked to answer separately without consulting each other), and the third part collects the information about the couples' joint choices. After answering SP questions, the couples reported how they made joint SP decisions [(1) almost always dominated by husband, (2) more or less dominated by husband, (3) equally, (4) more or less dominated by wife, (5) almost always dominated by wife]; satisfaction with final SP answers; and dominant decision makers of previous move to the current residence, purchases of durable electronic equipment, and household asset management, respectively.

The SP choice set includes five alternatives: one is current residential location (i.e., no move), and the other four alternatives are combinations of two residential areas near ASTRAMLINE stations and two commuting modes (car and ASTRAMLINE) of household heads (mainly husbands) for each assumed residential area. Household heads' workplaces were assumed to be at the city center.

- *Alternative 1* (6 km + CAR): living at a residential area located 6 km away from the city center and commuting by car
- *Alternative 2* (6 km + ASTRAMLINE): living at a residential area located 6 km away from city center and commuting by the ASTRAMLINE
- *Alternative 3* (12 km + CAR): living at a residential area located 12 km away from city center and commuting by car
- *Alternative 4* (12 km + ASTRAMLINE): living at a residential area located 12 km away from city center and commuting by the ASTRAMLINE

- *Alternative 5* (No move): no intention to move (i.e., stay at current residential location)

The option of taking the bus was excluded from the choice set, because, in the assumed areas, buses were not a competing mode with cars and the ASTRAMLINE. When a choice environment is made complex, respondents may choose the strategy to delay choice, seek new alternatives, or even revert to the status quo option. To avoid such situations, current residential location was taken as a reference for choices. In reality, the respondents selected in this survey may not want to move to a place along the ASTRAMLINE. For the sake of simplifying the discussion, they were asked to move to the selected residential areas nearby ASTRAMLINE stations.

Each residential area was assumed to have a high-rise condominium, which was characterized by housing price, distances to major neighboring urban facilities, including school, hospital, supermarket, and ASTRAMLINE stations. House prices were calculated based on the assumed floor area and the market price at the time of the survey. Travel time and cost (gasoline) to workplace (i.e., the city center) and parking fee at the city center were selected as car commuting attributes. The attributes and levels assumed in the SP survey are shown in Table 7.1. The attributes of the ASTRAMLINE were fixed to be its values at the time of the survey, together with car commuting travel costs (gasoline and parking fee).

An orthogonal fractional factorial design was applied to the SP survey, and 27 SP profiles were derived. Excluding unrealistic profiles, 20 profiles were finally selected. For mitigating SP biases caused by answering too many SP questions, the 20 profiles (or SP cards) were randomly grouped into five blocks. Each respondent was randomly assigned to answer just one block with four SP questions. Each SP card was designed in a visual format that displayed all the attributes in an easily comprehensible way. An example of an SP card refers to Zhang and Fujiwara (2009).

When a couple was asked to make independent choices, the four SP questions were shown to husband and wife in different orders at random, while in making joint decisions, husband and wife jointly answered the same four randomly ordered SP questions.

7.2.4 Survey implementation

To obtain more reliable survey data, it was necessary to select respondents who would possibly move to the areas along ASTRAMLINE. Such respondents may come from households living in a rental house, with

Table 7.1 Attributes and levels in stated preference (SP) survey.

Attributes	Level 1	Level 2	Level 3
Residential attributes			
Distance to elementary/secondary school (m)	500	1000	–
Distance to neighboring hospital (m)	500	1000	–
Distance to neighboring supermarket (m)	200	500	1000
Distance to ASTRAMLINE station (m)	200	500	1000
Housing price (million Yen)	24	27	30
Commuting travel attributes *Travel time to the city center by car (min)*			
1. Residential area being 6 km far from the city center	20	25	30
2. Residential area being 12 km far from city center	30	35	40

Access travel mode to ASTRAMLINE station (fixed in SP survey)

Distance to ASTRAMLINE station	Access mode	Access time	Access cost
100 m	Walk	3 min	0 Yen
500 m	Walk	7 min	0 Yen
1000 m	Bus	5 min	100 Yen

household members working in Hiroshima City. In 2004 a privacy protection law came into effect in Japan. Since then, it has become more and more difficult to conduct face-to-face interviews. It has also become difficult to get a permission from apartment owners. Due to such difficulties, face-to-face interviews were not conducted, and a web survey was implemented instead.

Web surveys have some good features in comparison to conventional paper-based surveys. First, it is much easier to select respondents at random. Second, there are much fewer missing data, because a web survey can be designed to ask respondents to answer all the required questions before submitting their answers. Third, manual data input is not needed.

The SP survey was conducted with the assistance of a major web survey company in Japan. Since this study only discusses methodological issues, representativeness of the population was ignored, and only households with married couples were selected. One of the spouses was recruited from the members registered at the survey company and asked to invite his/her partner to participate in the survey. For this survey, 539 registered members were contacted, among whom 233 accessed the survey website. Of the 233 members, 24% gave up participation before completing the survey. As a result, 173 households provided complete and valid responses. Thus the collection rate was 32%. As a web survey, this rate is not low. The average answer time was about 12 minutes.

7.2.5 Aggregation analysis

Among the 173 households, 17 households did not have married couples and were excluded from this study. Thus the valid sample was 156 households, of which about 100 couples fell within the 30–39 years old category. The husbands of about 130 households were businessmen. Households with residential durations of 2–4 years accounted for the largest percentage (about 65 households). More than 70% of households expressed their intention to move in the future.

Concerning the SP results chosen separately and jointly by the couples, the share of households with no intention to move (*Alternative 5*) was the highest (about 40%). This suggests difficulty in promoting residence along transit systems. Among the 60% of households intending to move in the future, more households preferred to live closer to the city center and commute by the ASTRAMLINE. This may indirectly reflect the tendency for people to move back to city centers, which was observed in some cities of Japan at the time of the survey. It is further confirmed that more husbands tended to use cars, but more wives hoped that their husbands would use the ASTRAMLINE. Except the case of *Alternative 4* (12 km + ASTRAMLINE), on average, the couple's joint choices tended to fall between the two independent choices.

As for the change of SP choices before and after joint decisions, 40% of households changed their SP choices after joint decisions. As a result of joint decisions, the share of households giving a priority to husbands' decisions is almost the same as that of the wives (about 14%). About 5.6% of households changed their choices after joint decisions, even though the couples showed consistent SP choices independently. For both independent and joint decisions, 6.2% of households showed inconsistent SP choices. Thus it can be concluded that survey and analysis methods ignoring group decision-making mechanisms could result in seriously biased estimation results and misleading policies.

Focusing on joint decisions on major household matters, the relative influences of husbands and wives varied largely with decision contexts. For household asset management, wives showed the largest influence, as expected. Husbands and wives had almost the same relative influences in joint decisions on durable electronic equipment. However, concerning the decision on moving to the current residence, the relative influence of husbands was clearly higher than that of wives. Such relative influences remained almost unchanged in the SP choices.

7.3 Methodology

Here, two additional types of intrahousehold interaction models are presented: attribute-based model and decision outcome-based model. No matter what kinds of group decision rules are applied in reality, it is necessary to clarify how to transform each household member's individual preferences into a household joint decision by reflecting each member's relative influence in the joint decision.

- *Attribute-based intrahousehold interaction model*

  There are three ways to represent members' relative influences in joint decisions, as shown as follows, which correspond to three intrahousehold interaction models.

  - Model 1: Model with members' individual attributes (Attri_IA model)

    Members' individual attributes are used to represent their relative influences in joint decisions on each attribute of an alternative in a choice set.

  - Model 2: Model with members' independent choice results (Attri_CR model)

    Members' independent choice results of an alternative in a choice set are used to represent their relative influences in joint decisions on each attribute of the alternative.

  - Model 3: Model with members' independent choice probabilities (Attri_CP model)

    Members' independent choice probabilities of an alternative in a choice set are used to represent their relative influences in joint decisions on each attribute of the alternative.

  - *Outcome-based intrahousehold interaction model*
    - Model 4: Weighted probability model (Weight_CP model)

Members' independent choice probabilities of an alternative in a choice set are linearly weighted by members' individual attributes to represent a joint decision on the alternative.

7.3.1 Attribute-based intrahousehold interaction models

This type of model assumes that different members attach different importance to different attributes of choice alternatives. Trade-offs between members' relative influences on each attribute may be completely compensatory on the one hand, while joint decisions might be made by members with the weakest or strongest relative influence on the other. To represent such heterogeneities in a general way, the concept of the generalized mean is adopted to accommodate the attribute-based joint decision. In this case, the utility function can be specified as follows:

$$U_{hj} = V_{hj} + \varepsilon_{hj} = \sum_k \beta_k \left(\sum_n \left[\frac{w_{hnj,k}{}^{\alpha_k}}{N} \right] \right)^{1/\alpha_k} x_{hj,k} + \varepsilon_{hj} \tag{7.1}$$

$$w_{hnj,k} = \frac{\exp(v_{hnj,k})}{\sum_{n'} \exp(v_{hn'j,k})} \tag{7.2}$$

where U_{hj} is the utility of household h's alternative j, V_{hj} is the nonstochastic part of U_{hj}, ε_{hj} is an error term of U_{hj}, $x_{hj,k}$ is the kth attribute of household h's alternative j, β_k is a parameter of $x_{hj,k}$, $w_{hnj,k}$ is a weight parameter (or relative influence parameter) that household member n has in a joint decision on attribute $x_{hj,k}$, $v_{hnj,k}$ is the nonstochastic term to describe $w_{hnj,k}$, α_k is a parameter dealing with noncompensatory trade-off relationships between members' relative influences in a joint decision on the kth attribute $w_{hnj,k}$ of alternative j, and N is the number of household members.

In the case that $\alpha_k \to -\infty$, there is no substitution relation between members' relative influences in a joint decision, and the member with minimum relative influence makes the joint decision on a target alternative-specific attribute. When $-\infty < \alpha_k < +1$, the substitution relation becomes incomplete, where a member with a weaker relative influence is more likely to make the final joint decision. When $\alpha_k = +1$, a complete substitution relation is observed. When $+1 < \alpha_k \to +\infty$, a different type of incomplete substitution relation occurs, where a member with a larger relative influence is more likely to make the final joint decision. When $\alpha_k \to +\infty$, no substitution relation exists and a member with maximum relative influence becomes the final joint decision maker.

In summary, when α_k approaches 1, there is a larger substitution relation between members' relative influences in a joint decision, and on the contrary, when α_k goes away from 1, the substitution relation become smaller.

Depending on how to specify the nonstochastic term $v_{hnj,k}$ of the weight parameter $w_{hnj,k}$, Eq. (7.2) can be further transformed into the following three special cases:

$$v_{hnj,k} = \sum\nolimits_s \gamma_{k,s} z_{hn,s} \tag{7.3}$$

$$v_{hnj,k} = \gamma_y y_{hnj} \tag{7.4}$$

$$v_{hnj,k} = \gamma_p p_{hnj} \tag{7.5}$$

where $z_{hn,s}$ is the sth attribute of household member n; y_{hnj} is a choice result of alternative j that household member n makes in the independent choice experiment (it is equal to 1 when j is chosen, otherwise 0); p_{hnj} is a choice probability of alternative j that household member n makes in the independent choice experiment; and $\gamma_{k,s}$, γ_y, and γ_p are the parameters of the above three variables.

Eq. (7.3) corresponds to the Attri_IA model, Eq. (7.4) indicates the Attri_CR model, and Eq. (7.5) is the Attri_CP model. By assuming that error terms (ε_{hj}) follow an independent and identical Gumbel distribution, a logit-type model can be derived. Different from the traditional logit model, the models based on the preceding equations capture intermember interactions, which take the form of intrahousehold interactions in this case.

7.3.2 Outcome-based intrahousehold interaction model

The most popular form of outcome-based group choice models is the following weighted probability model. It is assumed that group choice probability is a probability by weighing members' independent choice probabilities. This model can be described as follows.

$$P_{hj} = \sum\nolimits_n w_{hnj} p_{hnj} \tag{7.6}$$

$$w_{hnj} = \frac{\exp(v_{hnj})}{\sum_{n'} \exp(v_{hn'j})} \tag{7.7}$$

In this case, Eq. (7.7) can be only defined by using members' individual attributes,

$$v_{hnj} = \sum\nolimits_{s} \gamma_s z_{hn,s} \tag{7.8}$$

where $z_{hn,s}$ is the sth attribute of household member n; p_{hnj} is a choice probability of alternative j that household member n makes in the independent choice experiment; and γ_s is the parameter of $z_{hn,s}$.

Even though the preceding two types of choice models are built to represent intrahousehold interactions, they can be applied to more general group decisions.

7.4 Model analysis

In total, four choice models were estimated in this study, that is, Attri_IA, Attri_CR, Attri_CP, and Weight_PR. Explanatory variables introduced into utility functions of all five alternatives will be explained later. Concerning the weight parameters representing husband's and wife's relative influences in a joint decision, the following individual attributes of husband and wife are adopted: age, job type (self-employed or not), number of children under middle/high school, number of household members over 60 years old, and the major decision makers of electric household appliances, household budget management, and previous move, respectively. Even though each respondent answered SP questions four times, the obtained SP responses are regarded as responses from different respondents to simplify the discussion. As a result, 588 valid SP responses are used as the samples for this study.

7.4.1 Results of attribute-based intrahousehold interaction models

The following five types of alternative-specific variables are used to explain household residential and commuting mode choices.

- Alternative-specific log-transformed distance variables (m) are introduced to utility functions of the five alternatives (1–5), respectively, including *ln* (distance to supermarket), *ln* (distance to hospital), *ln* (distance to school), and *ln* (distance to station).
- Travel time to workplace (minute, one-way) is introduced to utility functions of the five alternatives (1–5).

- Travel cost to workplace (Yen, one-way) is introduced to utility functions of the five alternatives (1−5).
- Housing price (10,000 Yen)/household income (10,000 Yen) is introduced to utility functions of alternatives 1−4.
- Dummy variable of access travel mode to station (1: walk; 0: otherwise) is introduced to alternatives 2 and 4.

Estimation results of Attri_IA, Attri_CR, and Attri_CP models are shown in Tables 7.2 and 7.3. The adjusted McFadden's Rho-squared values are 0.1250 for Attri_IA, 0.5787 for Attri_CR, and 0.1877 for Attri_CP. It is obvious that Attri_CR model has the highest model accuracy, suggesting that knowing household members' independent choice results is most important to predict joint choices. The model with members' independent choice probabilities is the second best in terms of model accuracy. In reality, it is not easy to observe members' independent choices. In this sense, the accuracy of Attri_IA is also acceptable.

As for the generalized mean parameters [α in Eq. (7.1)], none of them was found to be statistically significant in Attri_CR and Attri_CP models. In contrast, the parameters related to distance, travel time, travel cost, and access travel mode to station are all significant. Thus if independent choices are known in advance, household members may not be sensitive to attributes of each alternative. However, in the case that independent choices are unknown before joint decision, household members are more likely to become sensitive to the alternative attributes.

If husband and wife make negotiations at the attribute level, husband or wife with weaker relative influence is more likely to become the major influencer for distances, housing price (relative to income), and travel time to workplace, because the relevant generalized mean parameters are smaller than one [distances: −1.095, housing price (relative to income): 0.137, and travel time to workplace: 0.249]. Regarding travel cost to workplace and access travel mode to station, their generalized mean parameters are larger than one (travel cost to workplace: 9.740, access travel mode to station: 40.247), indicating that husband or wife with stronger relative influence is more likely to become the major influencer for these two variables. Then, how about husband's and wife's relative influences in the joint decision? The model estimation results suggest that stronger influencers are older husband/wife, self-employed husband, wife without a job, major decision maker of household budget management, and husband/wife who was not the major decision maker of previous move.

Table 7.2 Estimation results of the Attri_IA model.

Model	Attri_IA	
Explanatory variable	**Parameter**	***t*-Score**
Alternative-specific variable		
ln (distance to supermarket) [1–5]	− 2.531	− 0.966
ln (distance to hospital) [1–5]	− 23.534	− 3.783***
ln (distance to school) [1–5]	4.532	0.916
ln (distance to station) [1–5]	1.770	0.373
Housing price/household income [1–4]	− 0.012	− 3.288***
Travel time to workplace [1–5]	− 0.183	− 2.453**
Travel cost to workplace [1–5]	0.000	− 0.766
Access travel mode to station [2,4]	0.630	3.504***
Alternative-specific constant term		
Constant term [1,3]	− 0.505	− 1.616
Constant term [2,4]	− 0.524	− 1.650
Parameter of generalized mean		
ln (distance) [1–5]	− 1.095	− 5.427***
Housing price/household income [1–4]	0.137	0.866
Travel time to workplace [1–5]	0.249	2.091**
Travel cost to workplace [1–5]	9.740	2.913***
Access travel mode to station [2,4]	40.247	3.903***
Weight parameter to represent members' relative influence in joint decision		
Age: husband and wife	0.204	3.330***
Job type (self-employed: 1, yes; 0, no): husband	0.121	1.983**
Job type (business person, agriculture, forestry, or self-employed: 1, yes; 0, no): wife	− 2.772	− 4.820***
Number of children under middle/high school: wife	− 0.100	− 0.168
Number of members aged 60 + years old: wife	0.002	0.011
Major decision maker of electric household appliance (1: yes, 0: no): husband and wife	1.897	1.660
Major decision maker of household budget management (1: yes, 0: no): husband and wife	1.120	2.458***
Major decision maker of previous move (1: yes, 0: no): husband and wife	− 1.563	− 5.589***
Constant term: husband	− 0.093	− 0.346
Age: husband and wife	2.283	2.703***
Sample size	588	
Initial log-likelihood	− 946.350	
Final log-likelihood	− 819.227	
McFadden's Rho-squared	0.1343	
Adjusted McFadden's Rho-squared	0.1250	

Note: * significant at 10% level, ** at 5% level, *** at 1% level; numbers 1–5 in the parentheses refer to the *Alternatives 1–5* in the SP survey, as shown in Section 7.2.3.

Table 7.3 Estimation results of Attri_CR and Attri_CP models.

Model	Attri_CR		Attri_CP	
Explanatory variable	**Parameter**	***t*-Score**	**Parameter**	***t*-Score**
Alternative-specific variables				
ln (distance to supermarket) [1–5]	− 0.876	− 2.067**	− 1.605	− 1.952*
ln (distance to hospital) [1–5]	− 0.404	− 0.929	0.092	0.108
ln (distance to school) [1–5]	− 0.538	− 1.454	0.239	0.225
ln (distance to station) [1–5]	− 0.186	− 0.439	− 0.914	− 0.986
Housing price/household income [1–4]	0.005	0.855	0.000	− 0.655
Travel time to workplace [1–5]	− 0.054	− 1.138	− 0.479	− 2.010**
Travel cost to workplace [1–5]	− 0.014	− 1.046	− 0.052	− 1.900*
Access travel mode to station [2,4]	0.940	2.160**	10.838	3.139***
Alternative-specific constant term				
Constant term [1,3]	− 0.564	− 2.413**	0.350	1.057
Constant term [2,4]	− 0.870	− 3.486***	0.146	0.531
Parameter of generalized mean				
ln (distance) [1–5]	0.167	1.522	0.138	0.771
Housing price/Household income [1–4]	− 5.811	− 0.622	5.936	0.870
Travel time to workplace [1–5]	0.242	0.785	− 3.317	− 0.826
Travel cost to workplace [1–5]	− 8.486	− 1.115	− 0.655	− 0.443
Access travel mode to station [2,4]	1.588	1.072	− 1.197	− 1.305
Weight parameter to represent members' relative influence in joint decision				
Husband's independent choice result/probability	− 2.005	− 5.478***	− 5.432	− 2.100**
Wife's independent choice result/probability	2.981	4.403***	5.253	2.240**
Constant term: husband	− 1.405	− 1.413	− 3.149	− 6.736***
Sample size	588			
Initial log-likelihood	− 946.350		− 946.350	
Final log-likelihood	− 405.342		− 762.682	
McFadden's Rho-squared	0.5719		0.1939	
Adjusted McFadden's Rho-squared	0.5686		0.1877	

Note: * significant at 10% level, ** at 5% level, *** at 1% level; numbers 1–5 in the parentheses refer to the *Alternatives 1–5* in the SP survey, as shown in Section 7.2.3.

If husband and wife negotiate with respect to their independent choices, husband's independent choice weakens his role in the joint decision (relevant parameters are −2.005 in Attri_CR and −5.432 in Attri_CP), while wife's independent choice strengthens her role in the joint decision (relevant parameters are 2.981 in Attri_CR and 5.253 in Attri_CP). This means that once wife decided her choice, it is difficult to change it in the joint decision, and as a result, husband has to compromise.

7.4.2 Results of outcome-based intrahousehold interaction model

For estimating this model, husband's and wife's independent choice probabilities are first calculated based on the independent SP choice experiments and then introduced into the household choice model as explanatory variables. The weight parameters for describing husband's and wife's relative influences in the joint decision are defined by using the same set of individual attributes as mentioned in the previous subsection. The estimation results are shown in Table 7.4. It is found that the model accuracy is acceptable with the adjusted McFadden's Rho-squared of 0.2089.

Except for the number of household members aged 60 + years old, the other variables explaining the weight parameters are all statistically significant. As the couple age, their relative influences in the joint decision become weaker. Self-employed husbands are more likely to show stronger influence in the joint decision, while wives with a job (business person, working in agriculture and forestry, or self-employed) tend to play a smaller role. In the households with more children, wives' relative influences in the joint decision become weaker. Husbands or wives who are the major decision makers of household budget management and previous move are more likely to show stronger influence in the joint decision in the SP experiment. Meanwhile, husbands or wives who are the major decision makers of electric household appliance tend to play a smaller role in the joint decision. Finally, the constant term indicates that omitted factors from the model are associated with weaker influence of husband in the joint decision.

7.4.3 Energy consumption implications of model estimation results

In this study, even though energy consumption is not directly modeled, it can be interpreted that living in a residential area farther from the city center and commuting by car lead to more energy consumption by car.

Table 7.4 Estimation results of the Weight_CP model.

Explanatory variable	Parameter	*t*-Score
Weight parameter to represent members' relative influence in joint decision		
Age: husband and wife	− 1.202	− 8.282***
Job type (self-employed: 1, yes; 0, no): husband	18.961	6.297***
Job type (business person, agriculture, forestry, or self-employed: 1, yes; 0, no): wife	− 21.843	− 1.747*
Number of children under middle/high school: wife	− 22.501	− 7.428***
Number of members aged 60 + years old: wife	− 0.470	− 0.469
Major decision maker of electric household appliance (1: yes, 0: no): husband and wife	− 4.457	− 2.482**
Major decision maker of household budget management (1: yes, 0: no): husband and wife	2.982	1.835*
Major decision maker of previous move (1: yes, 0: no): husband and wife	18.351	6.794***
Constant term: husband	− 13.253	− 7.523***
Sample size	588	
Initial log-likelihood	− 946.350	
Final log-likelihood	− 745.779	
McFadden's Rho-squared	0.2119	
Adjusted McFadden's Rho-squared	0.2089	

Note: * significant at 10% level, ** at 5% level, *** at 1% level.

Here, energy consumption implications of the estimation results of three attribute-based household choice models are discussed.

Factors affecting residential location and commuting mode choices differ depending on what husband and wife negotiate for the joint decision. In the case of attribute-based negotiation, distance to hospital, housing price (divided by household income), travel time to workplace, and access travel mode to station are influential, while in the case of outcome-based negotiation, distance to supermarket, access travel mode to station, and travel time/cost to workplace (only in the Attri_CP model) become statistically significant.

Thus only access travel mode to station is identified to be influential by all three attribute-based household choice models. Its parameter signs suggest that access to station by walking is associated with more use of public transport, that is, less use of car and as a result, less energy consumption by car. Actually, the choice share of *Alternative 2* (6 km + ASTRAMLINE) is more than twice higher than that of *Alternative 4* (12 km + ASTRAMLINE), indicating that access to station by walking could encourage more residence choices closer to the city center. Overall, making the station closer to residence could encourage more walking to the station, and as a result, less energy consumption by car could be expected.

Except for the preceding access travel mode to station, joint decisions negotiated in terms of alternative attributes and choice outcomes lead to different influential factors. If the negotiation is made at the attribute level, shorter distance to hospital and shorter travel time to workplace as well as lower housing price (relative to household income) are associated with more move from the current residence. Because distance to hospital and travel time to workplace are introduced, with common parameters, to all five alternatives in this study, the reported larger choice shares of residing in a residential area located 6 km away from the city center (in comparison with the area located 12 km away from the center) imply that shorter distance to hospital and shorter travel time to workplace may lead to more residence closer to the city center, and as a result, less energy consumption could also be expected, similar to the effects of encouraging more walking to the station. On the other hand, if the negotiation is made with respect to independent choice outcomes, less energy consumption by car could be expected by shortening the distance to supermarket and travel time/cost to workplace.

Finally, what about the influence of intrahousehold interaction on energy consumption by car? This can be answered by observing generalized mean parameters and weight parameters. Major influencers on distances and travel time to workplace are younger husband/wife, husband who is not self-employed, wife with a job, major decision maker of previous move, and husband/wife who was not the major decision maker of household budget management. Meanwhile, major influencers on travel cost to workplace and access travel mode to station are husbands and wives who are older and major decision makers of household budget management, but not major decision makers of previous move, husbands who are self-employed, and wives without a job. Thus reducing car-related energy consumption via various policies should pay more attention to major influencers on those policy-related factors. Successful transport-related energy policies need effective communication with trip makers, whose residential location and commuting mode choices are affected by intrahousehold interaction.

The preceding discussion suggests that more effective communication with trip makers needs to properly identify who are actual influencers on each policy variable. It is not easy to do. For this, more research accumulation is required in the future. The preceding discussion implies that policy communication is required to be made at the household level, rather than at the individual level.

7.5 Conclusions

This paper has presented new ways of modeling intrahousehold interactions, which are associated with transport and energy research. Residential location and commuting mode choices are examined in the context of Japan. This study has advanced transport and energy research by proposing attribute-based and outcome-based modeling approaches for intrahousehold interactions. These approaches can also be applied to represent intermember interactions in general group decisions. To quantitatively capture group decision-making mechanisms, an SP survey with both independent and joint choices was implemented. The higher modeling accuracy of outcome-based model suggests that households may make final joint decisions by making negotiations with respect to each member's independent choices made before a joint decision. It is found that wives are less likely to modify their independent choices for reaching a joint decision with their partners. It is further confirmed that attribute-based and outcome-based joint decisions lead to different influential factors to husbands' and wives' residential location and commuting mode choices. Husbands and wives impose different influences on different alternative attributes in joint decisions. Modeling analyses imply that transport and energy policymaking should pay more attention to the household context by differentiating between members' different roles in joint decisions.

There are some limitations of this study. First, this study only asked a couple to jointly choose their residential location and the household head's commuting mode. At least in Japan, more and more wives have entered the employment market, and dual-worker households have been increasing. Accordingly, the wife's workplace will become a joint decision target. Furthermore, if a child is present at a household, his/her influence is not negligible, because his/her parent needs to consider the ease of access from residence to locations of his/her daily activities (especially, schools). If there are more household members, residential location and commuting mode choices may not be the only joint decision target. Other choices may need to be decided simultaneously: for example, school location, locations of major daily activities, and social network members. It is therefore necessary to figure out a better modeling framework of representing multiple members' multiple joint choices.

Second, even though household members were asked to first answer the SP questions separately and then to answer jointly, we had no way to check whether they followed such survey instructions or not. Therefore it is important to further develop more reliable survey methods to investigate group decision-making mechanisms/rules.

Third, there are still other types of negotiation strategies in group decisions (Corfman and Gupta, 1993) that have not been examined in this study. As argued by Corfman and Gupta (1993), group decision-making mechanisms/rules may differ from context to context. Accordingly, it is worth clarifying whether to accommodate various decision-making mechanisms/rules in a unified modeling framework or to deal with them separately in different frameworks. This should be further examined by using different types of group choice data.

Fourth, it is important to clarify how to incorporate group decision-making mechanisms/rules into a city-wide or region-wide model to effectively support transportation and energy policy decisions.

Fifth, energy consumption has not been directly analyzed. In this regard, decisions on ownership and usage of vehicles and in-home energy-consuming end uses may involve more complicated intrahousehold interactions, because the decisions contain multiple discrete and continuous choices, and one discrete choice may correspond to two or more continuous choices. For example, once a car is chosen, decisions on how often to use it, how much household expenditure is assigned to it, and how long to own it will follow. Last but not least, transport and energy policies need to be better communicated with various stakeholders, including households, for achieving long-lasting consensus building. For this, more general group choice and social choice theories should be developed in the future.

Acknowledgment

This study was fully supported by the Grants-in-Aid for Scientific Research (A), Japan Society for the Promotion of Science (JSPS), no. 15H02271 (Principal researcher: Junyi Zhang, Hiroshima University).

References

Bhat, C.R., Pendyala, R.M., 2005. Modeling intra-household interactions and group decision-making. Transportation 32, 443–448.

Borgers, A., Timmermans, H., 1993. Transport facilities and residential choice behavior: a model of multi-person choice processes. Pap. Reg. Sci. 72 (1), 45–61.
Corfman, K.P., Gupta, S., 1993. Mathematical models of group choice and negotiations. Handb. Oper. Res. Manage. Sci. 5, 83–142.
Kim, C., Parent, O., 2016. Modeling individual travel behaviors based on intra-household interactions. Reg. Sci. Urban Econ. 57, 1–11.
Timmermans, H.J.P., Zhang, J., 2009. Modeling household activity travel behavior: examples of state of the art modeling approaches and research agenda. Transp. Res., B 43, 187–190.
Timmermans, H., Borgers, A., van Dijk, J., Oppewal, H., 1992. Residential choice behaviour of dual-earner households: a decompositional joint choice model. Environ. Plann. A 24, 517–533.
Weiss, A., Habib, K.N., 2018. A generalized parallel constrained choice model for intra-household escort decision of high school students. Transp. Res., B 114, 26–38.
Yu, B., Zhang, J., Fujiwara, A., 2011. Representing in-home and out-of-home energy consumption behavior in Beijing. Energy Policy 39 (7), 4168–4177.
Yu, B., Zhang, J., Fujiwara, A., 2012. Analysis of the residential location choice and household energy consumption behavior by incorporating multiple self-selection effects. Energy Policy 46, 319–334.
Yu, B., Zhang, J., Fujiwara, A., 2013. A household time use and energy consumption model with multiple behavioral interactions and zero-consumption. Environ. Plann. B 40 (2), 330–349.
Zhang, J., Fujiwara, A., 2009. Intrahousehold interaction in transit-oriented residential choice behavior represented in stated preference approach. Transp. Res. Rec. 2134, 73–81.
Zhang, J., Kuwano, M., Lee, B., Fujiwara, A., 2009. Modeling household discrete choice behavior incorporating heterogeneous group decision-making mechanisms. Transp. Res., B 43, 230–250.
Zhang, J., Timmermans, H.J.P., Borgers, A., 2002. A utility-maximizing model of household time use for independent, shared and allocated activities incorporating group decision mechanisms. Transp. Res. Rec. 1807, 1–8.
Zhang, J., Timmermans, H.J.P., Borgers, A., Wang, D., 2004. Modeling traveler choice behavior using the concepts of relative utility and relative interest. Transp. Res., B 38 (3), 215–234.
Zhang, J., Timmermans, H.J.P., Borgers, A., 2005. A model of household task allocation and time use. Transp. Res., B 39, 81–95.

CHAPTER 8

Battery electric vehicles in China: ownership and usage

Enjian Yao*, Chunfu Shao, Fanglei Jin, Long Pan and Rui Zhang
Beijing Jiaotong University, Beijing, P.R. China

8.1 Introduction

With the rapid development of national economy and the increasing improvements of people's living standards, the automobile industry in China has ushered in the golden period of development. China has become the largest automobile producer and the largest automobile consumer in the world (South China Morning Post, 2017). Undeniably, this fact will promote the development of important industries such as steel, energy, materials, and machinery. At the same time, it will also bring a series of challenges in resources and the environment, such as energy dependence and environmental pollution. Seeking a sustainable development pattern that takes both social and economic development and resource and environmental protection into account is the main way to solve the preceding problems. In particular, developing new-energy vehicles (NEVs), characterized by lower energy consumption, lower emissions, and lower pollution, is becoming a new trend worldwide, where electric vehicles (EVs) are the dominant player (Hu and Yuan, 2018).

Since EVs came on the scene in the 19th century, the development of EVs had been staying stuck for a while due to the high production costs and electricity supply shortages. Since the second half of the 20th century, EVs has been attached great importance again with the increasingly severe problems such as energy crisis and environmental pollution.

Some developed countries, such as Germany, Japan, and the United States, have drawn up various strategies and plans to ensure the development of EV industry. In 2009 the German government planned to have at least one million EVs [including the battery EV (BEV) and the plug-in hybrid EVs (or PHEVs)] registered by 2020 (Bubeck et al., 2016).

*Corresponding author.

Transport and Energy Research
DOI: https://doi.org/10.1016/B978-0-12-815965-1.00008-9

Meanwhile, the National Electromobility Development Plan was issued to support the research and development (R&D) investment, market early-stage preparations, and latter introduction for EV industry. In 2010 the Next Generation Auto Strategy 2010 was published by the Ministry of Economy of Japan. It expounds the strategic steps of EV R&D from many aspects. According to this strategy, the next-generation automobiles including PHEVs and EVs will account for 15%–20% of the new-car market in 2020 (Wu and Zhang, 2017). In 2011 the US government claimed to make the United States to be the first country in the world having more than a million EVs in 2015. Meanwhile, a series of measures have been taken to ensure the smooth implementation of this target (Yang et al., 2011).

In China the central government has paid more attention to the development of NEVs, especially the EVs, which is viewed as the important strategic direction for China's economic structural changes. According to Ministry of Science and Technology (2009), the Outline of National Mid-and-Long-Term Science and Technology Development Plan (2006–20) has positioned the R&D of low energy–consumption vehicles and NEVs as the priority theme in key development areas. Since 2009, a series of documents [e.g., Notice on the Launch of Demonstration Pilot Projects in Energy-efficient and New Energy Vehicles (Ministry of Finance, 2009), Notice on the Expansion of Energy-efficient and New Energy Vehicles Demonstration and Promotion Work in Public Service Areas (Ministry of Finance, 2010a)] have been released to encourage the promotion and application of EVs in public service areas, including public transportation, taxi, public affairs, sanitation, and postal services and so on. In 2010 the Notice on the Subsidy Pilot Promotion of Private Purchase New Energy Vehicles opened a new chapter about the EV pilot work in the private sector (Ministry of Finance, 2010b). In 2012 the Energy-efficient and New Energy Vehicles Automotive Industry Plan (2012–20) proposed the target that the production capacity of PHEVs and EVs will be up to 2.0 million, and the cumulative production and sales will exceed 5.0 million (Sohu, 2012). Released later in the same year, the 12th Five-Year Plan for the Development of Electric Vehicle Technology has also pointed out that it is an important science and technology strategic demand and strategic focus to develop EVs (Ministry of Science and Technology, 2012a).

Specifically, in order to ensure the smooth implementation of the preceding national strategic targets, a series of specific policies have also been issued to encourage citizens to buy and use EVs. It is believed that the adoption of EVs will be stimulated to some degree by giving a subsidy and tax preference to EV users (Zhang et al., 2017). To promote the adoption of NEVs in private areas the Notice on the Subsidy Pilot Promotion of Private Purchase New-energy Vehicles (Ministry of Finance, 2010b) stated to provide the subsidy for automobile manufacturers and battery lease companies. And the acquisition expenses of EVs for private users can be cut down significantly. The Notice on Further Finishing the Pilot Work of Demonstrating and Extending Energy-efficient and New-energy Vehicles (Ministry of Science and Technology, 2011) elaborated the essential requirements for pilot cities and manufacturing enterprises. According to this document, local governments are required to make corresponding supportive policies (e.g., freeing restrictions on registration plates, lottery draw of plates, on-road usage, industry access, subsidies) and build up special agencies for better coordination. Moreover, it should be noted that from 2013 to 2016, the subsidy range was expanded from the 28 promotion cities or city groups to the entire country (Ministry of Finance, 2015; Ministry of Industry and Information Technology, 2015).

Perfect EV charging infrastructures will provide more confidence to EV users or EV potential users. Motivated by such a consideration, a series of infrastructure promotion policies have been published by the Chinese central government since 2010. It was expected that a network-based power supply system consisting of 400,000 charging poles (CPs) and 2000 charging and swapping power stations (CSPSs) would be built up in more than 20 pilot cities in 2015 (Ministry of Science and Technology, 2012b). And it is planned that the business model dominated by pure electric drive will be determined and the integration degree of vehicles and network will be improved by 2020. The National Energy Administration issued a specialized development guideline for promoting the EVs' infrastructure establishment in 2015 (National Energy Administration, 2015). According to this document, the construction objective of EV charging infrastructure is determined based on the principle of appropriately advanced development. Specifically, new centralized CSPSs will exceed 12,000 and distributed CPs will exceed 4,800,000 in 2020 to fulfill the charging demand of 5,000,000 EVs nationwide. In order to accelerate the establishment of charging infrastructures for NEVs and foster a good application environment, the Ministry of Finance will continue to provide the economic subsidy to local governments for the charging infrastructure construction and operation during the period of 2016–20 (Ministry of Finance, 2016). According to this policy, the reward subsidy shall be used exclusively in related areas including construction and operation, transformation, and upgrading for charging infrastructure, as well as operation monitoring system construction for charging and swapping service network.

Faced with the risks resulting from the oil dependence and the requirements for energy saving and emission reduction, many countries have begun to push viable replacements of conventional gasoline-powered vehicles (GVs) and diesel-powered vehicles. Recently, some countries have made a clear statement on the timetable for banning the conventional GVs. For example, the deadlines for Norway, India, the United Kingdom, and France are 2025, 2030, 2040, and 2040, respectively, mainly concentrating on the time range of 2025–50. In China the research on the timetable for stopping production and sales of GVs has been carried out by the Ministry of Industry and Information Technology. This measurement will further promote the development of EV industry and push the transformation of urban transportation structure in China.

According to China Association of Automobile Manufacturers, NEVs were sold around 180,000 by the end of 2017 in China, accounting for more than half of the global market share, where the EVs played a dominant role (Peoples Daily, 2018). However, compared to the sales of total vehicles in China (i.e., 28,878,900 in 2017), the market share of EVs is only 2.7% (He et al., 2018). In other words, it is still a long way to go for the massive expansion of EVs.

It is noted that EV's cruising range confined by the battery capacity is still less than GVs (Woodjack et al., 2012). For example, the cruising range of a BAIC E150 is about 75 miles under actual traffic conditions, and a BYD Qin of its 2014 version can run up to 120 miles with full charge. But for a GV with 40–50 L fuel, its cruising range can be more than 310 miles. The worry of batteries running out of power en route, normally referring to range anxiety (Pearre et al., 2011), inevitably affects EV drivers' travel choices such as vehicle purchase intention, charging decision-making, charging station, and route choices.

Concerning the EV ownership behavior, a series of factors have been examined by focusing on purchase intention. These factors are categorized into three levels (Bjerkan et al., 2016; He et al., 2018; Sierzchula et al., 2014): (1) technological level including vehicle purchase cost, battery performance characteristics (e.g., driving range and charging time);

(2) consumer level including individual attributes (e.g., gender, age, income, and education level) and personality variables (e.g., lifestyle orientations, environmental beliefs, risk-taking propensity); and (3) context level including fuel price, electricity costs, charging infrastructure availability and some related government policies (e.g., vehicle purchase restriction, driving restriction, and other subsidies for EV purchase).

As for the EV usage behavior, it usually consists of two aspects: charging choice behavior and route choice behavior. So far, there are three kinds of charging modes, including fast charging, slow charging, and battery swapping. Due to the lack of standards of battery specifications and ownership of battery, it is difficult to construct battery swapping stations. Slow charging and fast charging are carried out in accordance with charging time. Particularly, the time spent in slow charging and fast charging is about 6–8 h and half an hour, respectively (Botsford and Szczepanek, 2009). Slow charging is more suitable for long-time stay locations, such as residential and office areas during night and/or working times. However, because of forgetting to charge or being occupied by other EVs, the state of charge (SOC) of EV is not always 100% before starting a travel, causing a higher possibility of charging during travel for completing the entire trip. Under such situations, fast charging, even harmful to battery, is needed to satisfy the charging demand. This behavior is similar to that of a GV user who can refuel the vehicle at any gas station and at any time (Bae and Kwasinski, 2012). However, compared to the refuel behavior of GV, the charging behavior of EV drivers during travel occurs more frequently and needs more time. Therefore EV drivers' travel behavior modeling should be different from that for GV drivers, specifically in the case when charging behavior occurs during travel. Obviously, the charging decision-making and charging station choice are not needed in GV drivers' travel behavior modeling.

Particularly, the main differences of route choice models between EV and GV drivers are shown as follows:

1. Variables: The route choice variables included in traditional route choice models for GV drivers usually are limited to some basic level of service (LOS) attributes of the alternative routes, such as travel time, fare, road grade, and route direction (De Dios Ortúzar and Willumsen, 2011; Raveau et al., 2011). In addition to the aforementioned attributes the vehicle variables (e.g., SOC of EV, energy consumption) may also have important impacts on EV route choices. Particularly, the service level and location of charging station cannot

be ignored when there is a charging demand. Therefore it is a new challenge to quantify the preceding variables and analyze their impacts on EV route choices.

2. Route choice set: The alternative routes of EV drivers can be classified into two categories: routes without charging and routes with charging. Obviously, GV drivers do not need to consider such charging. Therefore it is necessary to represent the EV route choice set in a different way.

In summary, due to the unique energy consumption characteristics, the EV users are bound to behave differently from the GV users, in terms of both ownership and usage. In order to reduce energy consumption from the passenger transport sector, it is necessary to put off the doubts and fears for the current EV users or future potential users. It is, therefore, becoming important to study the behavioral mechanisms related to EV ownership and usage.

8.2 Battery electric vehicles ownership

According to Global EV Outlook 2018, the global stock of electric passenger cars reached 3.1 million in 2017, 40% of which were in China. Promotion of alternative-fuel vehicles (AFVs), including EVs, has become a primary goal of the Chinese government. The number of AFVs reached 1.7 million by the end of 2017, and it is expected to reach 5.0 million by the end of the 13th Five-Year Plan in 2020. However, to increase ownership of EVs in China, various barriers should be removed, such as the cruising range and charging methods, and impacts of policy contexts, culture, lifestyles, and so on have also remained unknown.

8.2.1 Ownership-related studies

Related to EV ownership, existing studies mainly examined influences of technical features of EVs against GVs, infrastructure attributes, travel patterns, and policies as well as personal attributes (Liao et al., 2017), and the effects of various factors affecting EV ownership, such as cruising range and vehicle performance (Egbue and Long, 2012; Lieven et al., 2011; Moons and De Pelsmacker, 2012), charging time and charging opportunities (Carley et al., 2013; Graham-Rowe et al., 2012), and ownership cost (Dumortier et al., 2015; Jensen et al., 2013). Existing studies also evaluated the impacts of household travel patterns (e.g., using daily driving distances derived from GPS-logged movements to investigate the feasibility of BEV substitution) (Jakobsson et al., 2016; Karlsson, 2017) and discussed how socioeconomic factors influence the purchase intention (Carley et al., 2013; Glerum et al., 2011). Besides, some studies investigated how attitudinal variables such as environmental consciousness influence people's decision-making process of vehicle purchasing (Lane and Potter, 2007; Larson et al., 2014; Liu et al., 2015).

Because of the recent development of EVs in China, relevant studies about EV purchase behavior of Chinese people have gradually accumulated. It is shown that older consumers have a more optimistic attitude to EVs than younger generations, more consumers are interested in adopting EVs as the second family car, experienced drivers are more concerned about the cost of battery and maintenance, and consumers who are more concerned about the environment are more likely to adopt EVs (She et al., 2017). Focusing on vehicle technical features and infrastructure attributes (Cen et al., 2018), it is revealed that driving range alone is not a major obstacle for EVs adoption, and due to relatively short inner-city trips, the detour cost and charging cost negatively impact the EV penetration rate; EV drivers tend to choose routes with a longer distance but a shorter travel time, due to the lower electricity cost of EVs as compared to the fuel cost of GVs, and drivers with a longer commute distance have a higher propensity to choose EVs due to their comparatively lower electricity cost.

Apart from the preceding common factors explored in other countries, China has some special characteristics that play important roles in the EV ownership, such as policy, culture, and economic development. For cultural factors, it is clarified that the Chinese cultural values are key to understand Chinese consumers' intentions to adopt EVs, and it is suggested that public policy and social marketing efforts should pay more attention to the role of cultural values (Qian and Yin, 2017). For urban economic development, it is observed that Chinese consumers in lower tier cities (defined by the population size) are generally sensitive to monetary attributes, charging service, and driving range of EVs (Huang and Qian, 2018). When comparing the differences in second-tier versus third-tier cities, it is found that consumers in the third-tier cities are more sensitive to purchase price, subsidy of purchase, and coverage of charging stations than those in the second-tier cities.

Despite the preceding findings, there is still lack of studies in the context of China (Hao et al., 2014), especially lack of case studies about cities such as Beijing, where unique policy environment exists. However, studies on China have failed to thoroughly analyze the impacts of complex policy environment or design various policy scenarios. Like the study by Helveston et al. (2015), they removed the Beijing data due to the unavailability of a part of the data, leading to a lack of specific preference for BEV purchase in Beijing, and they also neglected the external policy environments of Shanghai, Chengdu, and Shenzhen. Hao et al. (2014) summarized the subsidy and nonsubsidy incentive policies of main cities in China, including license plate lottery exemption for BEV in Beijing; however, they did not reveal the quantified compensation relationship between subsidy and other policies. Sun and Xu (2018) did a comparative study on the impacts of subsidies on Chinese consumer preferences for AFVs without considering policy environments of local cities. It is, however, essential to include the external policies as variables to explain the BEV purchase intention.

In addition, some studies integrated the vehicle ownership and usage and intended to investigate whether EVs, as a supplementary vehicle in family, are feasible to cover daily travel demand or cover the demand that cannot be fulfilled by GVs. Karlsson (2017) tested the potential for BEVs to replace one of the GVs in two-car households considering the households' driving demand. Different vehicle substitution and use strategies were considered in their analyses, and the results showed that households with multiple GVs could be ideal candidates to enhance BEV adoptions beyond the early adopters. Pan et al. (2017) considered the BEV charging demand when planning charging stations. When modeling the demand, it was assumed that travelers keep the same activity plan and time budget as possible as they can. As a result, if a BEV is feasible to replace a GV, it has to guarantee the maximum fulfillment of daily travel demand. Therefore from the angle of BEV travel demand fulfillment, the related research is rare.

8.2.2 Attitude and purchase intention to battery electric vehicles: a case study of Beijing

To study the attitude and purchase intention to BEV in China, the China-unique policy environment cannot be neglected. To be specific, Beijing began restricting the vehicle purchase via a quota system in 2011, differing from an auction system (Yang et al., 2014). In addition to purchase restriction, Beijing has implemented administrative measures to restrict the use of vehicles, which require specific vehicles barred within the fifth Ring Roads for the period from 07:00 to 20:00 one weekday per week, depending on the last digit of the license plate number. Very likely, these vehicle restrictions will affect consumers' attitudes and purchase intentions to any vehicle type. Especially from 2018, the quota for BEVs has remained 60,000 vehicles, but GVs have been reduced from 90,000 to 40,000, while there are 2,800,927 efficient encodings with staircase success rates. The attitudes and intentions under the preceding restrictions are probably no longer about simple preferences for vehicle technologies but they have to be related to the opportunity to purchase vehicles.

Thus a questionnaire survey was designed and implemented in the early 2018 to collect Beijing respondents' personal attributes, attitudes, and purchase intentions under various scenarios, by targeting BEVs. And some results about the attitudes and purchase intentions are shown in the following subsections.

8.2.2.1 Battery electric vehicles attitude

A five-point Likert scale was used to gather respondents' attitudes to BEVs in Beijing. By reflecting comments from mass media and existing research, 31 questions were included to capture BEV features and impacts as many as possible. Each statement was measured ranging from 1 (strongly disagree) to 5 (strongly agree). Finally, 604 valid responses were collected. Table 8.1 shows all the questions together with their means and standard deviations.

The first nine questions ask respondents to express their attitudes toward potential benefits in owning a BEV. The question gaining a great awareness is the relationship between driving restrictions and BEV purchase, followed by cheaper operating cost, environmental consciousness, and subsidy (B2, B3, B8, and B9). The other 22 questions represent negative attitudes to BEVs and have been scored inversely. What respondents worry more about are range, low battery, and charging problem (ETG1-ETG4). As for the questions about ownership cost (ETG8-ETG10), respondents still have a strong disagreement that owning a BEV can save money, implying that potential consumers should be informed more about BEV energy cost efficiency over the whole lifecycle, according to the results obtained by Conti et al. (2015).

In order to figure out a valid structure an exploratory factor analysis (EFA) was carried out based on half of the samples, and the EFA results are listed in Table 8.2. The Kaiser–Meyer–Olkin (KMO) measure of sampling adequacy (KMO = 0.869) and the Bartlett test of sphericity

Table 8.1 Attitudinal questions and their statistical characteristics.

Index	Question	Mean	Standard deviation
B1	BEV can satisfy my travel demand	3.67	1.074
B2	BEV has subsidies for purchase	3.75	1.099
B3	Electric cost is cheaper than fuel cost	3.77	1.122
B4	BEV is less noisy in driving than GV	3.49	1.120
B5	BEV has better acceleration than GV	3.17	1.149
B6	License plate lottery is not needed because I have greater opportunity to purchase a BEV within 3 years	3.67	1.159
B7	I prefer to apply for purchase a BEV and wait, because now the waiting will not take long time and the application is relatively easy	3.69	1.130
B8	BEV has no driving restrictions	3.86	1.077
B9	BEV is environment friendly	3.76	1.136
ETG1	I worry about the insufficient cruising range	1.88	1.103
ETG2	I worry about a low battery capacity when using air-condition	1.99	1.020
ETG3	It is not possible for me to drive a BEV for a long distance	1.84	1.098
ETG4	I worry about that it is hard to find charging facilities	1.86	1.061
ETG5	BEV is not friendly to users who are living in a rental apartment, because changing residence needs to consider the availability of chargers	2.49	1.192
ETG6	I worry about that it is not easy to install a charger for me	2.13	1.126
ETG7	GVs always occupy the parking space with a charger	2.64	1.129
ETG8	BEV price is unacceptable, even considering subsidies	2.25	1.102
ETG9	Even though BEV energy cost is cheaper, BEV is still expensive considering its purchase price	2.34	1.069
ETG10	The maintain cost of BEV is too high, especially for battery	2.15	1.077
ETG11	The value of BEV does not last as long as a GV does	2.46	1.143
ETG12	It is better to purchase a BEV a few years later since technologies have been developed very fast	2.19	1.084
ETG13	I worry about that BEV has a greater failure rate	2.32	1.118

(*Continued*)

Table 8.1 (Continued)

Index	Question	Mean	Standard deviation
ETG14	The battery in BEV brings pollution to soil	2.89	1.124
ETG15	I worry about that the after-sales service of BEV is worse than GVs	2.48	1.046
DE1	The driving in BEV is too quiet to be noticed	3.17	1.276
DE2	Available brands and models of BEVs are fewer than GV	3.22	1.249
DE3	There is less fun to reform vehicle in the case of BEVs	3.64	1.303
DE4	I feel that driving a BEV does not look like driving an automobile	2.84	1.194
PI1	My family and friends do not want me to buy a BEV	3.42	1.203
PI2	Purchasing a BEV is regarded to be unwise	3.31	1.306
PI3	Purchasing a BEV will put one's image down	3.43	1.327

BEV, Battery electric vehicles; *GV*, gasoline-powered vehicles.

(with $P = .000$) were calculated. These values indicate that the factor analysis is appropriate. As a result, four latent factors were identified. Finally, there are 27 questions that were considered, and 4 questions were omitted due to low factor loadings on every latent factor. These four latent factors are energy-technology gap (11 indexes), BEV benefits (11 indexes), personal image (3 indexes), and driving experience (4 indexes). The Cronbach's α values indicate that all the question items in each factor are reliable.

Since a valid factor structure is verified, a multiple indicators multiple causes (MIMIC) model is applied to analyze the relationship between latent factors and various observed variables using the other half of the sample. The observed variables considered in the model are described in Table 8.3. Among them, daily travel distance, daily travel frequency, years of driving, age, annual income, and education are set to be ordinal variables, and others are continuous variables. Note that, for ease of understanding, the question statements related to energy-technology gap, driving experience, and personal image represent a negative attitude and respondents were asked about how much they agree with the negative aspects of BEVs. Intending to make sure all the attitude data representing the same kind of attitude in the model, the points of the three categories are inverted into points representing positive attitude.

Table 8.2 Exploratory factor analysis results.

Index	Factor loading			
	Factor 1	Factor 2	Factor 3	Factor 4
	Energy–technology gap	BEV benefits	Personal image	Driving experience
ETG2	0.774			
ETG3	0.741			
ETG4	0.724			
ETG1	0.717			
ETG10	0.662			
ETG6	0.634			
ETG13	0.618			
ETG9	0.594			
ETG12	0.546			
ETG11	0.497			
ETG15	0.434			
B2		0.790		
B1		0.712		
B3		0.674		
B9		0.634		
B6		0.628		
B8		0.601		
B4		0.597		
B7		0.558		
B5		0.496		
PI1			0.812	
PI3			0.796	
PI2			0.754	
DE2				0.801
DE1				0.622
DE3			0.440	0.622
DE4				0.416
Cronbach α	0.884	0.860	0.880	0.788

Note: (1) Extraction method: maximum likelihood. (2) Rotation method: Varimax with Kaiser normalization. (3) KMO: 0.869, Bartlett test: $P = .000$. (4) Factor loadings under 0.4 are omitted. *KMO*, Kaiser–Meyer–Olkin.

Looking at the test results in Table 8.4, where the minimum chi-square/degree of freedom (CMIN/DF), root mean square error of approximation (RMSEA), comparative fit index (CFI), and standardized root mean square error residual (SRMR) are calculated. Comparing with

Table 8.3 Description of observed variables.

Observed variables	Description
Daily travel distance (km)	(1) <10; (2) 10–20; (3) 20–40; (4) 40–60; (5) 60–100; (6) >100
Daily travel frequency	(1) ≤1 round trip; (2) 2 round trips; (3) ≥3 round trips
Years of driving	(1) <2; (2) 3–6; (3): 7–10; (4) >10
Number of GVs	Continuous variable
Number of driving licenses in family	Continuous variable
Number of GV applicants in family	Continuous variable
Number of BEV applicants in family	Continuous variable
Age (years old)	(1) ≤18; (2) 18–25; (3) 26–30; (4) 31–40; (5) 41–50; (6) 51–60; (7) >60
Annual income (RMB)	(1) <50,000; (2) 50,000–100,000; (3) 100,000–150,000; (4) 150,000–250,000; (5) 250,000–350,000; (6) ≥350,000
Education	(1) Middle school; (2) high school; (3) undergraduate; (4) postgraduate
Number of family members	Continuous variable
Number of children	Continuous variable

BEV, Battery electric vehicles; *GV*, gasoline-powered vehicles.

Table 8.4 Test results of multiple indicators multiple causes model.

	CMIN/DF	RMSEA	CFI	SRMR
Test results	1.56	0.043	0.902	0.062
Acceptable level	<3	<0.06	>0.9	<0.08

the acceptable levels, all the test results suggest that the model is a valid model.

Based on Table 8.5, up to 21 observed variables were tested in the MIMIC model to explore their relationships with the four latent factors, trying to explain how these factors affect respondents' attitudes to BEVs. Finally, after screening the variables with acceptable significance, 17 observed variables were selected to describe the four latent factors simultaneously, and all the variables are significant, at least at 1% level.

Table 8.5 Estimation results of multiple indicators multiple causes model.

	B	ETG	DE	PI
Daily travel distance		− 0.072 (−1.870)	0.091 (1.738)	0.103 (1.875)
Daily travel frequency			− 0.183 (−2.387)	
Years of driving			− 0.124 (−2.151)	
Number of GVs	0.125 (1.734)			
Number of driving licenses in family		− 0.093 (−1.702)		− 0.176 (−2.682)
Number of GV applicants in family	− 0.107 (−1.930)		− 0.110 (−1.980)	
Number of BEV applicants in family	0.257 (3.136)	0.228 (3.122)	0.247 (3.209)	
Age			0.122 (2.230)	
Annual income			− 0.077 (−2.087)	
Education		− 0.130 (−1.695)		
Number of family members				0.126 (2.592)
Number of children				− 0.205 (−2.219)

Note: The values in the parentheses of the last four columns are t values. *GV*, Gasoline-powered vehicles.

In the model the more the points of questions are, the more positive attitude to BEV each question represents. As a result, all the parameters are estimated to measure their relationships with the positive attitudes to BEVs in various aspects. Number of GVs, number of BEV applicants, and number of GV applicants in households are significantly correlated to the latent factor "BEV benefits," indicating that families who believe more in BEV benefits will have more people to buy BEVs and less people to buy GVs. The reason why the number of GVs is correlated with BEV benefits is probably when they consider buying an extra vehicle, since there are almost enough GVs to cover daily travel demand, they are very likely to consider BEVs positively as a supplementary mobility tool, resulting in a positive attitude to BEV benefits.

Regarding the latent factor "energy-technology gap," it is significantly correlated with daily travel distance, number of driving licenses, number of BEV applicants, and education. Not surprisingly, people traveling a long distance have less confidence in BEV energy technology, probably due to the poor cruising range and the insufficient charging facilities. As for number of driving licenses, which has a negative value, the families with more driving licenses show a great travel demand of vehicles in general so that the required cruising range is as long as possible. Thus under this requirement, families with more driving licenses are reasonable to have a negative attitude to BEV energy technology. Besides, it is interesting to find that people having higher level of education are more likely to show a negative attitude to BEV energy technology, implying that well-educated people have more knowledge about the deficiency of BEV, and as a result, they are more sensitive to the negative effects than positive ones.

Concerning the latent factor "driving experience," it is found that people with a lower travel frequency and a longer travel distance are more likely to have a positive attitude to BEV. When driving BEV frequently, the perception about driving experience difference between BEV and GV becomes quite obvious due to the preference of GVs. However, people driving a long distance indicate their dependence on vehicles. Such dependence makes them less sensitive to the difference between BEV and GV. In addition, it is easy to understand that people with more years of driving show a negative preference for BEV, probably because BEV is newly deployed in the market and driving a GV has become a habit for people with more years of driving. It is also found that younger and higher income people have a negative attitude to BEV driving experience. It is interesting to find that daily travel distance and number of family members positively affect BEV's personal image, while number of driving licenses and number of children show a negative effect. Among them, the number of children plays the most important role in measuring the BEV personal image, indicating that parents with children care more about the personal image.

All these findings present useful information for policy making and market strategies.

8.2.2.2 Purchase intention

Due to the unique restriction policies, the decision-making mechanisms about vehicle purchase in Beijing should be captured in a different way from other cities. It is necessary to model the choice as a vehicle application behavior, rather than just a pure choice of a vehicle technology. Here, a binary logit (BL) model is applied to analyze the BEV adoption intention. To emphasize the impacts of policies on BEV choice, this study examines several existing policies and future potential policies for revealing how policies affect vehicle purchase.

Table 8.6 shows the variables to be introduced in the model. As for policy variables, we focus on purchase restriction, drive restriction, purchase tax, purchase subsidy, and two potential policies, that is, daily use subsidy and the permission to use bus line for BEVs.

The parameter estimation of the BL model was done using the maximum likelihood method, and the results are shown in Table 8.7. To intuitively compare the effects of variables on BEV purchase intention, the equivalent purchase subsidy coefficients (*EPSC*) are calculated based on parameter estimation.

For the vehicle technology parameters, both *CC* and *CR* values are positive, indicating that the convenience of charging and the improvement of cruising range have a strong promotion effect on the purchase of BEV. This reveals that the technological development related to these two aspects in the future is crucial for users to purchase a BEV.

For the restriction policy parameters, both *PR* and *DR* values are negative, meaning that the purchase restriction and driving restriction policy for BEVs have a strong negative effect on the choice probability of BEV. This may be due to the fact that people's increasing demand for vehicles cannot be directly met. In addition, the *EPSC* value of the *PR* equals to -6.99 that is the largest among all the relevant values. This indicates that

Table 8.6 Explanatory variables to describe battery electric vehicles (BEV) adoption intention.

Variables	Description
Basic variables	
• Personal attributes	
AGE1	Dummy for 18–25 years old {1: yes, 0: no}
AGE2	Dummy for 25–40 years old {1: yes, 0: no}
AGE3	Dummy for 40–60 years old {1: yes, 0: no}
EDU	Dummy for educational level being a postgraduate degree {1: yes, 0: no}
AHI	Dummy for annual household income being more than $29,560 {1: yes, 0: no}
• Travel attributes	
VO	Dummy for owning a vehicle in the present {1: yes, 0: no}
VPD	Dummy for urgent vehicle purchase demand{1: yes, 0: no}
VPB	Dummy for vehicle purchase budget being more than $29,560 {1: yes, 0: no}
DTD	Dummy for daily travel distance being more than 40 km {1: yes, 0: no}
Scenario variables	
• Vehicle technological attributes	
CC	Dummy for convenient charging of BEV {1: yes, 0: no}
CR	Dummy for BEV's technological level of cruising range {1: same as GV, 0: otherwise}
• Restriction policy variables	
PR	Dummy for BEV's purchase restriction policy {1: same as GV, 0: otherwise}
DR	Dummy for BEV's driving restriction policy {1: same as GV, 0: otherwise}
• Subsidy policy variables	
PT	BEV's purchase tax rate (%)
PS	BEV's purchase subsidy (US dollar)
• Substitutional policy attributes	
DUS	Subsidy for BEV's daily use (US dollar)
BLP	Dummy for permitting use of bus lines by BEVs {1: yes, 0: no}

GV, Gasoline-powered vehicles.

the GV purchase restriction is the most influential factor to promoting BEV purchases in the current status.

For the subsidy policy parameters, the *PT* and the *PS* value are negative and positive, respectively, indicating that as the BEV purchase tax rate increases or subsidy decreases, the probability of BEV purchase will decline. The EPSC of *PT* is −0.50, which means that a 1% increase in

Table 8.7 Parameter estimation results for binary logit model.

Explanatory variables	Parameter	*t* Value	*EPSC*
Scenario variables			
• Vehicle technological attributes			
CC	− 1.23***	− 7.24	− 4.07
CR	− 1.11***	− 6.75	− 3.68
• Restriction policy attributes			
PR	− 2.11***	− 12.14	− 6.99
DR	− 1.67***	− 10.69	− 5.53
• Subsidy policy attributes			
PT (%)	− 0.15***	− 10.95	− 0.50
PS ($1000)	− 0.30***	− 11.75	− 1.00
• Substitutional policy attributes			
DUS ($1000/year)	− 1.98***	− 8.16	− 6.56
BLP	− 1.72***	− 11.2	− 5.73
Basic variables			
• Personal attributes			
AGE1	− 1.54***	− 3.27	− 5.10
AGE2	− 1.16**	− 2.47	− 3.84
AGE3	− 0.90*	− 1.88	− 2.97
EDU	− 0.19**	− 2.22	− 0.63
AHI	− 0.10**	− 2.28	− 0.32
• Travel attributes			
DTD	− 0.30***	− 3.04	− 1.00
VO	− 0.55***	− 7.66	− 1.82
VPD	− 0.51***	− 6.27	− 1.68
VPB	− 0.23***	− 2.92	− 0.76
Constant term	− 5.68***	− 8.17	− 18.81

Significant at *10% level, **5% level, ***1% level; $\rho^2 = 0.237$; sample size: 8730.

the BEV purchase tax rate is equivalent to a reduction of $500 in BEV purchase subsidy. This value may be due to the difference between the two policies. The purchase tax is linearly related to the purchase price, while purchase subsidies are less relevant. The purchase tax is a direct exemption and does not need the purchaser to pay, while the purchase subsidy requires the purchaser to pay first and then to be subsidized.

For the substitutional policy parameters, the *DUS* value is positive, indicating that the BEV daily use subsidy increases the BEV choice probability. The *EPSC* of *DUS* is 6.56, which means that the daily use subsidy increases by $1000 per year equivalent to purchase subsidy of

$6560. The reason for this value may be the difference between the purchase subsidy and the daily use subsidy. The purchase subsidy is a one-time subsidy when purchasing the vehicle and the daily use subsidy is subannual subsidy, and the annual subsidy amount is related to the intensity of daily use, and the subsidy period is also determined depending on actual conditions. The *EPSC* of the *BLP* is 5.73, meaning that when BEV is allowed to use bus lanes, it is equivalent to a purchase subsidy of $5730. This shows that the use of bus lane has a strong stimulating effect on users. The reason may be related to traffic congestion during peak hours.

For the personal attribute parameters, all the *AGE* values are positive, indicating that the user under the age of 60 are more likely to purchase a BEV, and *AGE1* > *AGE2* > *AGE3*, meaning that younger persons are more likely to purchase a BEV. The *EDU* value indicates that people with postgraduate qualifications are more inclined to purchase a BEV. And the *AHI* value is negative, indicating that people with annual household income under $29,560 are more likely to purchase a BEV. Increasing the adoption rate of BEVs should target the above-mentioned people.

For the travel attribute parameters, the *DTD* value is negative, indicating that people who travel less than 40 km per day are more likely to purchase a BEV. This reveals that BEV is more suitable for meeting short-distance travel needs, and the cruising range and the convenience of charging are the factors to be improved. The value of *VO* shows that people who already own a car are more likely to buy a BEV as a supplement to meet travel needs. The reason for this phenomenon may be that the majority of the vehicle ownership in this sample is GV, whose purchase restriction policy has a stimulating effect on BEVs. The *VPD* value is positive, which suggests that people with an urgent vehicle purchase demand are more likely to buy a BEV, which is largely due to the extremely difficult GV lottery policy. The negative *VPB* value means that people with a vehicle purchase budget less than $29,560 are more inclined to buy a BEV. This is most likely due to the significant incentives of BEV purchase subsidy and tax exemption policy for people with a low purchase budget. Therefore it can be concluded that the current BEV's technological characteristics and incentive policies are more beneficial to people who meet the conditions of "short-distance travel demand," "owning a vehicle in the present," "urgent vehicle purchase demand," and "low purchase budget."

8.2.3 Summary

In this section, a brief literature review is demonstrated, and two case studies on China BEV ownership are carried out based on a stated preference (SP) survey. EFA, exploratory factor analysis, and MIMIC are applied to study the attitudes to BEVs. Four latent factors, that is, energy–technology gap, BEV benefits, personal image, and driving experience, are identified. It is found that travel pattern, current vehicle ownership, and personal/family characteristics have great impacts on the attitudes to BEVs. Besides, a BL model is estimated to analyze the intention to purchase a BEV, where local policies including the purchase restriction policy and driving restriction policy in Beijing are examined, together with other policies such as purchase tax and one-time subsidy, as well as substitutional policies. The results show that the purchase restriction policy and driving restriction policy have a positive impact on the adoption rate of BEVs, indicating the current policy environment encourages the BEV ownership.

8.3 Battery electric vehicles usage

Although BEVs have a better environmental performance than GVs, the limited cruising range of BEVs has still to some extent hindered BEV promotion. Therefore further understanding of the Chinese BEV usage behavior could help improve the user experience of BEVs and their charging systems. Focusing on BEVs, this section first investigates Chinese BEV usage behavior, particularly the route and charging choice behaviors, by establishing a BEV drivers' charging and route choice behavior model with heterogeneity and an angular cost. Then, based on an SP data, a BEV charging choice model is developed to estimate the charging demand, which is incorporated into a charging station location model for locating the public charging stations with the objective of maximally maintaining the daily trips of BEV drivers in the context of Beijing.

8.3.1 Battery electric vehicles route and charging choice behavior

8.3.1.1 Introduction

EVs have received wide attention of researchers, policy makers, consumers, and industry for less-pollution during use, low noise, and high energy efficiency (Jung and Jayakrishnan, 2011). However, confined by the battery capacity, the BEV's cruising range is still far less than GV, causing BEV drivers' range anxiety (Pearre et al., 2011) and inevitably affecting their travel behaviors. To overcome the limited cruising range and reduce range anxiety, charging has become an essential part in the BEV drivers' travel decision-making process.

Compared to the refuel behavior of GV drivers, the charging behavior of BEV drivers during travel occurs more frequently and needs more time. As mentioned in Section 8.1, effects of range anxiety and effects of charging behavior are two main differences between GV and BEV charging behavior.

This section is devoted to analyzing Chinese BEV drivers' charging and route choice behaviors based on a nested logit (NL) model, with an additional consideration of the influences imposed by vehicle variables of BEVs and service levels of charging stations. Meanwhile, considering the preference heterogeneity across individuals, a latent class model is estimated to examine differences between different groups of BEV drivers.

8.3.1.2 Battery electric vehicles route and charging choice modeling

8.3.1.2.1 Preference heterogeneity: a latent class model

The latent class model (Hess, 2009) is commonly used to identify preference heterogeneity across individuals by estimating class-specific sets of parameters. It makes an assumption that there are a certain number of latent classes among individuals, and the choice preferences are homogeneous within each class but heterogeneous across classes. Given a finite and fixed number of S classes, and given that individual i belongs to class s ($s = 1, 2, \ldots, S$), the utility function of s for any alternative k can be expressed as

$$U_{ik|s} = \beta_s' X_{ik} + \varepsilon_{ik|s} \tag{8.1}$$

where β_s' is a vector of the utility parameters for class s, X_{ik} is a vector of independent variables, and $\varepsilon_{ik|s}$ is the error term of the utility function.

The probability that individual i chooses alternative k in the latent class model is given as follows:

$$P_i(k) = \sum_{s=1}^{S} P_i(k|s) \cdot H_i(s) \tag{8.2}$$

Thus the choice probability is calculated by two types of probabilities: one is the conditional probability $P_i(k|s)$ that individual i chooses alternative k within class s, and the other is the class membership probability $H_i(s)$, which represents the probability that individual i belongs to class s.

Both $P_i(k|s)$ and $H_i(s)$ adopt a multinomial logit (MNL) formulation for the standard latent class model, namely, the LCMNL model:

$$P_i(k|s) = \frac{\exp\left(\beta_s' X_{ik}\right)}{\sum_{k' \in K_i^s} \exp\left(\beta_s' X_{ik'}\right)} \tag{8.3}$$

$$H_i(s) = \frac{\exp\left(w_s' z_i\right)}{\sum_{s'=1}^{S} \exp\left(w_{s'}' z_i\right)} \tag{8.4}$$

where K_i is the choice set for individual i, w_s' is a vector of unknown parameters for class s, and z_i is a vector of variables to explain each class membership function, which usually includes individual characteristics.

The class membership function may contain a set of class-specific constant terms, but the parameters of one class should be set to zero for parameter identification. The inclusion of individual characteristics in the class membership functions allows to identify typical member profiles for each class.

8.3.1.2.2 Choice set: the nested logit model

Compared to GV drivers, BEV drivers' route choice behaviors are significantly influenced by charging behavior when charging is needed. Depending on whether detouring to charge is needed or not, the BEV drivers' route choice sets could be classified into two categories: routes without charging and routes with charging. Considering the difference in variances of error terms between these two categories, an NL formulation is more appropriate than MNL to describe the choice situation. The nested structure is shown in Fig. 8.1. The upper level indicates the decision of charging or not, while the lower level shows the route choice combinations for no-charging and charging. Furthermore, to overcome the overlapping problem of routing, a path size logit (PSL) structure is applied to route choice without charging in the lower level. The correction term of PSL is given by the following equation:

$$\mathrm{PS}_k = \sum_{a \in \Gamma_k} \frac{l_a}{L_k} \frac{1}{\sum_{j \in K_{rs}} \delta_{aj}} \tag{8.5}$$

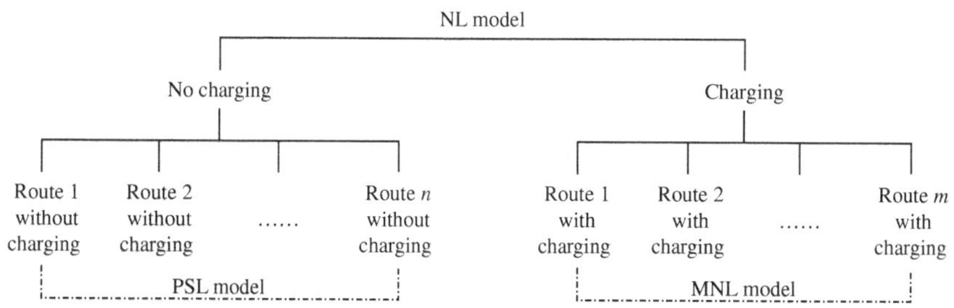

Figure 8.1 NL model structure of BEV drivers' charging and route choices.

where l_a is the length of link a, L_k is the length of route k, and δ_{aj} is equal to 1 for link a on route k and 0 otherwise.

The MNL structure is applied to route choice with charging in the lower level, since the routes with different charging stations usually have less overlapped links, which are less likely to violate the independence of irrelevant alternatives (IIA) property.

8.3.1.2.3 The latent class nested logit model

As shown above, here, a latent class NL (LCNL) model is adopted to describe the charging and route choice behaviors of BEV drivers. It simultaneously captures the flexible correlation structure between the choice alternatives and preference heterogeneity across individuals.

The probability formulation of the LCNL model (a two-level NL model) can be expressed as follows (Kamakura et al., 1996):

$$P_i(k) = \sum_{s=1}^{S} [P_i(k|n,s) \cdot P_i(n|s)] \cdot H_i(s) \tag{8.6}$$

$$P_i(k|n,s) = \frac{\exp\left(\beta_s' X_{ik} / \mu_n^s\right)}{\sum_{k' \in K_i} \exp\left(\beta_s' X_{ik'} / \mu_n^s\right)} \tag{8.7}$$

$$P_i(n|s) = \frac{\exp\left(\mu_n^s I_{in}^s\right)}{\sum_{n'} \exp\left(\mu_{n'}^s I_{in'}^s\right)} \tag{8.8}$$

$$I_{in}^s = \ln \sum_{k' \in K_{in}^s} \left[\exp\left(\beta_s' X_{ik} / \mu_n^s\right)\right] \tag{8.9}$$

$$H_i(s) = \frac{\exp\left(w'_s z_i\right)}{\sum_{s'=1}^{S} \exp\left(w'_{s'} z_i\right)} \tag{8.10}$$

Within class s, $P_i(k|n, s)$ is the probability that individual i chooses route k in nest n, $P_i(n|s)$ is the probability that individual i is in nest n, K_{in}^s is the route choice set for individual i in nest n, I_{in}^s is the logsum function or inclusive value of nest n for individual I, and μ_n^s is the parameter of I_{in}^s for nest n. The parameter μ_n^s should fall into [0, 1] for meeting the principle of random utility maximization.

8.3.1.3 Explanatory variables

Here, route attributes, vehicle variables, and charging station attributes are introduced as explanatory variables.

In the group of route attributes, travel time and travel cost from origin to destination are used to describe individuals' preferences for the levels of services provided by each route alternative. To reflect drivers' preferences for high-grade roads, the total ratio of expressway and arterial road is also included. Given that drivers prefer the most direct routes from origin to destination, an angular cost variable is added to represent the directionality of routes (Raveau et al., 2011).

Considering the characteristics of BEVs, three variables are adopted in the group of vehicle variables, including initial SOC at origin, average electric consumption (AEC) factor, and SOC at destination. The initial SOC at origin is a crucial reference to evaluate the cruising range of BEV before travel. AEC, suggested by Yao et al. (2013) [see Eqs. (8.11) and (8.12)], represents the AEC for the whole route. And the SOC at destination is used as an indicator reflecting the impact of drivers' range anxiety on charging and route choices. That is, the higher SOC the BEV has at destination, the less range anxiety the driver feels during travel. And the value of SOC at destination can be calculated by Eq. (8.13).

$$\mathrm{EC} = \mathrm{EF}/U = 3.576/v - 7.895 \times 10^{-3} \cdot v + 7.845 \times 10^{-5} \cdot v^2 + 0.574 \tag{8.11}$$

$$\mathrm{AEC} = \sum_i \frac{l_i}{L} \mathrm{EC}(v_i) \tag{8.12}$$

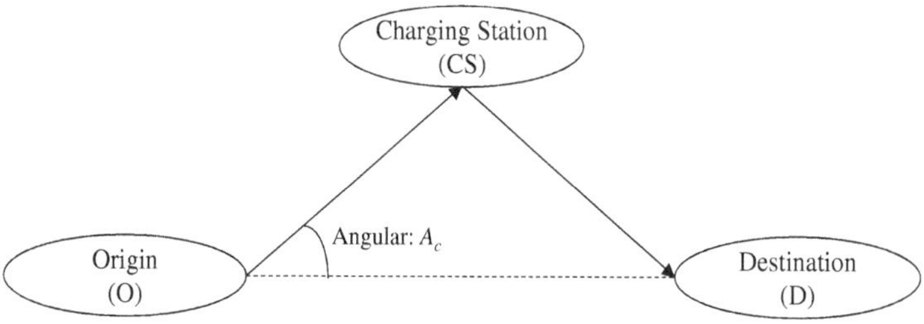

Figure 8.2 Diagram of angular formed by origin, charging station, and destination.

$$SOC_d = SOC_0 - \frac{AEC \cdot L}{Q} \times 100\% \tag{8.13}$$

where EC is the electric consumption factor (A h/km), which is a function of average speed v; U is the output voltage of power battery (V); U is determined to be 380 V in this study based on observation of modeling data; l_i is length of the ith link of the route (m); L is the length of the route (m); v_i is the average speed of the ith link; SOC_d is SOC at destination (%); SOC_0 is the initial SOC at origin (%); and Q is the battery capacity of BEV (A h).

When any two variables among three vehicle variables are determined, the other one is automatically obtained by Eq. (8.13). Therefore only two vehicle variables can be used for model estimation.

As for charging station attributes, the charging time, which is the sum of waiting time and charging service time, is adopted to reflect the LOS attribute of charging station. And the location of charging station, two variables reflecting factors related to the angular formed by origin, charging station, and destination (hereafter A_c: see Fig. 8.2), and the distance from origin to charging station (hereafter d_{OC}) are also included into this group of attributes.

As shown in Fig. 8.2, A_c is adopted to reflect deviation of charging station with travel direction, and its value ranges from 0 to π. It is assumed that travelers prefer the charging station that is closest to the travel direction. To reflect this preference, and to facilitate the evaluation of the effects of charging station location, A_c is defined as a penalty indicator. Three forms of A_c variable are presented: (1) $\sin(A_c/2)$, (2) $\tan(A_c/4)$, and (3) A_c/π, which are described in Fig. 8.3. These forms all satisfy the

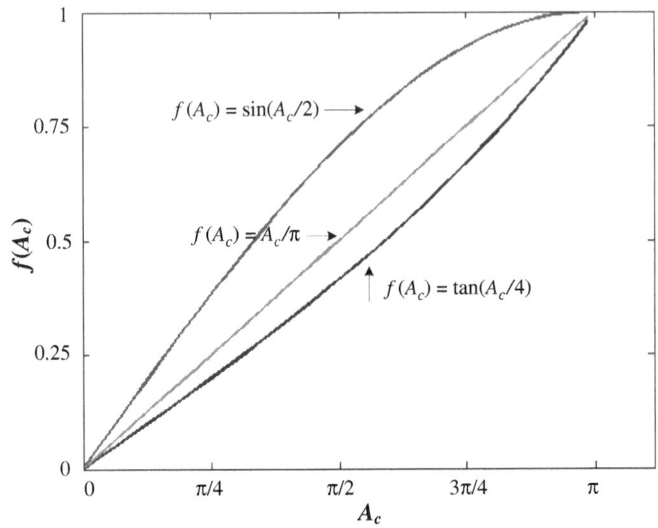

Figure 8.3 Three forms of A_c variables.

condition that the penalty is minimum when A_c is 0 and maximum when A_c is π. Form (3) is a base form, meaning that the variation rate of A_c perception is constant. Compared to form (3), the variation rate of A_c perception in form (1) decreases, while that in form (2) increases. The goal of setting the three forms is to reflect three kinds of drivers in terms of the perception of A_c: having the same amount of A_c, drivers of form (1) are impacted more than those of form (3), while drivers of form (2) are impacted less than those of form (3).

8.3.1.4 Estimation results and discussions

Since there is no widespread adoption of BEVs in Beijing at present, an SP survey was conducted to collect BEV charging and route choice behavior data from drivers. In order to ensure that the respondents' preference extracted from the questionnaire survey conforms to the reality, some basic features of BEVs, such energy consumption characteristics, cruising range, were first explained to the respondents before showing the route choice scenarios, which were given based on the route alternatives' information such as travel time, travel cost, and electric consumption.

The SP survey was implemented in the Beijing, China, from February to June in 2014. In total, 237 respondents provided valid answers. All respondents are assumed to own a BEV of BAIC E150, and each respondent was asked to make a decision of charging and route choice for 12 scenarios as the combination of four different origin-destination (OD) pairs and three different levels of initial SOC. Therefore the number of observations used in the model estimation was 1404 by eliminating the invalid data.

An MNL model containing only traditional LOS variables (i.e., without vehicle variables and charging station attributes) is used for comparison (MNL model 1). An extended MNL model with three groups of variables, namely, route attributes, vehicle variables, and charging station attributes, is further estimated (MNL model 2). Considering that the initial SOC is identical to each route, AEC and the SOC at destination are involved into the utility function of MNL model 2. An NL model is built based on the nested structure shown in Fig. 8.1. Compared to MNL models, initial SOC at origin are more intuitive than AEC to reflect the impacts of vehicle attributes on the decision on charging. Therefore the initial SOC is included in the NL model.

Table 8.8 presents the estimation results of MNL and NL models, which are estimated by using GAUSS software. The result of MNL model 1 indicates that the traditional LOS variables have negative signs, as expected, and they were significantly different from zero at the 5% level. However, MNL model 1 does not have a satisfactory goodness-of-fit, because the likelihood ratio is only 0.0307. Therefore the special variables, including vehicle variables and charging station attributes, were included in utility function. The results of MNL model 2 show that the parameters of these variables have logical signs, and all are statistically significant at the 5% level. Meanwhile, the goodness-of-fit of MNL model 2 are considerably increased.

Regarding the NL model, the logsum parameters are significantly different from both zero and one at the 5% level. In addition, the adjusted ρ^2 values demonstrate that the NL model outperforms the MNL models. In other words the NL model structure is appropriate to explain BEV charging and route choice behaviors.

Focusing on the results of NL model, in addition to the logsum parameters, the variables in the upper level including initial SOC at origin and constant term are used to explain the decision-making of charging. The initial SOC at origin, as a variable to determine the level of range anxiety, is estimated to be significant and positive, meaning that higher possibility of charging results from lower initial SOC. Alternative-specific constant for no-charging in the upper lever is estimated to be significantly negative, indicating that drivers are inclined to charging to relax the range anxiety.

Table 8.8 Estimation results for MNL and nested logit (NL) models.

Explanatory variables	MNL model 1	MNL model 2	NL model
Generic variables for route choice			
Travel time (min)	− 0.109 (− 10.915)	− 0.086 (− 6.028)	− 0.071 (− 4.039)
Travel cost (RMB)	− 0.146 (− 5.739)	− 0.136 (− 5.010)	− 0.111 (− 4.263)
Total ratio of expressway and arterial road (%)	0.304 (2.600)	0.580 (4.574)	0.638 (6.216)
Variables for route without a charging facility			
Angular cost	− 0.107 (− 4.793)	− 0.340 (− 9.214)	− 0.079 (− 1.493)
SOC at destination (%)	–	11.880 (50.507)	12.582 (7.438)
AEC (A h/km)	–	− 4.871 (− 10.261)	–
ln (PS)	–	–	1.890 (2.814)
Variables for route with a charging facility			
Charging time (min)	− 0.018 (− 4.641)	− 0.065 (− 5.754)	− 0.056 (− 3.706)
d_{OC} (km)	–	− 0.126 (− 6.268)	− 0.091 (− 3.641)
A_c/π	–	− 0.299 (− 3.729)	− 0.066 (− 2.295)
Logsum			
The nest of routes without a charging facility	–	–	0.453 (5.254)
The nest of routes with a charging facility	–	–	1.000 (fixed)
Upper level (charging demand)			
Initial SOC at origin (%)	–	–	1.698 (0.488)
Constant term	–	–	− 10.767 (− 6.495)
Sample size	1404	1404	1404
$LL(0)$	− 2515.631	− 2515.631	− 2515.631
$LL(\beta)$	− 2438.495	− 2064.428	− 2030.465
ρ^2	0.0307	0.1794	0.1929
Adjusted ρ^2	0.0306	0.1782	0.1912

Note: The values in the parentheses of the last three columns are t values. *AEC*, Average electric consumption; *PS*, path size; *SOC*, state of charge.

The generic variables in the utility functions of route choices without charging and route choices with charging are estimated to be reasonably significant. Among these variables, the parameters of travel time and travel cost are negative, and the parameter of total ratio of expressway and arterial road is positive. It is consistent with previous studies that drivers prefer to choose a route with less travel time, less travel cost, and more expressway and arterial road.

The parameter of the angular cost is only significant in the utility functions of routes without charging. And it is negative, which is consistent with previous studies (e.g., Raveau et al., 2011). This phenomenon reveals that BEV drivers pay more attention to the route direction in the route choice without charging. And SOC at destination and AEC, reflecting the energy consumption, are only added in the utility functions of routes without charging. This is because the energy consumption is paid less attention by BEV drivers due to the full charging on the way.

The logarithm value of the path size variable is estimated to be 1.890. It is significantly different from 1.0, which would be the expected value if the path-size parameter captures only the statistical error caused by the IIA property of MNL model. It has been suggested that the path-size parameter should not be arbitrarily fixed to 1.0, since it may have a meaningful behavioral interpretation (Frejinger and Bierlaire, 2007).

By considering the charging behavior, the charging station variables, including service level variable and location variable, are tested as specific variables in the utility functions of routes with charging stations. The parameter for the charging time is significantly negative, expressing the drivers' willingness for reducing the charging time. The location of charging station is reflected by A_c: the distance from origin to charging station in the estimated model. By testing three forms of A_c variables in utility functions, A_c/π is adopted in the utility function and estimated with a significant and negative value, which is consistent with the above assumption. It reveals that penalty utility of A_c variable for drivers is linear under quantitative information provision. The parameter for d_{OC} is also significantly negative, indicating that drivers prefer to charge as soon as possible for reducing the range anxiety in the route choice with charging. The above phenomenon indicates that the charging station should be located close to the areas with high travel demand and the direction of OD with more traffic volume.

Table 8.9 shows the parameter estimates of the two-class LCNL model. This model statistically outperforms the NL model (see Table 8.1).

Table 8.9 Estimation results of the two-class latent class nested logit model.

Explanatory variables	Class 1	Class 2
Generic variables for route choice		
Travel time (min)	− 0.015 (−0.414)	− 0.127 (−2.780)
Travel cost (RMB)	− 0.138 (−1.842)	− 0.148 (−2.261)
Total ratio of expressway and arterial road (%)	− 0.057 (−0.157)	2.306 (5.026)
Variables for routes without a charging facility		
Angular cost	− 1.430 (−1.487)	− 0.069 (0.485)
SOC at destination (%)	30.548 (1.728)	21.093 (1.771)
log (PS)	13.506 (1.683)	4.850 (2.602)
Variables for routes with a charging facility		
Charging time (min)	− 0.025 (−0.881)	− 0.097 (−2.079)
d_{OC} (km)	− 0.152 (−2.540)	− 0.010 (−0.177)
A_c/π	− 0.225 (−2.746)	0.410 (1.793)
Logsum		
The nest of routes without a charging facility	0.346 (2.464)	0.976 (3.855)
The nest of routes with a charging facility	1.000 (–)	1.000 (–)
Variables for the upper level (charging demand: no charge)		
Initial SOC at origin (%)	− 0.608 (−0.045)	2.171 (0.288)
Constant term	− 8.415 (−1.996)	− 7.281 (−4.729)
Class membership variables		
Constant term	0.542 (1.211)	
Female (1: female, 0: male)		− 0.702 (−2.550)
Driving years >5 years		0.981 (2.743)
Working for private companies, individual operation, or foreign company		1.197 (2.576)
Income >10,000 RMB per month		12.320 (0.191)
Having a master or PhD degree		0.363 (0.982)
Sample size	1404	
Average probability of each class (%)	37	63
$LL(0)$	− 2515.631	
$LL(\beta)$	− 1957.471	
ρ^2	0.2219	
Adjusted ρ^2	0.2171	

Note: The values in the parentheses of the last two columns are *t* values. *PS*, Path size; *SOC*, state of charge.

And the logsum parameters in two classes are statistically significant, supporting the adoption of the nested model structure. Based on preliminary testing results, five attributes of drivers (all dummy variables) are introduced in the membership function of Class 2's membership function and estimated to be statistically significant (see Table 8.2). The constant term parameter of membership function for Class 2 is set to zero for identification. The results of membership variables show that the female novice drivers with lower personal income and lower education level are more likely to belong to Class 1.

Class 1, with an average probability of 37%, has significant values for angular cost and distance from origin to charging station. Except for the total ratio of expressway and arterial road and initial SOC at origin, the parameters of the other variables have the same signs as those in the NL model. Drivers in Class 1 are inclined to choose the direct route and the charging station close to origin and pay more attention to SOC at destination rather than initial SOC at origin under quantitative information provision, because the parameter of initial SOC at origin is insignificant.

In Class 2, with an average probability of 63%, travel time and charging time are two significant variables. Except for A_c, the other variables have the same signs as those in the NL model. Compared to the drivers in Class 1, these time-sensitive drivers in Class 2 prefer not to charge. This is probably because that drivers in Class 2 have higher income and incline to take the risk of batteries running out of power en route. In addition, the parameter of A_c/π is positive in Class 2, expressing that time variables, including travel time and charging time, are core factors in choice decision-making of drivers in Class 2, and the influence of location variables can be ignored.

The value of time for a travel choice model can be used for various purposes in practice. It can be calculated by using the parameters of time and cost

$$\text{VOT} = \frac{\rho_t}{\rho_c} \tag{8.14}$$

where ρ_t is the parameter of travel time or charging time and ρ_c is the parameter of travel cost.

Table 8.10 shows the values of travel time and charging time in NL and two-class LCNL models. The value of travel time in the NL model is similar to the average hourly wage in Beijing (37 RMB/h), even though it is not necessarily the same as that of working time. And the NL model

Table 8.10 Values of travel time and charging time.

	NL model	LCNL model	
		Class 1	Class 2
Travel time	38.1	6.3	51.6
Charging time	29.9	10.7	39.4

estimated that the value of travel time is higher than that of charging time. Despite the charging time is extra time for travel, the time spent in charging station can be used for work affairs or rest. The LCNL model reveals that Class 1 highly evaluates charging time, while Class 2 attaches a higher value to charging time. Thus the NL model cannot capture such heterogeneous evaluations about different types of time.

8.3.2 Battery electric vehicles charging behavior–based deployment of charging infrastructures

8.3.2.1 Introduction

EVs have been widely considered as a potential alternative to replace internal combustion engine vehicles, which usually cause heavy air pollution. One significant barrier to a wider adoption of BEVs is, however, the limited driving range, leading to the range anxiety (Franke and Krems, 2013). In this regard, deploying more charging infrastructures could be helpful to mitigate the rage anxiety and maintain BEV drivers' daily travel demand. In line with such a consideration, Beijing has built a charging network of 92,300 public and private chargers (Yin, 2017) and also made a plan to further promote a charging network of 435,000 chargers by 2020 (Beijing Municipal Commission of Development and Reform, 2016).

One of the obstacles to develop a sufficient private charging network [i.e., every BEV owner has a home charger (HC)] is that the land sources are scarce in Beijing, leading to the lack of private parking lots. Existing studies show that the ownership rate of HCs (ORH) is 70%, meaning that 30% of BEV drivers do not have an HC (Yin, 2017), and these drivers have to charge their BEVs at public chargers. Due to this, Beijing government has mainly focused on the construction of public charging facilities, for allowing more BEV drivers to charge their BEVs in a similar convenience such as GVs.

Even though it seems that the charging demand exceeds the public charging supply, the current utilization rate of public chargers is extremely lower than expected, which is less than 15% in terms of occupied time (Jiang, 2018). What causes this result is that the current distribution of public chargers cannot match the charging demand spatially and temporally.

This section uses a public charging location model to figure out how to maximally maintain the existing activity patterns of BEV drivers. First, a BEV charging decision model is built by considering the charging choice behaviors of both BEV drivers with an HC (HC BEV drivers) and BEV drivers without an HC (NHC BEV drivers) in Beijing. Then, a location choice model for public charging stations is constructed for maximally maintaining the BEV drivers' daily trips.

8.3.2.2 Battery electric vehicles charging decision process considering charging choice behavior

BEV drivers' charging decision process is affected by factors such as vehicle condition, destination condition, and the remaining travel distance. BEV drivers have to decide whether to charge or not by considering these aforementioned factors.

8.3.2.2.1 Charging decision process

The BEV charging decision process considering charging choice behavior is shown in Fig. 8.4.

In Fig. 8.4, when a BEV driver arrives at a destination with a charging facility, the driver will first make a judgment about whether the remaining range is affordable to the remaining trips based on the SOC of BEV and the distance of the remaining trips. The remaining trips stand for the trips within this trip and the next destination with a charging facility. If the remaining range cannot afford the remaining trips, the driver chooses to charge. If the remaining range could afford the remaining trips, the driver decides whether to charge or not through the charging decision process, in which the core is to use the charging choice model for estimating the charging probability. In the charging decision process, it is assumed that if the charging probability calculated by the charging choice model is larger than 0.5, the driver will charge at this destination.

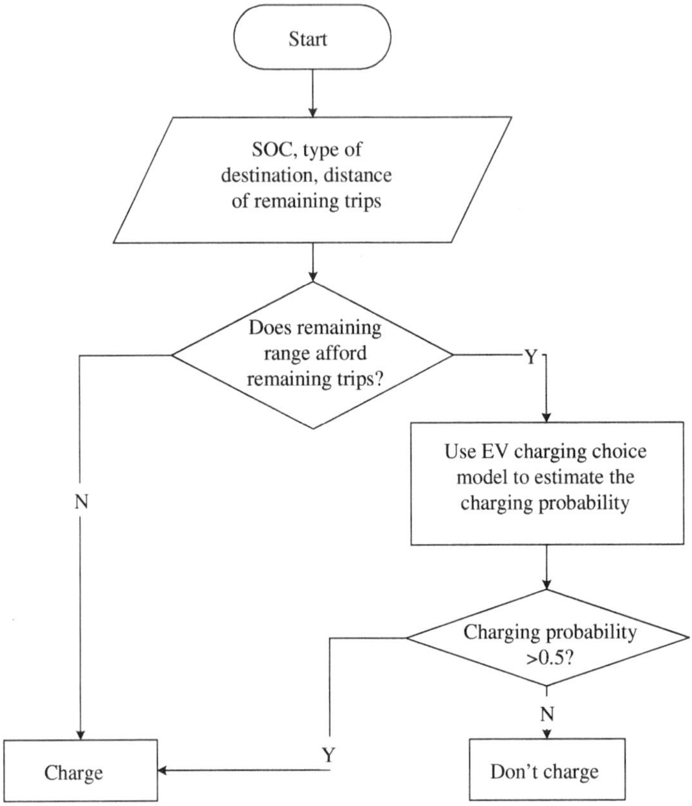

Figure 8.4 Charging decision process considering charging choice behavior.

8.3.2.2.2 Battery electric vehicles charging choice model

A BEV charging choice model with a charging demand estimation and locations of chargers is developed. In this model, factors such as SOC at destination, excess range, charging at home (1, if charge is done at home; 0, otherwise) are examined. We excluded the socio-demographic variables from the choice model, because the input data of the charging demand estimation could not provide these variables. The model is a BL model.

The data used for estimating the BL model was obtained by a web-based SP survey conducted in February 2018, where 160 BEV owners were recruited. The survey contains three parts: experimental scenarios, general attitudinal questions, and socio-demographics. Here, data about the answer of the experimental scenarios are extracted to build the charging choice model. In the experimental scenarios, SOC at destination, distance of the remaining trips, and type of the destination (home,

Table 8.11 Estimation results of battery electric vehicles charging choice model.

Explanatory variables	Parameter	*t* Value
Constant term	2.400	12.32
SOC	− 4.460	− 10.22
Excess range (100 km)	− 1.200	− 2.10
Charging at home	0.832	3.61
Sample size	800	
ρ^2	0.227	
Adjusted ρ^2	0.220	

SOC, State of charge.

workplace, or shopping mall) are used to define the scenarios. As a result, 800 valid samples were obtained.

The estimation results of the BEV charging choice model are shown in Table 8.11.

The *t* value of each variable shows that these explanatory variables are statistically significant for BEV driver's charging choice behavior. The results show that BEV drivers tend to charge when SOC is lower, the excess range is lower, and the destination is home.

8.3.2.3 Battery electric vehicles public charging station location model

Building the charging facilities is costly in terms of use of land sources and construction cost. Therefore a crucial issue of locating the charging facilities is to maximize the BEV drivers' daily trips when the sources are limited. To this end, a BEV public charging station location model (Pan et al., 2018) is applied with the objective of maximally maintaining the daily trips of BEV drivers.

The objective can be quantified by BEV drivers' missed trips, which is calculated by comparing the completed trips under a certain amount of charging facilities with the planned trips. The control variables of the model are 0−1 variables, in which each variable stands for whether a candidate site builds a public charging facility or not. BEV drivers all have their planned successive trips. When they arrive at a destination with a charging facility, they decide whether to charge or not based on the aforementioned charging decision process.

Sets:

I—set of candidate regions for building public charging stations.

J—set of the BEV drivers.

Parameters:

U_j—battery voltage of driver j's BEV (V).

Q_j—battery capacity of driver j's BEV (A h).

P_i—pubic charging power (kW) at candidate site i: if no charging station is built, then $P_i = 0$.

P_j^h—HC power (kW) of driver j.

$d_j(k)$—distance of driver j's trip k (km).

EF_j—energy consumption factor of driver j's BEV (kW h/km).

K_j—number of trips of driver j.

$t_j(k)$—parking time at trip k's destination of driver j (h).

p—number of public charging stations to be built.

Intermediate variables:

Y_j—missed trips of driver j: refers to Dong et al. (2014). If the remaining range cannot support the BEV to finish the trip, this trip and the subsequent trips are considered as a missed trip.

$h_j(k)$—HC accessibility variable: If the destination of driver j's trip k is at home and driver j has an HC, $h_j(k)$ equals one; otherwise, $h_j(k)$ equals zero.

$E_j(k)$—the energy increase from recharging at the destination of driver j's trip k (kW h).

$\mathrm{SOC}_j(k)$—SOC at the destination of driver j's trip k (%).

$\mathrm{EC}_j(k)$—energy consumption of driver j's trip k (kW h).

$y_j(k)$—a binary decision variable: If driver j's trip k is a missed trip, $y_j(k) = 1$; otherwise, $y_j(k) = 0$. When $\mathrm{SOC}_j(k) < 0$, indicating that driver j cannot complete trip k and the subsequent trips, $y_j(k) = 1$, and $y_j(k+1), \ldots, y_j(K_j)$ also equals one.

Decision variables:

x_i—decision variable in the model: If a charging station is built at candidate site i, $x_i = 1$; otherwise, $x_i = 0$.

Based on the deterministic process in Section 8.3.1, the BEV public charging station location model is given by Eqs. (8.14)–(8.19).

$$\min f(x) = \sum_j Y_j \tag{8.14}$$

subject to

$$\sum_{i \in I} x_i = p \tag{8.15}$$

$$E_j(k) = \begin{cases} \min\left((1 - \mathrm{SOC}_j(k)\right) \cdot U_j Q_j, P_j^h t_j(k)) \cdot \; \textit{if } h_j(k) = 1 \\ \min\left((1 - \mathrm{SOC}_j(k)\right) \cdot U_j Q_j, P_i t_j(k)) \cdot \; \textit{if } h_j(k) = 0 \end{cases} \tag{8.16}$$

$$\mathrm{SOC}_j(k) = \mathrm{SOC}_j(k-1) + \frac{10^3 \cdot \left(E_j(k-1) - EC_j(k)\right)}{U_j Q_j} \tag{8.17}$$

$$\mathrm{EC}_j(k) = d_j(k) \cdot \mathrm{EF}_j \tag{8.18}$$

$$Y_j = \sum_{k=1}^{K_j} y_j(k) \tag{8.19}$$

The objective function [Eq. (8.14)] is to minimize the missed trips on the condition of a fixed number of public charging stations and ORH.

The constraint [Eq. (8.15)] determines the total number of public charging stations (p). The constraint [Eq. (8.16)] defines the calculation of the energy increase from recharging. BEV drivers will utilize the parking time maximally. If the time is sufficient, BEVs will be fully charged; otherwise, the energy increase can be calculated by the corresponding charging power and the parking time. The BEV drivers can use the HC when parking at home and having an HC (if $h_j(k) = 1$). Regarding $h_j(k)$, since ORH is fixed before calculating, we randomly choose the BEV drivers who have an HC. The constraint [Eq. (8.17)] is an SOC balance equation. The constraint [Eq. (8.18)] determines the calculation of energy consumption of a trip, which is estimated by the distance and the energy consumption factor. The constraint [Eq. (8.19)] is used to calculate the number of missed trips of driver j, which is the sum of $y_j(k)$.

A genetic algorithm (GA)–based optimization model is applied to minimize the objective function of the location model subject to the constraints. GA can find globally optimal solutions for complex optimization problems (Holland, 1975) and has been applied to many charging station location problems.

8.3.2.4 Charging demand estimation and charging station location choice

The data used for this part of study was extracted from a database obtained from Beijing Household Survey in 2014, including 40,000 families with 180,000 people. In the survey, daily trips of the respondents were recorded with basic information, such as the start and end time, locations and types of origin and destination, travel distance, trip purpose, and traffic mode. The regions targeted in the survey are divided into traffic analysis zones (TAZs), and therefore the locations (e.g., origins and destinations of trips and residences) corresponded to the TAZs.

The central area of Beijing (within the fourth ring road) was selected as the study area, which is divided into 395 TAZs. To estimate the charging demand, 5182 drivers with a destination in the study area (totally 12,117 trips) were extracted. As of the year 2017, the number of BEVs in Beijing reached 120,000, and the number of private vehicles in Beijing reached 4.67 million. Therefore the penetration rate of BEVs was 2.5%. Considering the sample rate of the household survey was 0.5%, we totally obtained 25,910 BEV drivers (hereinafter referred to as "the BEV drivers") in the study area.

Since the travel survey only collected the trips of the respondents on a randomly selected day, to better understand the multiday charging demand, it is assumed that daily trips and activities of the BEV drivers are the same on weekdays.

8.3.2.4.1 Charging demand estimation

Slow charging (3.3 kW) at home and fast charging (40 kW) at public destinations are two charging modes included in the charging demand estimation and charging location model. It is also assumed that all the BEV drivers' BEVs are full of power at the beginning of the week.

We estimated the charging demand by assuming that all the candidate sites have charging facilities. The distribution of the charging demand and the completion rate of BEV drivers' five-weekday trips are summarized in Table 8.12.

Table 8.12 Trip completion performance at different ownership rates of home chargers.

ORH (%)	Number of missed trips	Number of completed trips	Completion rate (%)
50	9195	208,780	95.78
60	7605	210,370	96.51
70	6365	211,610	97.08
80	5880	212,095	97.30
90	4660	213,315	97.86
100	3035	214,940	98.61

ORH, Ownership rate of home charger.

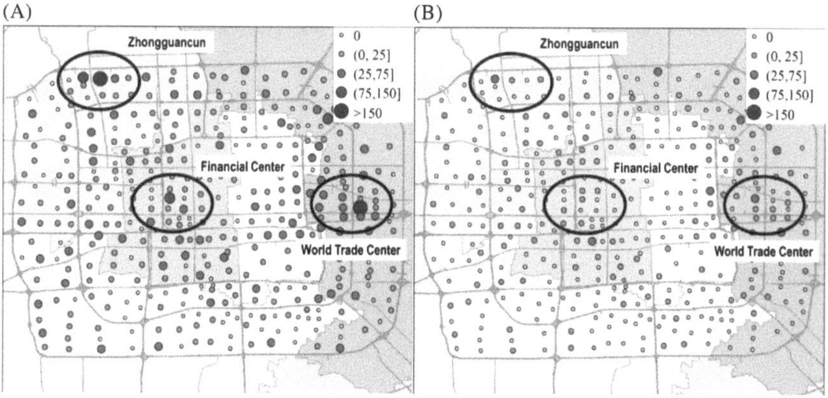

Figure 8.5 Distribution of charging demand based on different ownership rates of home chargers: ORH is (A) 70% and (B) 100%. *ORH*, Ownership rate of home charger.

In Table 8.12, when the ownership rate of HCs is 70%, about 97% of the planned trips can be completed based on a sufficient public charging network, indicating that most of the BEV drivers can maintain their daily trips without adjusting them by relying on a public charging network.

In Fig. 8.5 the distribution of charging demand with an ORH of 70% and 100% are displayed as examples, in which the magnitude of the point stands for the frequency of public charging. The charging demand is mainly observed at Zhongguancun, the National Trade Center, and the Financial Street, indicating that the BEV drivers' activities in these areas last relatively longer, which is suitable for charging. When the ORH increases, the overall charging demand decreases. For example, the charging demand at Zhongguancun and the National Trade Center decreases when the ORH increases to 100%.

8.3.2.4.2 Location of public charging stations

This section simulates the scenarios with different numbers of TAZs with public charging facilities (20, 50, 100, 200, and 300) and different ORHs (50%, 70%, and 100%). The capacity of the facility is not considered, meaning that the TAZs with charging facilities could entirely fulfill the charging demand. Moreover, the HC BEV drivers are selected randomly from all the BEV drivers based on the ORH, and other BEV drivers are set to be the NHC BEV drivers.

The fulfilling effects of the charging network could be reflected by the number of missed trips. A lower number of missed trips means that more

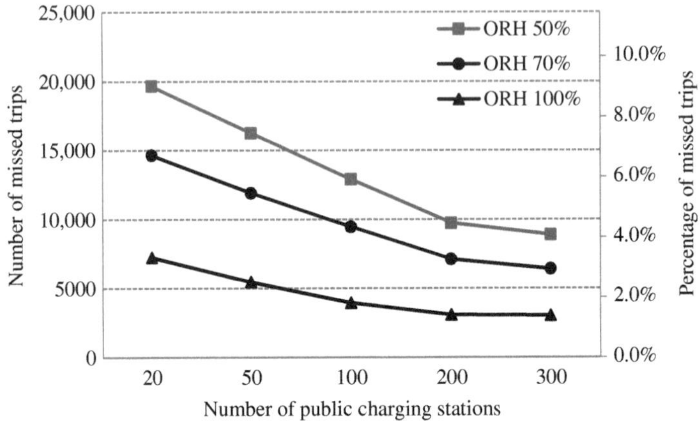

Figure 8.6 Variations of missed trips under different scenarios.

charging demand is fulfilled by the charging network, indicating the fulfilling effect is greater.

The variations of missed trips under various scenarios are shown in Fig. 8.6. The increase of public charging stations can significantly affect the number of missed trips. For example, when the ORH is 70%, the number of missed trips decreases from 14,635 (20 charging stations) to 6415 (300 charging stations), where 97.1% of the planned trips are fulfilled. The result also shows the fulfilling effect obeys the diminishing returns: when the number of public charging stations exceeds 200, the decrease of the number of missed trips starts to be slower.

Fig. 8.7 shows the spatial distributions of public charging stations when the ORH is 70%, and the number of public charging stations are 50, 100, and 200 as the examples. In this figure, the solid points represent the TAZs with a public charging station, and the hollow points represent the TAZs without a public charging station. Under each scenario, some main areas, including Zhongguancun, Financial Street, and Beijing World Trade Center, are covered by public charging stations. When the number of public charging stations increases, the locations of the charging stations gradually expand to other areas. Comparing Fig. 8.7 with Fig. 8.5, it is found that the location of public charging stations matches the charging demand spatially.

8.3.3 Summary

Based on an SP survey, Chinese BEV drivers' charging and route choice behaviors are analyzed based on an NL model, with additional

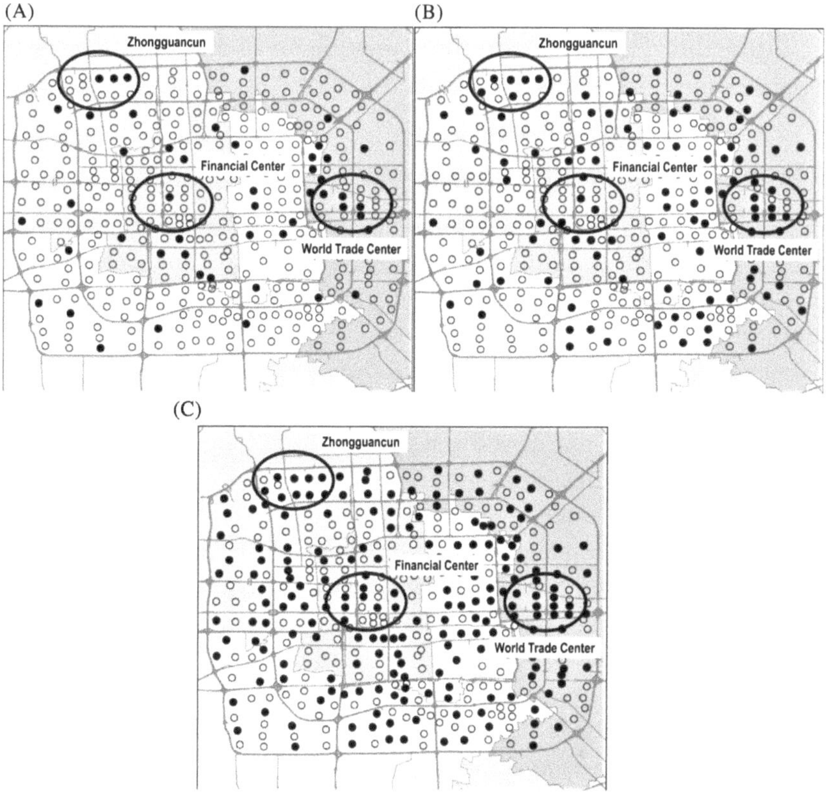

Figure 8.7 Locations of public charging infrastructures under various scenarios when ownership rate of home chargers is 70%: (A) 50, (B) 100, and (C) 200 public charging stations.

consideration of the influences imposed by vehicle variables of BEVs and service levels of charging stations. Meanwhile, to capture the preference heterogeneity across individuals, a two-class LCNL model is estimated to examine differences between segments of BEV drivers. The results show that drivers in Class 1 (the average membership probability is 37%) are inclined to choose the direct route and the charging station closer to origin and pay more attention to SOC at destination rather than initial SOC at origin under quantitative information provision, while in Class 2, drivers pay more attention to travel time and charging time.

In addition, a charging choice model with the charging demand estimation and locations of chargers is developed based on an SP survey of BEV drivers. This choice model is then applied to a case in Beijing to estimate charging demand for locating public charging stations. The results show that the charging demand is mainly observed at financial or commercial centers in Beijing.

8.4 Conclusion and future research

EVs are expected to resolve the environmental issues in the transport sector in a dramatic way, within which, China will play a key role. Such an expectation, however, depends on whether people like to own EVs or not, and how to use them. Taking Beijing as an example, this chapter makes use of SP approach and discrete choice models to clarify what factors affect the ownership and usage of BEVs by paying special attention to the role of unique policies and the influence of charging infrastructures. Even though the sample sizes used for the studies in this chapter are not very large and it is therefore difficult to make any general conclusions, various useful and unique observations to policy makers have been derived in terms of unique BEV promotion policies (subsidies and charging infrastructure construction, etc.), technological development, trip attributes, and trip makers' personal attributes and attitudes.

Future research should incorporate actual trip making patterns based on large-scale travel diary data and various Big Data (GPS data, smart card data, etc.). BEV ownership and usage should be jointly analyzed by combining SP data and actual travel behavior data. More types of EVs (not only BEV, but also PHEV and HEV) should be analyzed in a unified framework, by further considering the future development of privately owned, connected, and shared autonomous vehicles. Charging station locations and EV usage should be optimized by considering the whole transportation networks and the progress of various smart technologies. Not only behavior itself but also behavioral change should be explored in a dynamic way, because travel behavior and other related choice behaviors as well as lifestyles usually change over time. Various subsidy policies in China have largely helped the deployment of EVs in the market; however, various future uncertainties (due to changes in both domestic and international policymaking environment) still remain, which are worth being carefully explored. Finally, more research on more cities in China and other countries should be accumulated by considering various city- and country-specific policies and cultures in order to derive scientifically sound and practically feasible and effective policies for realizing environmentally sustainable transportation.

References

Bae, S., Kwasinski, A., 2012. Spatial and temporal model of electric vehicle charging demand. IEEE Trans. Smart Grid 3, 394–403.

Beijing Municipal Commission of Development and Reform, 2016. Sub-Plan for Beijing Electric Vehicle Charging Infrastructure [WWW Document]. <http://www.chinaequip.gov.cn/2016-04/23/c_135333592.htm> (in Chinese; accessed 10.06.19.).

Bjerkan, K.Y., Nørbech, T.E., Nordtømme, M.E., 2016. Incentives for promoting Battery Electric Vehicle (BEV) adoption in Norway. Transp. Res., D 43, 169–180.

Botsford, C., Szczepanek, A., 2009. Fast charging vs. slow charging: pros and cons for the new age of electric vehicles. In: EVS24 International Battery, Hybrid and Fuel Cell Electric Vehicle Symposium, pp. 1–9.

Bubeck, S., Tomaschek, J., Fahl, U., 2016. Perspectives of electric mobility: total cost of ownership of electric vehicles in Germany. Transp. Policy 50, 63–77.

Carley, S., Krause, R.M., Lane, B.W., Graham, J.D., 2013. Intent to purchase a plug-in electric vehicle: a survey of early impressions in large US cites. Transp. Res., D 18, 39–45.

Cen, X., Lo, H.K., Li, L., Lee, E., 2018. Modeling electric vehicles adoption for urban commute trips. Transp. Res., B 117, 431–454.

Conti, M., Kotter, R., Putrus, G., 2015. Energy efficiency in electric and plug-in hybrid electric vehicles and its impact on total cost of ownership. In: Beeton, D., Meyer, G. (Eds.), Electric Vehicle Business Models. Lecture Notes in Mobility. Springer, Cham.

De Dios Ortúzar, J., Willumsen, L.G., 2011. Modelling Transport. John Wiley & Sons.

Dong, J., Liu, C., Lin, Z., 2014. Charging infrastructure planning for promoting battery electric vehicles: An activity-based approach using multiday travel data. Transp. Res., C 38, 44–55.

Dumortier, J., Siddiki, S., Carley, S., Cisney, J., Krause, R.M., Lane, B.W., et al., 2015. Effects of providing total cost of ownership information on consumers' intent to purchase a hybrid or plug-in electric vehicle. Transp. Res., A 72, 71–86.

Egbue, O., Long, S., 2012. Barriers to widespread adoption of electric vehicles: an analysis of consumer attitudes and perceptions. Energy Policy 48, 717–729.

Franke, T., Krems, J.F., 2013. Understanding charging behaviour of electric vehicle users. Transp. Res., F 21, 75–89.

Frejinger, E., Bierlaire, M., 2007. Capturing correlation with subnetworks in route choice models. Transp. Res., B 41, 363–378.

Glerum, A., Thémans, M., Bierlaire, M., 2011. Modeling demand for electric vehicles: the effect of car users' attitudes and perceptions. In: Proceedings of the Second International Choice Modeling Conference, Leeds, UK, July 4–6.

Graham-Rowe, E., Gardner, B., Abraham, C., Skippon, S., Dittmar, H., Hutchins, R., et al., 2012. Mainstream consumers driving plug-in battery-electric and plug-in hybrid electric cars: a qualitative analysis of responses and evaluations. Transp. Res., A 46, 140–153.

Hao, H., Ou, X., Du, J., Wang, H., Ouyang, M., 2014. China's electric vehicle subsidy scheme: rationale and impacts. Energy Policy 73, 722–732.

He, X., Zhan, W., Hu, Y., 2018. Consumer purchase intention of electric vehicles in China: the roles of perception and personality. J. Cleaner Prod. 204, 1060–1069.

Helveston, J.P., Liu, Y., Feit, E.M., Fuchs, E., Klampfl, E., Michalek, J.J., 2015. Will subsidies drive electric vehicle adoption? Measuring consumer preferences in the U.S. and China. Transp. Res., A 73, 96–112.

Hess, S., 2009. Latent class structures: taste heterogeneity and beyond. In: Hess, S., Andrew, D. (Eds.), Handbook of Choice Modelling. Edward Elgar Publishing, pp. 311–330.

Holland, J.H., 1975. Adaptation in Natural and Artificial Systems: An Introductory Analysis With Applications to Biology, Control, and Artificial Intelligence. The University of Michigan Press.

Hu, Z., Yuan, J., 2018. China's NEV market development and its capability of enabling premium NEV: Referencing from the NEV market performance of BMW and Mercedes in China. Transp. Res., A 118, 545–555.

Huang, Y., Qian, L., 2018. Consumer preferences for electric vehicles in lower tier cities of China: Evidences from south Jiangsu region. Transp. Res., D 63, 482–497.

Jakobsson, N., Gnann, T., Plötz, P., Sprei, F., Karlsson, S., 2016. Are multi-car households better suited for battery electric vehicles? – Driving patterns and economics in Sweden and Germany. Transp. Res., C 65, 1–15.

Jensen, A.F., Cherchi, E., Mabit, S.L., 2013. On the stability of preferences and attitudes before and after experiencing an electric vehicle. Transp. Res., D 25, 24–32.

Jiang, Y., 2018. National Energy Administration: Utilization Rates of Public Chargers Are Less Than 15% [WWW Document]. <http://finance.cnr.cn/txcj/20180122/t20180122_524107201.shtml> (in Chinese; accessed 05.06.19.).

Jung, J., Jayakrishnan, R., 2011. High-coverage point-to-point transit: study of path-based vehicle routing through multiple hubs. Transp. Res. Rec. 2218 (1), 78–87.

Kamakura, W.A., Kim, B.-D., Lee, J., 1996. Modeling preference and structural heterogeneity in consumer choice. Market. Sci. 15, 152–172.

Karlsson, S., 2017. What are the value and implications of two-car households for the electric car? Transp. Res., C 81, 1–17.

Lane, B., Potter, S., 2007. The adoption of cleaner vehicles in the UK: exploring the consumer attitude–action gap. J. Cleaner Prod. 15, 1085–1092.

Larson, P.D., Viáfara, J., Parsons, R.V., Elias, A., 2014. Consumer attitudes about electric cars: pricing analysis and policy implications. Transp. Res., A 69, 299–314.

Liao, F., Molin, E., van Wee, B., 2017. Consumer preferences for electric vehicles: a literature review. Transp. Rev. 37, 252–275.

Lieven, T., Mühlmeier, S., Henkel, S., Waller, J.F., 2011. Who will buy electric cars? An empirical study in Germany. Transp. Res., D 16, 236–243.

Liu, H., Sato, H., Morikawa, T., 2015. Influences of environmental consciousness and attitudes to transportation on electric vehicle purchase intentions. Asian Transp. Stud. 3, 430–446.

Ministry of Finance, 2009. Notice on the Launch of Demonstration Pilot Projects in Energy-Efficient and New Energy Vehicles [WWW Document]. <http://www.mof.gov.cn/zhengwuxinxi/caizhengwengao/2009niancaizhengbuwengao/caizhengwengao2009dierqi/200904/t20090413_132178.html> (in Chinese; accessed 05.06.19.).

Ministry of Finance, 2010a. Notice on the Expansion of Energy-Efficient and New Energy Vehicles Demonstration and Promotion Work in Public Service Areas [WWW Document]. <http://www.ndrc.gov.cn/rdzt/jsjyxsh/201006/t20100603_352046.html> (in Chinese; accessed 05.06.19.).

Ministry of Finance, 2010b. Notice on the Subsidy Pilot Promotion of Private Purchase New Energy Vehicles [WWW Document]. <http://www.mof.gov.cn/gp/xxgkml/jjjss/201006/t20100602_2499641.html> (in Chinese; accessed 05.06.19.).

Ministry of Finance, 2015. Notice on the Finance Support Policy of NEV Promotion and Adoption During 2016 to 2020 [WWW Document]. <http://jjs.mof.gov.cn/zhengwuxinxi/zhengcefagui/201504/t20150429_1224515.html> (in Chinese; accessed 05.06.19.).

Ministry of Finance, 2016. Notice on NEV Charging Infrastructure Incentive Policy of "Thirteen Five" and Strengthening the Promotion and Application of NEVs [WWW Document]. <http://www.mof.gov.cn/mofhome/mof/zhengwuxinxi/caizhengwengao/wg2016/wg201603/201607/t20160705_2344732.html> (in Chinese; accessed 05.06.19.).

Ministry of Industry and Information Technology, 2015. Four Ministries Determine the First Batch of NEV Promotion and Adoption Cities or Urban List [WWW Document]. <http://www.miit.gov.cn/n1146285/n1146352/n3054355/n3057585/n3057592/c3610711/content.html> (in Chinese; accessed 05.06.19.).

Ministry of Science and Technology, 2009. The Outline of National Mid-and-long-term Science and Technology Development Plan (2006−2020) [WWW Document]. <http://www.most.gov.cn/mostinfo/xinxifenlei/gjkjgh/200811/t20081129_65774.htm> (in Chinese; accessed 05.06.19.).

Ministry of Science and Technology, 2011. Notice on Further Finishing the Pilot Work of Demonstrating and Extending NEVs [WWW Document]. <http://www.most.gov.cn/tztg/201111/t20111110_90811.htm> (in Chinese; accessed 05.06.19.).

Ministry of Science and Technology, 2012a. The 12th Five-Year Plan for the Development of Electric Vehicle Technology [WWW Document]. <http://www.most.gov.cn/tztg/201204/t20120420_93807.htm> (in Chinese; accessed 05.06.19.).

Ministry of Science and Technology, 2012b. Notice on the Issuance of EV Science and Technology Development the Twelfth Five-Year Special Plan [WWW Document]. <http://www.most.gov.cn/tztg/201204/t20120420_93807.htm> (in Chinese; accessed 05.06.19.).

Moons, I., De Pelsmacker, P., 2012. Emotions as determinants of electric car usage intention. J. Market. Manage. 28, 195−237.

National Energy Administration, 2015. Notice on the Issuance of EV Charging Infrastructure Development Guide (2015−2020) [WWW Document]. <http://www.nea.gov.cn/2015-11/18/c_134828653.htm> (in Chinese; accessed 05.06.19.).

Pan, L., Yao, E., Zhang, R., 2017. A location model of EV public charging station considering drivers' daily activities and range anxiety: Case study of Beijing. In: Proceedings of the 96th Annual Meeting of Transportation Research Board, Washington, DC, USA, January 8−12.

Pan, L., Yao, E., Zhang, R., 2018. A location model for evaluating the demand-fulfillment effect of public and home charging networks: A case study of Beijing. In: Proceedings of the 97th Annual Meeting of Transportation Research Board, Washington DC, USA, January 7−11.

Pearre, N.S., Kempton, W., Guensler, R.L., Elango, V.V., 2011. Electric vehicles: How much range is required for a day's driving? Transp. Res., C 19, 1171−1184.

People's Daily, 2018. China's new energy vehicle sales have the highest market share in the world for three consecutive years [WWW Document]. <http://auto.people.com.cn/n1/2018/0417/c1005-29930896.html> (in Chinese; accessed 05.06.19.).

Qian, L., Yin, J., 2017. Linking Chinese cultural values and the adoption of electric vehicles: The mediating role of ethical evaluation. Transp. Res., D 56, 175−188.

Raveau, S., Muñoz, J.C., de Grange, L., 2011. A topological route choice model for metro. Transp. Res., A 45, 138−147.

She, Z.Y., Sun, Q., Ma, J.J., Xie, B.C., 2017. What are the barriers to widespread adoption of battery electric vehicles? A survey of public perception in Tianjin, China. Transp. Policy 56, 29−40.

Sierzchula, W., Bakker, S., Maat, K., van Wee, B., 2014. The influence of financial incentives and other socio-economic factors on electric vehicle adoption. Energy Policy 68, 183−194.

Sohu, 2012. Energy-Efficient and New Energy Vehicles Automotive Industry Plan (2012−2020) [WWW Document]. <http://db.auto.sohu.com/baike/778.shtml> (in Chinese; accessed 05.06.19.).

South China Morning Post, 2017. China 2016 Car Sales Surge at Fastest Rate in Three Years. <https://www.scmp.com/business/china-business/article/2061642/china-2016-car-sales-surge-fastest-rate-three-years> (in Chinese; accessed 05.06.19.).

Sun, X., Xu, S., 2018. The impact of government subsidies on consumer preferences for alternative fuel vehicles. J. Dalian Univ. Technol. 39, 8–16.

Woodjack, J., Garas, D., Lentz, A., Turrentine, T., Tal, G., Nicholas, M., 2012. Consumer perceptions and use of driving distance of electric vehicles: changes over time through lifestyle learning process. Transp. Res. Rec. 2287 (1), 1–8.

Wu, Y., Zhang, L., 2017. Can the development of electric vehicles reduce the emission of air pollutants and greenhouse gases in developing countries? Transp. Res., D 51, 129–145.

Yang, F., Zhang, Y., Ge, X., 2011. Electric vehicle development tendency and characteristics in China, USA, and Japan. Energy Technol. Econ. 23, 40–44.

Yang, J., Liu, Y., Qin, P., Liu, A.A., 2014. A review of Beijing's vehicle registration lottery: short-term effects on vehicle growth and fuel consumption. Energy Policy 75, 157–166.

Yao, E., Yang, Z., Song, Y., Zuo, T., 2013. Comparison of electric vehicle's energy consumption factors for different road types. Discrete Dyn. Nat. Soc. 2013, 328757 (7 pages). Available from: https://doi.org/10.1155/2013/328757.

Yin, X., 2017. Number of BEV in Beijing is Larger Than Germany [WWW Document]. <http://bj.people.com.cn/n2/2017/1128/c233088-30973423.html> (in Chinese; accessed 05.06.19.).

Zhang, X., Liang, Y., Yu, E., Rao, R., Xie, J., 2017. Review of electric vehicle policies in China: content summary and effect analysis. Renew. Sustain. Energy Rev. 70, 698–714.

CHAPTER 9

Impacts of shared mobility on energy consumption and emissions in China

Biying Yu*, Xiaoyi Li and Meimei Xue
Beijing Institute of Technology, Beijing, P.R. China

9.1 Introduction

With the rapid development of the economy and the improvement of people's living standard, the annual growth rate of transportation energy consumption and CO_2 emission in China is as high as 9.7% and 9% in the past 10 years. In the transport sector the CO_2 emissions of urban road traffic account for more than 75% (ITF, 2010; IEA, 2011–2015). Domestic and international studies indicate that the environmental issue of urban road traffic will be further intensified.

New emerging shared travel modes such as "Zhuanche," "Kuaiche," and "Shunfengche" are developing very rapidly in China, which indicates that the transport sector in China has entered the "new normal" situation. In China, such kind of shared mobility, indicated as an app-based private taxi, has been promoted in more than 400 cities, and the number of registrants and orders has increased exponentially since 2015. Ride-sourcing services have reached 12% of the total number of taxi orders. Shared mobility, as a new travel mode in the era of "Internet Plus," has had great impacts on traditional travel modes. Although shared mobility only rose to popularity in the past few years, it has had significant influences on the existing transport system, overall car usage, and urban residents' lives. First of all, shared mobility service makes up for the travel demand which existing public transport modes cannot satisfy. Especially in megacities such as New York, Beijing, and Shanghai, during rush hours, the public transportation systems are usually in a state of extreme crowdedness. The shared mobility services offer more choices for people. To some extent, it

* Corresponding author.

Transport and Energy Research
DOI: https://doi.org/10.1016/B978-0-12-815965-1.00009-0

compensates for insufficient public transport capacity. However, on the other side, the public has raised many questions about its externality, such as whether the shared mobility will increase the user's travel demand, whether the emergence of shared mobility services will have negative effects on the environment. Therefore evaluating the environmental impact of shared travel and exploring how to realize the green development of the sharing mobility in different cities will provide reference and inspiration for the energy conservation and emission reduction for both the local and national governments and enterprises.

In China, there are five main app-based shared mobility services, including online taxi-hailing service, hitching service, regular ride-sourcing service, shared ride-sourcing service, and premium ride-sourcing service. The definition of each service is given in Table 9.1. Different from traditional taxi, online taxi-hailing service allows passengers to send a taxi request on mobile apps instead of hailing the taxi on the road which could save their waiting time and get a taxi easier. Hitching service is a kind of ridesharing service, which enables private car drivers and passengers who share a similar route to be matched and share the ride (Chen et al., 2017). Regular ride-sourcing service, shared ride-sourcing service, and the premium service are three types of ride-sourcing services which could help to meet people's tiered travel demands. Generally, comparing to other two ride-sourcing services, shared ride sourcing is the cheapest way to travel because passengers who share part of route could choose to split a ride and fare for sharing the route. On the other hand, premier ride-sourcing service offers higher end mobility experience with luxury cars at a relatively higher price.

It is thought that the emergence of shared mobility may have great impacts on people's travel behaviors. In the short term, passengers shift to shared mobility from bus, taxi, private car, metro, green modes, or others, what is to say, people's travel mode choice behavior has been changed due to the existence of shared mobility. Such kind of effect of shared mobility caused by the mode shift is called direct effect. While from the middle- or long-term viewpoint, shared mobility is likely to influence people's demand for private vehicles due to the convenient shared mobility service, which accordingly change their intention of purchasing cars (such as buying a new car or replacing the old car with new ones). Such kind of changes will result in the decline of energy consumption and emissions (CO_2 and air pollutant emissions) during the processes of vehicle manufacturing, transportation, and disposal. Here, we regard this effect as indirect environmental effect.

Table 9.1 Definition of shared mobility services in China.

Service type	Service	Definition
Taxi	Online taxi-hailing service	Passengers send a ride request through shared mobility app and a taxi will pick up the passengers no matter where they are.
Ridesharing	Hitching service (Shunfengche 顺风车)	Hitching matches private car drivers and passengers who share similar routes. Both the private car drivers and passengers could benefit from shared travel costs.
Ride sourcing	Regular ride-sourcing service (Kuaiche 快车)	Private car drivers are able to connect passengers with travel needs on shared mobility platform and provide a ride for them.
	Shared ride-sourcing service (KuaichePingche 快车拼车)	Sharing ride-sourcing service allows private car drivers picking up more than one group of passengers who share the similar destinations or part of route. Passengers are able to share the total costs.
	Premium ride-sourcing service (Zhuanche 专车)	Premium offers standardized, high-quality, and professional services with luxury vehicles. The price is relatively higher than other services.

Therefore it is not an easy question to answer the impacts of shared mobility on energy consumption and emissions. Under such circumstance, we conducted a joint research with DiDi Chuxing, a Chinese transportation network company offering the largest peer-to-peer platform for private vehicles in China. By employing the raw big data on sharing trips which include user and vehicle information provided by DiDi, we aim to quantify the environmental impacts of ridesharing and further answer how to realize more environmental benefits for sharing mobility.

9.2 Characteristics of sharing trips

The big data of shared mobility trips across China collected from Didi Chuxing shows specific spatial and temporal characteristics. Here, we take Beijing as an example to interpret. Fig. 9.1A and B shows the average

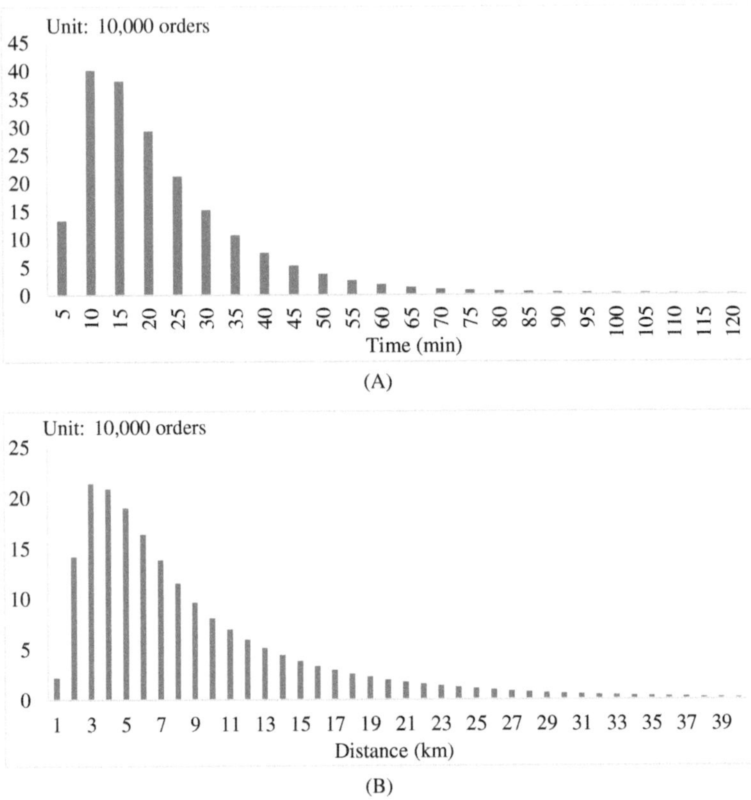

Figure 9.1 Distribution of (A) travel time and (B) distance for all sampled ride-sourcing trips (Xue et al., 2018).

travel time and trip distance for ride-sourcing trip in Beijing. The travel time includes two parts: pickup and service time. The former is the time before the driver picking up the passenger(s) once he or she accepts the order request and the latter means passenger(s)'s travel time from the origins to the destinations. The distance of ride-sourcing trips also includes pick-up distance and service distance. The average service time is about 20 minutes. The most frequent service time of a ride-souring trip in Beijing is around 10 minutes, followed by 15 minutes. The average service distance is about 8.5 km, which is close to the average trip distance (8.6 km) of taxi in Beijing (BTRC, 2015). This indicates a competitive but also complementary relationship between ride-sourcing travel and existing taxi service. Ride sourcing is much less attractive to trips longer than 20 km (service distance) and trips shorter than 1 km. The most

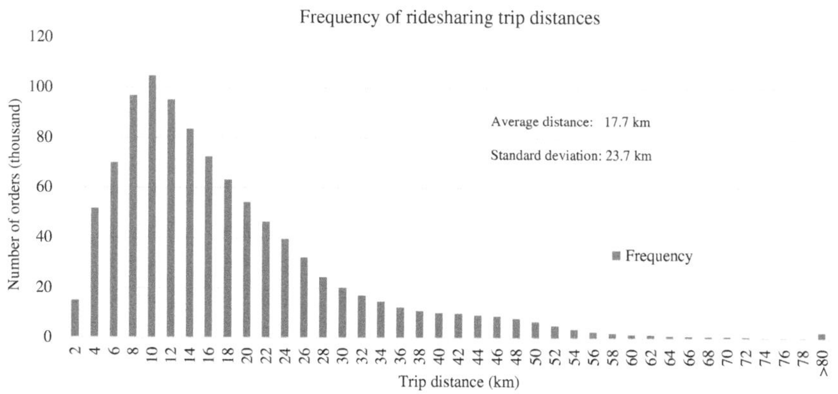

Figure 9.2 Frequency of ridesharing trip distances (Yu et al., 2017).

common service distance of a ride-souring trip in Beijing is between 2 and 7 km (Fig. 9.2).

Compared with the ride-sourcing trips, the average ridesharing trip distance is much longer, approximately 17.7 km in Beijing with the most frequent ridesharing trip distance ranging between 8 and 14 km, similar to the average travel distance of subway (17.8 km) in Beijing (BTRC, 2015), implying that ridesharing service may make up the current shortage of railway system.

9.3 Behavioral impacts of shared mobility

To better understand the phenomenon of shared mobility and the potential changes of shared mobility services on people's travel behaviors and motivations, we designed a questionnaire for the shared mobility frequent users (who use shared mobility service at least twice per week) and conducted a large-scale app-based survey through DIDI Chuxing platform during May to November 2016. Users received and filled the questionnaire from their DIDI Chuxing app on their mobile phones. This survey covered a total sample of 150,000 medium- and high-frequency users. After removing samples with missing and invalid answers a total of 9,762 valid survey feedbacks, which covered passenger from 10 main cities of China, were used for this analysis.

In the survey the users were asked to answer their alternative travel modes if they do not use shared mobility. In addition to the options of traditional travel modes, we gave another option "cancel the trip" so as to see whether shared modes trigger extratravel demand or not. The survey results show that only 0.3% (negligible) of passengers reported to cancel the trip, suggesting that the occurrence of shared mobility mainly has influence on the share of existing modes, and it does not increase trip demand. Fig. 9.3 shows that 39.0% of them chose taxi as the original travel mode if there was no sharing mobility service, 15.7% chose subway, 20.6% chose bus, and 16.0% chose their own private vehicle. It can be seen that shared mobility partakes certain travel demand from taxi, bus, and subway, which adds up to 75.3%. Shared mobility also attracted the users of private vehicles. If the cars used for shared mobility are more efficient than passengers' own vehicles, shifting to shared mobility will have positive effects on the environment for this part of travel demand. Furthermore, this will be more possible with the increasing prevalence of electric vehicles in the market.

It is thought that shared mobility service may influence passengers' vehicle ownership behavior, on the one hand due to its convenience and economic benefit and on the other hand due to the constraint of the vehicle purchase and the insufficient capacity of public transport as well as the difficulties of parking in the current situation. According to our survey results (Fig. 9.4), 50% of passengers reported that they will change or may change their willingness for buying new car or replacing their old car in the future if the shared mobility service is always available. While for the rest frequent users, only 18% of them said that the existence of shared mobility service will not change their car ownership behavior. Generally, the intention for buying private cars will decline attributing to the shared mobility service, resulting in the decreasing demand in the automotive manufacturing sector, and hence the energy consumption and emissions.

Considering the performances and triggers for people's travel mode change and car-purchasing behavior change might differ across individuals

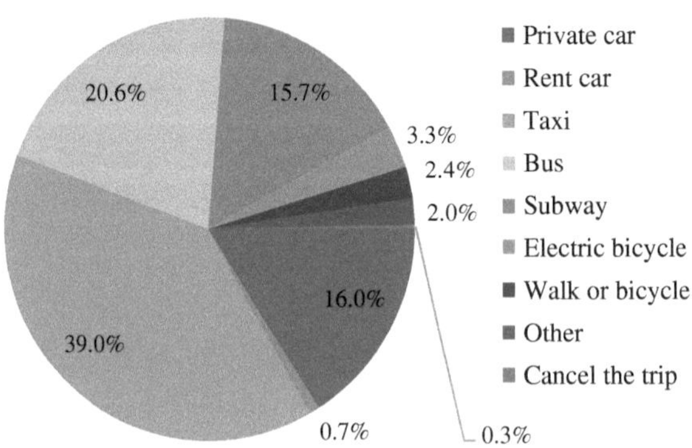

Figure 9.3 Alternative travel modes if not using sharing mobility.

given their diverse sociodemographic attributes, travel behaviors and motivations, residential environmental characteristics, and surrounding transportation infrastructures. Consequently, the multinomial logit model is used to investigate these factors that influence people who turned to using shared mobility services instead of using the original travel mode. And the binary logit model is utilized to represent the car-purchasing choice of shared mobility users. The framework is shown in Fig. 9.5. The results are displayed in Figs. 9.6 and 9.7. More details can be found in Tang et al. (2019).

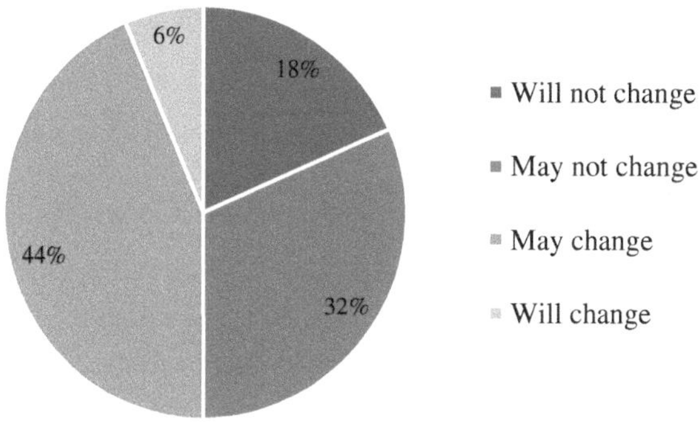

Figure 9.4 Willingness changes for buying a new car or replacing the old car when sharing mobility is available.

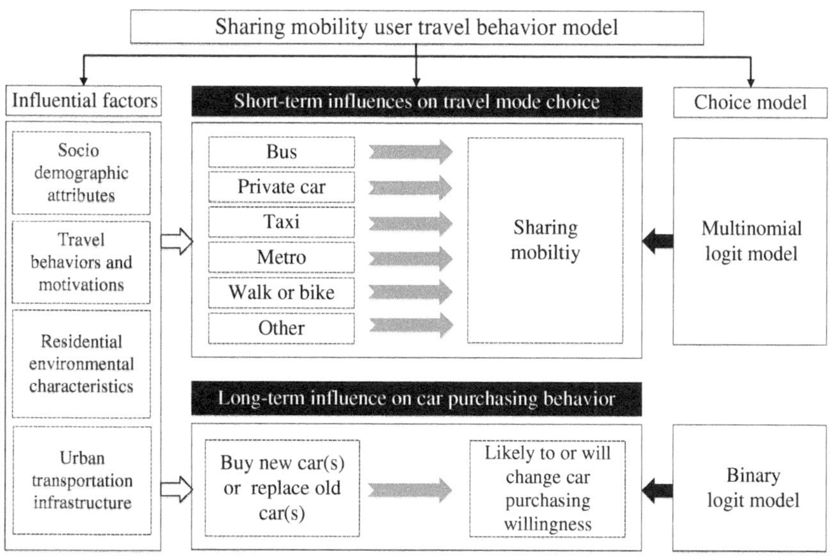

Figure 9.5 Framework of evaluating the impacts of shared mobility on travel behavior.

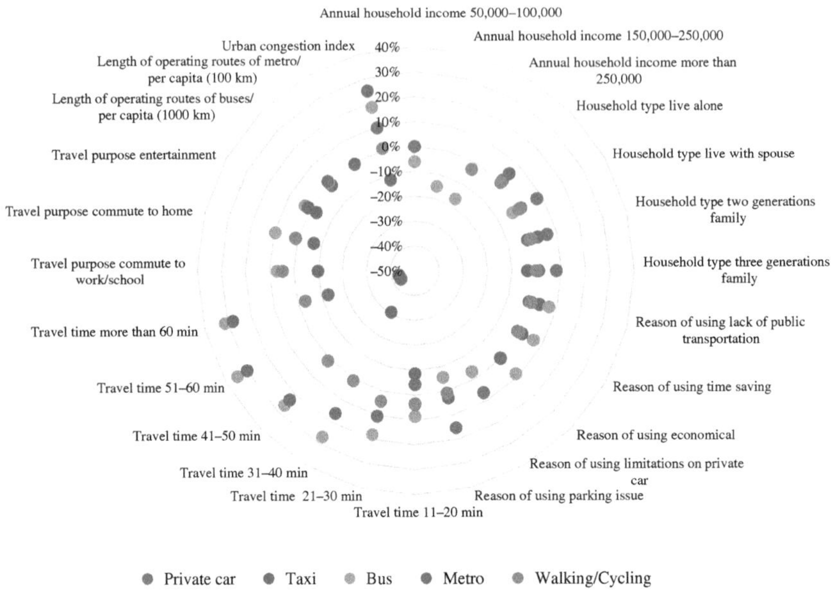

Figure 9.6 Impacts of influencing factors on the travel mode shift for sharing mobility users.

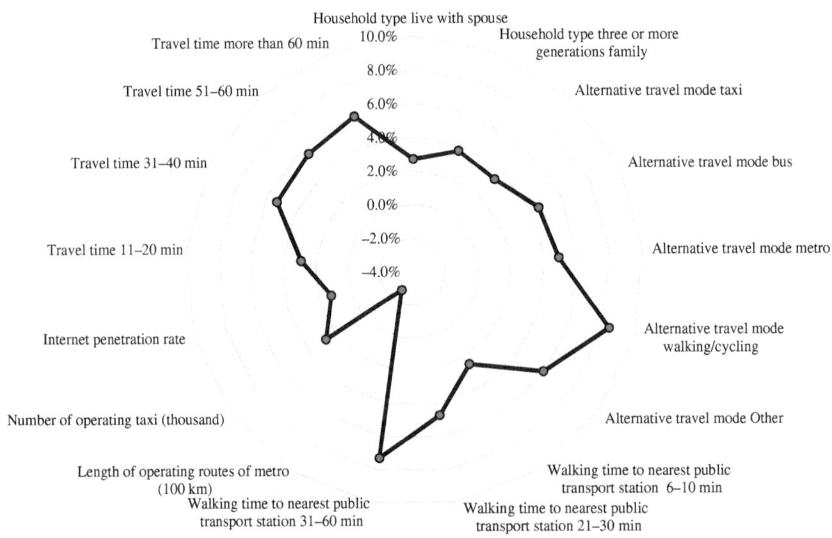

Figure 9.7 Impacts of influencing factors on the vehicle purchase behavior change for sharing mobility users.

The travel choice behavior of shared mobility users is related to their household type, income, travel time, reason, and purpose of using app-based ride hailing. Fig. 9.6 shows the marginal effect that demonstrates how these factors influence users' shift to app-based shared mobility from other travel modes. Original private car and taxi passengers with higher household income show higher probability of using app-based shared mobility. People who live with their spouse and with more than two-generation family are more likely to shift to app-based ride hailing from private car. This may be due to the fact that with the increase of family members, private cars might not satisfy everyone's travel demand. Parking issues are a major concern by private car drivers which lead them to give up their original travel modes. It reflects the common phenomenon of shortage of parking lots. With the increase of travel duration, original private car users and taxi passenger were less likely to shift to app-based ride hailing. On the contrary, original bus and metro passengers show higher probability of shifting to app-based ride hailing when the travel time increases. One plausible explanation is that public transport serves a limited area and is usually with transfers that worsen the experience.

In addition, residential environment also influences people's travel mode choices. Residents who lived in cities with longer operating routes of metro have lower probability to shift from metro to app-based shared mobility. The increase of urban congestion will increase the probability of original taxi passengers to shift to app-based ride hailing. By using ride-hailing platforms, passengers are able to call a cab easier and make full use of the waiting time. The original metro passengers are more likely to stick on taking underground transportation to avoid road congestion.

Car-purchasing decision of app-based shared mobility users is influenced by their household type, travel time, their alternative travel mode, and the availability of public transportation. Fig. 9.7 shows the marginal effect that demonstrates how these influential factors change users' car-purchasing decision. In terms of travel time, if longer journeys have to be taken by app-based ride hailing, the demand for private cars will decrease. Residents whose original travel mode is on foot or bike have the highest probability (8%) to change their mind. Original bus and metro passengers also show high probability to reconsider their car-purchasing decision. If travel needs of the people who take public transportation and bike or walk can be met by app-based ride hailing, then they are more likely to reconsider their car purchases. Walking time to reach nearest public transport station has a noticeable impact on people's car-purchasing willingness. If residents need to take a 21–60 minute walk to reach the nearest public transportation, and app-based ride hailing is an option of travel mode, they are more likely to change their car-purchasing decision.

Shared mobility users' car-purchasing willingness is influenced by the level of urban transportation infrastructure. One percent increase in metro-operating routes per capita will decrease the probability of changing car-purchasing willingness by 2%. This suggests that if app-based ride-hailing services increase in popularity in cities with shorter metro lines, people in these cities may change their mind on purchasing cars because of ride-hailing services. Residents who live in cities with more taxis show higher probability to change their car-purchasing decision. App-based ride hailing allows taxi drivers to use online hailing service to reach customers instead of aimlessly cruising on the road.

9.4 Environmental impacts of shared mobility

For different types of shared mobilities, their environmental impacts are quite different attributing to diverse influencing mechanisms. Here, we take two typical sharing modes as the examples to explain, including ride-sharing and ride-sourcing travels.

9.4.1 Energy-saving and emissions reductions of ridesharing service

Ridesharing (Hitching) services encourage private car drivers to share their vehicles and rides with travelers with similar destinations. On the basis of the existing car travel the hitch services undertake additional passenger travel needs that are partly diverted by cars, buses, subways, or other travel modes. Thus the energy consumption and emissions resulting from the use of alternative travel modes for hitching service's passengers are the energy-saving and emission reductions associated with the use of this service. Based on the life cycle assessment of transport fuel (Hawkins et al., 2013; Norman et al., 2006; Wang, 1996) and the survey results, we calculated the energy consumption and emissions generated by the alternative travel modes other than shared mobility service. The top 20 cities with the highest energy-saving and emission reductions in 2016 in China were shown in Fig. 9.8. The national cumulative energy saving reached up to 102,000 tce and 174,000 t of CO_2 which is equivalent to the CO_2 absorbed by 5.8 million trees in a year or the CO_2 emitted by 110,000 cars traveling 10,000 km a year. Beijing ranked at the top of the list. Shenzhen, Hangzhou, Dalian, Guangzhou, Wuhan, and other cities also have substantial contributions.

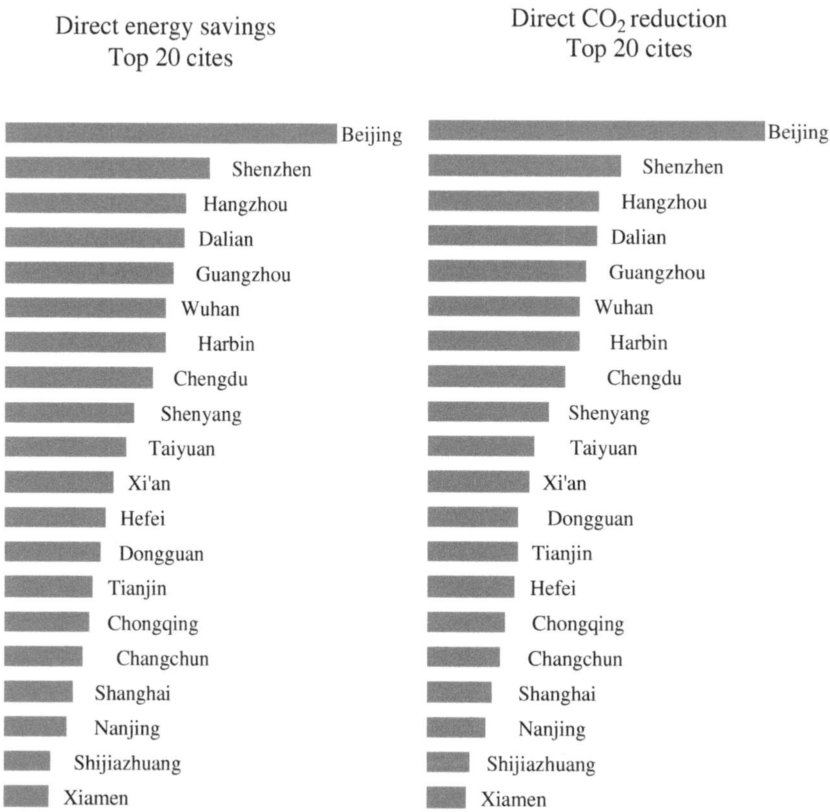

Figure 9.8 Direct energy-saving and CO_2 emission reductions in the top 20 cites of China.

9.4.2 Energy-saving and emission reductions potential of ride-sourcing service

Ride-sourcing services refer to as taxi-like services provided by private cars. From the supply side, ride-sourcing services can facilitate the development of sharing economy. On the basis of no additional infrastructure investment and practicing the sharing concept, ride-sourcing services expand the existing transport supply. From the demand side, ride-sourcing services have enriched residents' travel modes, providing more efficient, convenient, and comfortable services. To some extent, these services ease the difficulty of parking, car purchasing, and the insufficiency of public transportation facilities. However, the use of ride-sourcing services is also accompanied by extraenergy consumption and CO_2 emissions.

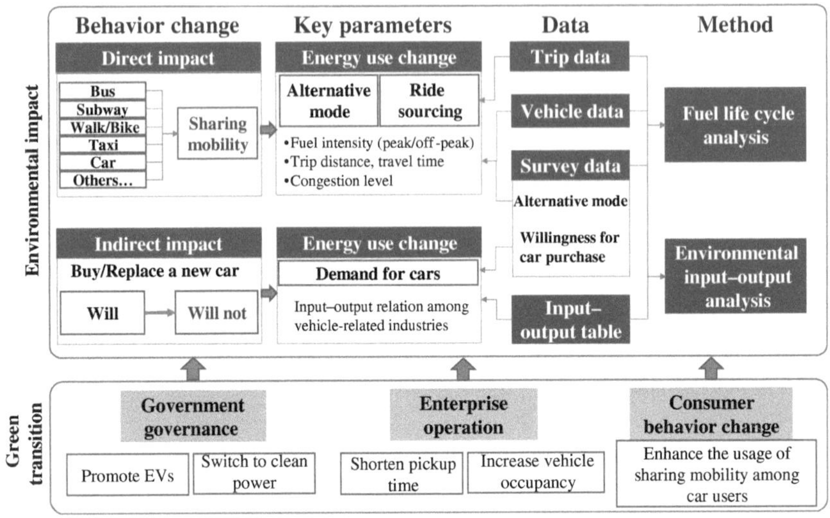

Figure 9.9 Analytical framework for evaluating the environmental impacts of ride-sourcing travel.

Fig. 9.9 shows the analytical framework for evaluating the environmental impacts of ride-sourcing travel.

The direct environmental impacts of ride sourcing on the complete road transport system are the difference between the energy consumption and emissions related to the original travel mode and that related to the ride-sourcing mode. Fig. 9.10 shows the additional energy consumption and CO_2 emissions due to the travel mode shift to ride-sourcing travel. Beijing, Chengdu, Shenzhen, and Guangzhou have much higher energy and emissions generated compared to other cities in China in 2016. Consequently, under the premise of meeting residents' travel demand, accelerating the green development of ride-sourcing service will play a positive role in promoting energy conservation and emission reduction in the transportation sector and even in the whole country. To comprehensively evaluate the environmental impacts of ride-sourcing travel, we further quantified the potential reduction amount of energy consumption and emissions due to the decreasing car purchase demand based on input–output method (Isard, 1951; Miller and Blair, 2009).

With the consideration of the issues long-term development plans for the country and enterprises, our research group conducted an in-depth evaluation of the potential energy-saving and emission reductions of ride-sourcing service, from the perspective of three agents involved in the shared mobility, that is, government, enterprises, and consumers. Among them,

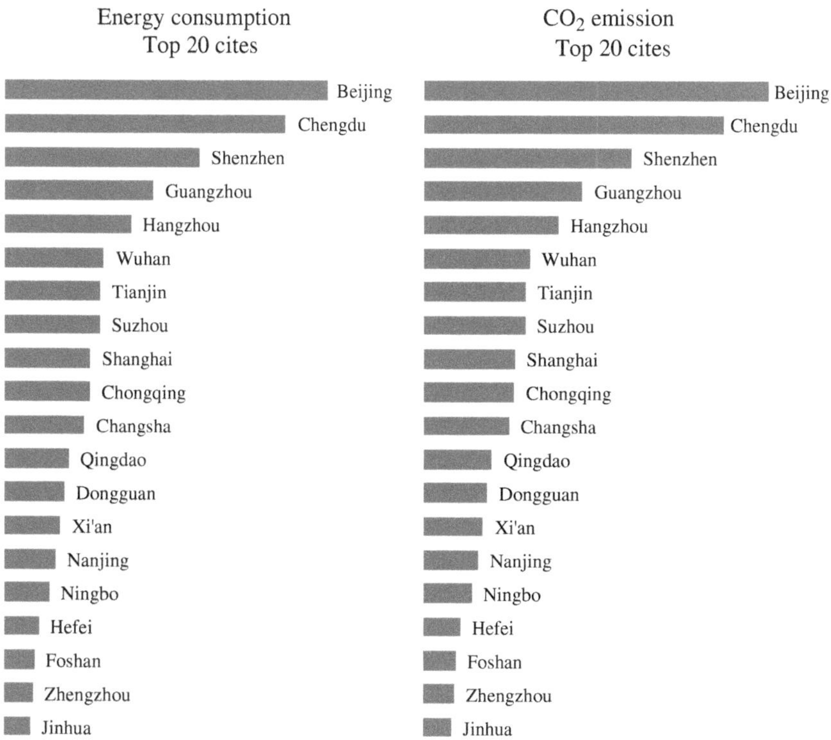

Figure 9.10 Additional energy consumption and CO_2 emissions for the top 20 cites in China.

government's actions include increasing the proportion of electric cars and promoting the usage of clean electricity for the electric cars; enterprises' actions include encouraging users to choose carpool services and shortening the time for matching the orders; and consumers' actions include increasing the use of share mobility instead of owning and or using private cars. By combining the efforts from different agents, we further analyze the energy-saving and emission reductions potentials for ride-sourcing travel (Fig. 9.9).

Joint efforts by governments, enterprises, and consumers will have the greatest potential for reducing emissions. In 2016 the major cities in China can achieve nearly 50% of energy-saving and emission reductions compared to the case without intervention (see Fig. 9.11). The energy-saving percentage of Changchun, Guiyang, Harbin, Shanghai, and Shijiazhuang is higher than national average level (57%). The CO_2 emission reduction percentage of Changchun, Guiyang, Harbin, Langfang, Shanghai, Shenyang, Shijiazhuang, and Zhengzhou is higher than the

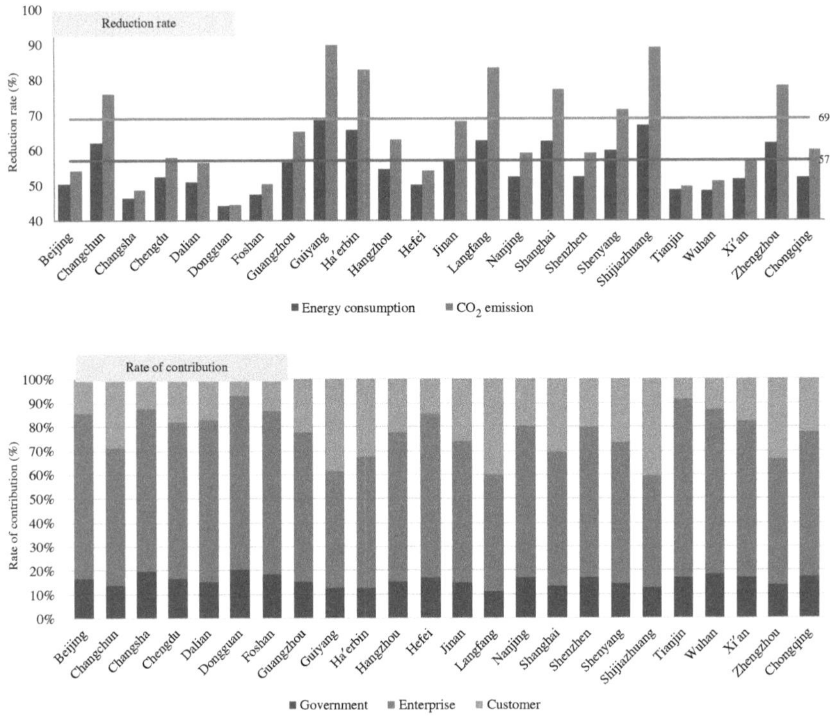

Figure 9.11 Impact of all-inclusive cooperation strategy.

national average level (69%). In comparison, Dongguan and Changsha do not show great emission reduction potential. Enterprises contribute the most to reduction under cooperation scenario. Especially in Beijing, Changsha, Chengdu, Dalian, Dongguan, Foshan, Tianjin, and Wuhan, the contribution of the enterprises is outstanding; the government contribution rates among cities are relatively stable and do not show great differences; and the residents' contribution rates vary substantially. In second-tier and third-tier cities such as Guiyang, Langfang, and Shijiazhuang, residents' behavior plays an important role in influencing the energy conservation and emission reduction.

Due to the implementation difficulties of the six types of policies, it is difficult to fully realize the assumed energy-saving and emission reductions under the all-inclusive cooperation strategy. In order to further investigate the energy-saving and emission reductions potential of each agent, as well as their cooperation effects, we further estimated the energy-saving and emission reductions of each agent alone and under the dual cooperation strategies between each two agents.

When the two agents cooperate to reduce emissions, cooperation of enterprise and consumer offers the greatest potential. When governments and enterprises cooperate to reduce emissions, the energy consumption will decrease by 39%–46% and the CO_2 emissions by 37%–43%. When governments and consumers cooperate to reduce emissions, the energy consumption will decrease by 16%–43% and the CO_2 emissions by 15%–69%. When enterprises cooperate with consumers' behavior change, the energy consumption will decrease by 36%–65% and the CO_2 emissions by 40%–94% (see Fig. 9.12).

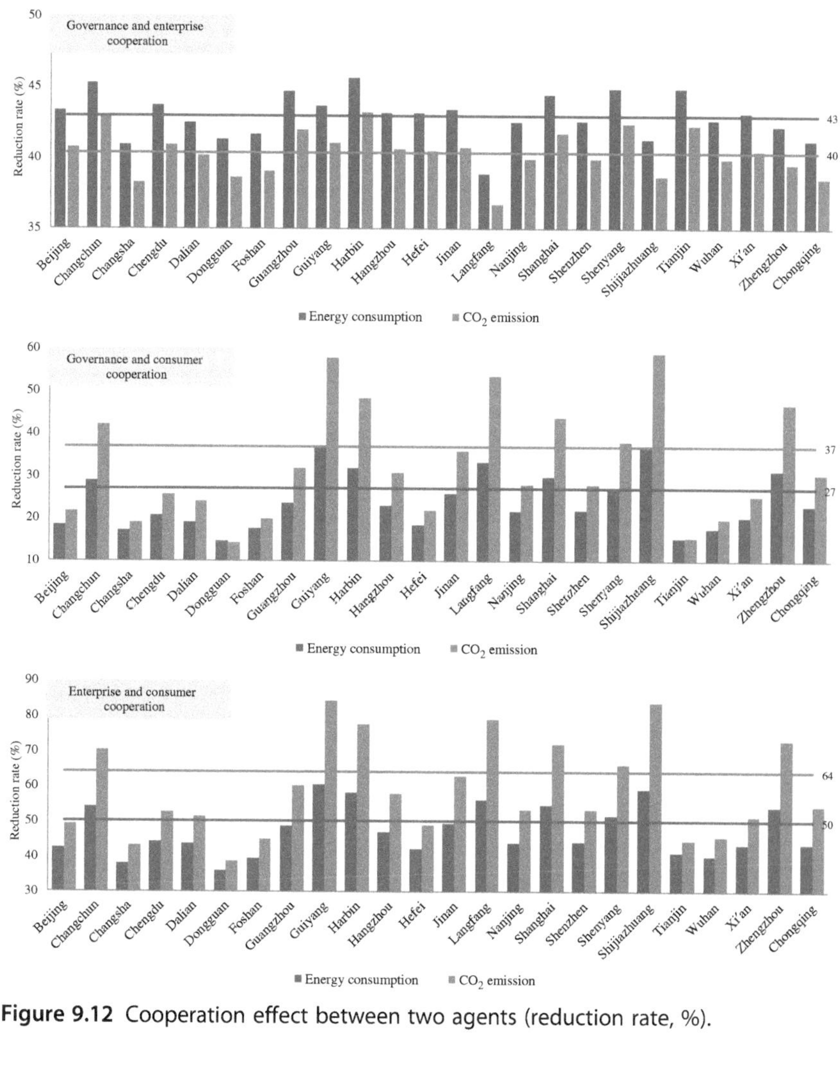

Figure 9.12 Cooperation effect between two agents (reduction rate, %).

Regarding the independent effect of each agent, on average, enterprises alone can achieve the largest CO_2 emissions reduction, ranging between 32% and 38%. Among the three emission reduction policies of enterprises, increasing vehicle occupancy to two persons is the best and most operational way to reduce emissions. The energy consumption and the CO_2 emissions will decrease by 20%–25% (see Fig. 9.13). When the pickup time is shortened to 3 minutes, the energy consumption and CO_2 emission will decrease by 20%. The northeastern regions, Shanghai and Guiyang, have much more obvious impacts. At present, carsharing service has not been fully promoted. If carsharing success rates could be increased, the energy consumption and CO_2 emission will decrease by 1%–4%. As carsharing services expand, their effectiveness will increase, similar to that of increasing vehicle occupancy.

Changing consumers' behavior is likely to have substantial impacts on emissions reduction as well, but with large differences among cities (see Fig. 9.14). The existence of shared mobility may change consumers' willingness to purchase new car or replace old car in the future. This will reduce the energy consumption and emissions during the motor vehicles production, manufacturing, transportation, and recycling process. Since changing residents' behavior is a medium- and long-term policy, we calculated the potential energy-saving and CO_2 emission reductions by changing consumers' behavior based on the survey results of consumers' car-purchasing willingness and distributed the energy-saving and CO_2 emission reductions into 10 years. As a result, if 60% of passengers use the ride-sourcing services in the first-tier cities, 3%–31% of energy

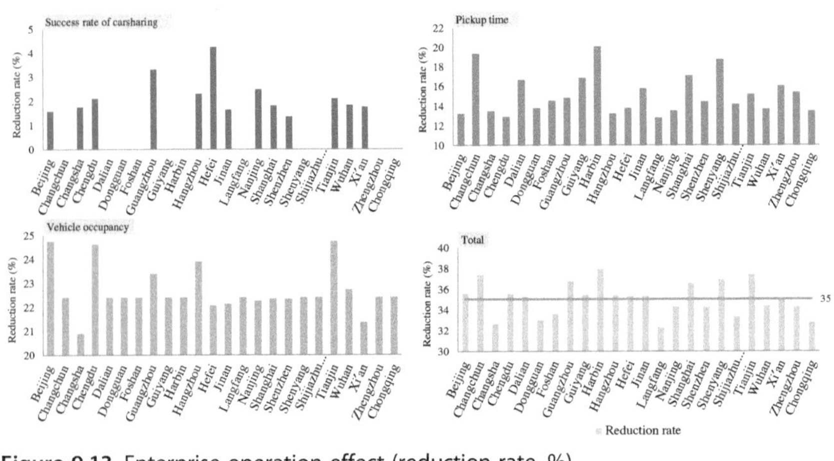

Figure 9.13 Enterprise operation effect (reduction rate, %).

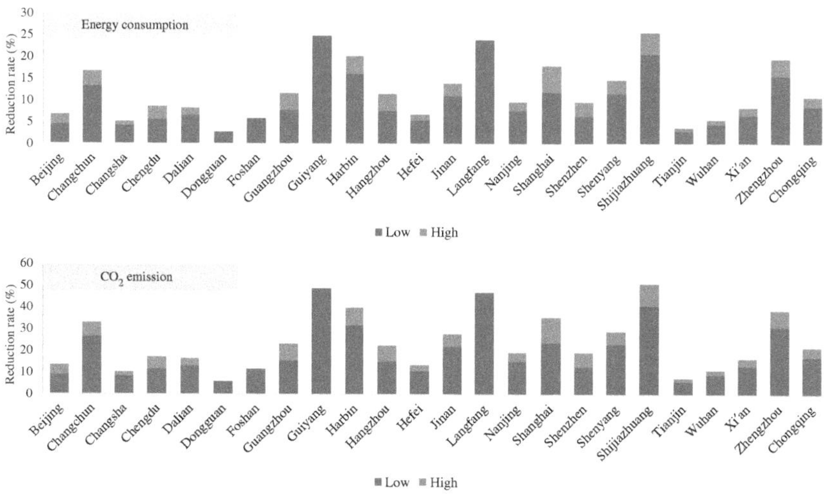

Figure 9.14 Consumer behavior change effect (reduction rate, %).

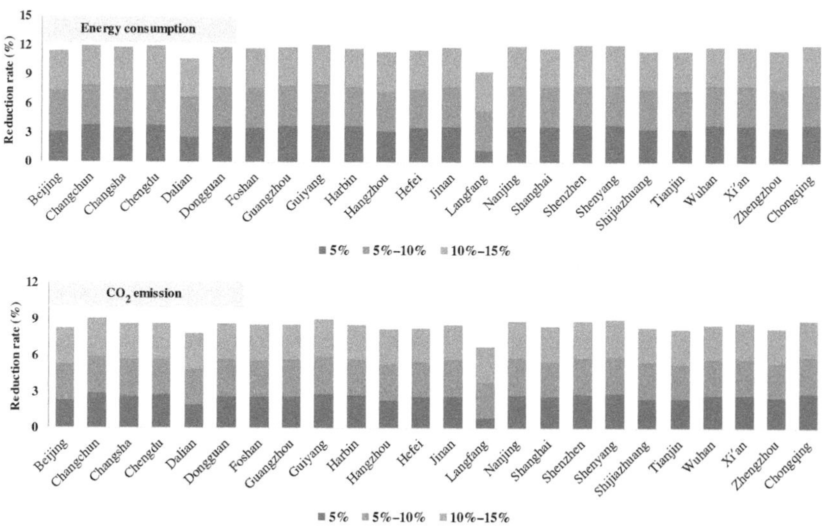

Figure 9.15 Governance and management effect (reduction rate, %).

consumption and 7%–61% of CO_2 emissions additionally generated by ride-sourcing trips can be reduced, with an average value of 15% and 29%, respectively.

The effectiveness of governmental governance on reducing emissions will rise significantly with the penetration of electric vehicles (see Fig. 9.15). When the proportion of electric vehicles in cities reached 15%, the energy consumption would be reduced by 12%, and the CO_2 emission would be reduced by 9%. The differences among cities are small.

9.5 Conclusions and policy implication

Shared mobility as a new emergent travel mode has a significant impact on people's existing travel patterns. However, the heavier pressure on the environment from the road transportation sector has called for a sustainable development of the shared mobility. This study has shed light on the environmental impacts of the ridesharing and ride-sourcing travel by taking into account the influence of shared mobility on consumers' current travel mode choices and the potential travel behavior related to car ownership. China is taken as the empirical context. The energy consumption change and emissions change caused by the ridesharing and ride-sourcing travel in the main Chinese cities are evaluated and ranked. Different policy packages are further designed for exploring the potentials of energy-saving and emission reductions for each city. Several conclusions can be drawn as follows:

1. Ride-sourcing services mainly contribute to meet the demand for the short- and middle-length trips, while ridesharing services are more welcomed among the long-trip users.
2. Sharing mobility has evident influence on residents' current travel model choices and their future attitudes toward purchasing new vehicles or replacing the old vehicles. Shared mobility partakes certain travel demand from taxi, bus, subway, and private vehicles. Fifty percent of passengers are likely to change their willingness for buying new car or replacing their old car in the future if the shared mobility services are always available.
3. Ridesharing services could significantly reduce the energy consumption and emissions. The national cumulative energy saving attributing to ridesharing travel reached up to 102,000 tce and 174,000 t of CO_2.
4. Travel mode shift to ride-sourcing travel caused additional energy consumption and CO_2 emissions. Beijing, Chengdu, Shenzhen, and Guangzhou have much higher energy and emissions generated compared to other cities in China in 2016. The strategy with the joint efforts from governments, enterprises, and consumers, which includes the measures of promoting the use of EVs, switching to clean electricity generation, shortening pickup time, increasing vehicle occupancy,

and enhancing the ride-sourcing usage among car users or potential car users would have substantial impacts for reducing emissions. In 2016 this strategy could make the major cities in China achieve nearly 50% of energy-saving and emission reductions.

Based on the conclusions, several policy implications are proposed here for promoting the shared mobility in light of energy saving.

1. The government may use the market or other financial measures to encourage shared mobility enterprises to vigorously develop ridesharing services. The government can incorporate the shared mobility enterprises into the carbon trading market. Through market incentives, enterprises can improve the operating platform and system, attracting more users to provide and use ridesharing services, so as to further increase the energy-saving and emission reductions from ridesharing service. In addition, according to the contribution of shared mobility enterprises in energy-saving and emission reductions, the government may appropriately reduce or exempt these enterprises from taxation to better incentive their actions.
2. The government should encourage more new energy vehicles to undertake shared mobility services. The government should focus on subsidizing the new energy vehicles that provide shared mobility services instead of implementing nontargeted promotion strategies. The government can increase the subsidy and exempt purchase tax to the new energy vehicles that provide shared mobility services. It would contribute to facilitating the use of new energy vehicles, maximizing the energy-saving and emission reductions potential of new energy vehicles and realizing the transition of the transport sector to the "new normal."
3. The technical and operational management modes of shared mobility enterprises should be further optimized. Shared mobility enterprises should continuously shorten the order processing time of ride-sourcing services by improving the efficiency of the vehicle-user matching platform. At the same time, users of shared mobility services can be guided to adopt carsharing travel with further incentive schemes, so as to improve carsharing success rate and vehicle occupancy rate.
4. Joint efforts by governments and enterprises to cultivate users' travel habits are needed. Governments and enterprises should cooperate to increase the publicity of the shared mobility and attract car users and potential car users to use shared mobility services. Our results show that because of the shared mobility services, nearly 50% users will or are likely to change their car-purchasing decisions. Therefore training

users to use shared mobility services will reduce the energy consumption and emissions of automotive industry during the production, manufacturing, transportation, recycling, and other processes.

References

BTRC, 2015. Beijing Transport Annul Report 2015. Beijing Transportation Research Center.

Chen, X., Zahiri, M., Zhang, S., 2017. Understanding ridesplitting behavior of on-demand ride services: an ensemble learning approach. Transp. Res. C: Emerg. Technol. 76, 51–70.

Hawkins, T.R., Singh, B., Majeau-Bettez, G., Strømman, A.H., 2013. Comparative environmental life cycle assessment of conventional and electric vehicles. J. Ind. Ecol. 17, 53–64.

International Energy Agency (IEA), 2011–2015. CO2 Emissions from Fuel Combustion Highlights. IEA.

International Transport Forum (ITF), 2010. Transport Greenhouse Gas Emissions: Country Data 2010. <http://www.itf-oecd.org/sites/default/files/docs/10ghgcountry.pdf> (accessed 16.11.16).

Isard, W., 1951. Interregional and regional input-output analysis: a model of a space-economy. Rev. Econ. Stat. 33, 318–328.

Miller, R.E., Blair, P.D., 2009. Input-Output Analysis: Foundations and Extensions. Cambridge University Press.

Norman, J., MacLean, H.L., Kennedy, C.A., 2006. Comparing high and low residential density: life-cycle analysis of energy use and greenhouse gas emissions. J. Urban Plann. Dev. 132, 10–21.

Tang, B.-J., Li, X.-Y., Yu, B., Wei, Y.-M., 2019. How app-based ride-hailing services influence travel behavior: an empirical study from China. Int. J. Sustain. Transp. Available from: https://doi.org/10.1080/15568318.2019.1584932.

Wang, M., 1996. Development and Use of the GREET Model to Estimate Fuel-Cycle Energy Use and Emissions of Various Transportation Technologies and Fuels. Energy Systems Div., Argonne National Lab., IL.

Xue, M., Yu, B., Du, Y., Wang, B., Tang, B., et al., 2018. Possible emission reductions from ride-sourcing travel in a global megacity: the case of Beijing. J. Environ. Dev. 27, 156–185.

Yu, B., Ma, Y., Xue, M., Tang, B., Wang, B., Yan, J., et al., 2017. Environmental benefits from ridesharing: a case of Beijing. Appl. Energy 191, 141–152.

CHAPTER 10

Rural migrant workers' energy consumption in passenger transport and domestic sectors of China: case study from a social perspective

Ying Jiang[1], Junyi Zhang[2,*] and Linghan Zhang[2]
[1]Hiroshima University, Hiroshima, Japan
[2]Hiroshima University, Higashi Hiroshima, Japan

10.1 Introduction

Research connecting transport and energy in a household context has mainly focused on urban residents. This is probably because urbanization is a global trend that has affected urban development in various positive and negative ways. Positive aspects mainly refer to its contribution to economic and social developments, while negative aspects include the resulting pressure on use of various limited resources (e.g., land, water, gas, fuel). In China, there are about 287 million rural migrant workers in cities, who account for 37.0% of the whole working population.[1] These rural migrant workers are also urban residents. In 2018 the urbanization rate in China was 59.6%[2]; however, the rate was just 42.4% if rural migrant workers in cities are excluded. Urbanization in China is largely due to rural–urban migration (Chen and Song, 2014). China established a special household registration system in the 1950s, called *hukou*, to control the flow of rural migrants into cities (Chan and Zhang, 1999). Under this *hukou* system, rural migrant workers (in cities) do not have equal access to various public services (e.g., medical treatment, social security,

* Corresponding author.

[1] <https://baijiahao.baidu.com/s?id = 1609553390005317019&wfr = spider&for = pc> (accessed 02.05.19.).

[2] <https://baijiahao.baidu.com/s?id = 1623601284313679808&wfr = spider&for = pc> (accessed 02.05.19.).

Transport and Energy Research
DOI: https://doi.org/10.1016/B978-0-12-815965-1.00010-7

education) as they do not have an urban *hukou*. More discussion about urbanization and *hukou* in China can be found in Jiang et al. (2017). Fortunately, since the implementation of the China's National New-type Urbanization Plan (2014–20),[3] the traditional *hukou* system has been gradually abolished, and more and more rural migrant workers are expected to become a truly urban resident. As a result, more and more rural population will migrate into cities. Accordingly, public policy makers need to prepare for such a new trend of urbanization in China, which is already the world's largest energy-consuming country. However, the transport and energy research on this special urban population group, rural migrant workers, is very limited. With the progress of urbanization and abolition of *hukou* system, rural migrant workers will gradually become a part of major consumers of in-home and out-of-home energy-consuming end uses in cities.

Motivated by the above background, this study attempts to present additional evidence on rural migrant workers' energy consumption and its influential factors. Both in-home and out-of-home energy consumption are targeted and complicated cause–effect relationships between its influential factors are reflected, which include ownership of vehicles and in-home energy-consuming end uses, residential behavior, the built environment attributes, and various life conditions (e.g., employment conditions, various difficulties encountered in daily life, barriers to survive in cities, and future concerns). Reflecting such life conditions into transport and energy research emphasizes a social perspective for the future transport and energy policymaking. In other words, this study argues that the transport and energy policies associated with rural migrant workers should not neglect the various social issues that have long been bothering this special population group in China. Even though this research focuses on China, its findings may also have various policy implications to other developing countries, urbanization of which has also been pushed up because of massive rural-to-urban population migration.

For the above research purpose, this study adopted a questionnaire survey data collected by the authors to 986 rural migrant workers in Dalian (a coastal city), Guiyang (an economically less-developed city), and Chongqing (one of the four municipalities directly under the central government) in 2014 and 2015. These cities were selected in consideration of

[3] <http://www.gov.cn/zhengce/2014-03/16/content_2640075.htm> (in Chinese; accessed 26.04.16.).

the fact that cities with different population sizes and locations are subjected to different requirements regarding the transformation of rural migrants to residents with an urban *hukou* in China.

10.2 Findings from two relevant existing studies

When conducting the aforementioned survey to rural migrant workers, the authors also implemented a survey to residents with an urban *hukou* in Dalian and Guiyang. For details, refer to Jiang et al. (2017, 2019). Here, these two studies are emphasized, not only because they provided an extensive literature review related to this study, but also because they presented comparative results. Details of the *hukou* system can be found by Jiang et al. (2017). These two papers further reviewed studies on migration, urbanization, and *hukou* in China, determinants of general energy consumption behavior (socio-demographics; lifestyle, attitude, belief, and consciousness; policies and technologies; the built environment), rural migrants' energy consumption behavior, factors of living conditions, and quality of life of rural migrant workers. Thus further review about existing studies will not be given. Instead, the above two studies are summarized as follows for bettering the position of this study in the literature.

Jiang et al. (2019) extracted data of Dalian and Guiyang (667 rural migrant workers) from the same data used in this study and compared energy consumption (electricity, gasoline, and gas) with residents with an urban *hukou* in the two cities, whose data (671 residents) were also collected in 2014 and 2015. They found that residents with an urban *hukou* consumed electricity 1.2−1.6 times higher, gasoline 2.2−4.4 times higher, gas 1.9−2.7 times higher than rural migrant workers did, respectively. Analyses based on a zero-inflated negative binomial model revealed that electricity consumption was mostly associated with residential behavior for rural migrant workers and the built environment for residents with an urban *hukou*. For both population groups, gasoline consumption was mostly affected by individual and household attributes, and gas consumption was mostly influenced by residential behavior. They further examined the influences of purchased insurances (pension, medical, unemployment, and employment injury insurances) and confirmed that the insurances were more influential to gasoline consumption by rural migrant workers, and electricity and gas consumption by residents with an urban *hukou*. Different types of purchased insurances did not impose consistent influences on the three types of energy consumption.

Jiang et al. (2019) made the first attempt to examine how insurance purchase and the built environment affect energy consumption between rural migrant workers and urban residents with a *hukou* in China. However, other factors of living conditions were ignored.

Differently, Jiang et al. (2017) examined factors affecting rural migrant workers' intentions to permanently reside in cities in the future and their associations with future energy consumption. They revealed that life satisfaction and viability (or survival ability) in cities were most influential to these workers' intention to reside permanently in cities, where the viability was measured in terms of skill level, job type (temporal or permanent), nonwork days per week, and income. They further identified that job type and nonwork days per week were two key factors to define the viability and rural migrant workers with a permanent job, and more days for rest were more likely to reside permanently in cities by obtaining an urban *hukou*. In the case of future energy consumption (future ownership of various end uses were used as a proxy variable), confidence and barriers to become a permanent urban resident mostly affected future energy consumption. Unexpectedly, the above intention did not impose any influence on future energy consumption. As for end uses, they found that vehicle ownership (car, motorcycle, and electric bike) was not a key end use for determining the future energy consumption preference, which was mostly characterized by washing machine, television (TV), and refrigerator. They concluded that "permanent residence in cities may not necessarily lead to the increase of energy consumption. Rather, rural migrants' concerns about their future life do matter to energy policymaking." However, Jiang et al. (2017) did not focus on rural migrant workers' current energy consumption.

In summary, this study will differentiate from existing studies in terms of examining how factors of living conditions affect rural migrant workers' energy consumption in a comprehensive way.

10.3 Survey and data

The questionnaire survey was conducted in Dalian in November 2014 (324 rural migrant workers), in Guiyang in June 2015 (343 rural migrant workers), and in Chongqing in August 2015 (333 rural migrant workers). The final sample size is 986 by excluding missing and invalid data. Details of survey implementation refer to Jiang et al. (2017).

The questionnaire consists of four parts: Part I investigates rural migrant workers' current lives that are the main contents for this study. Part II is about rural migrant workers' future lives, which are beyond the scope of this study (a study using this part of data was done by Jiang et al., 2017). Part III records rural migrant workers' employment histories (employment period, workplace, employment industry, firm type, firm size, monthly income, and contract). Part IV collects information about all household members, including those who did not live/work in the above cities (age, gender, relationship with the household head, education level, employment status, workplace, marital status, and annual income).

The contents of Part I are summarized in Table 10.1. Characteristics of the major contents are described as follows.

- Energy consumption
  It is investigated in terms of use of electricity, gas (liquefied gas and coal gas), fuel (gasoline, kerosene, and diesel), water, and so on across four seasons of a year. It was reported as monetary expenditure or amount of energy consumption. Here, its unit is unified to money (RMB/month). In other words, all the reported amounts of energy consumption were transformed into money. This study targets four major types of energy: electricity, gas, fuel, and water. Rural migrant workers in this case study spent 254 RMB/month on electricity, 177 RMB/month on fuel, 81 RMB/month on gas, and 81 RMB/month on water. Thus electricity and fuel are two energy sources for these workers' daily lives.
- Ownership of energy-consuming end uses
  Energy-consuming end uses include vehicles and in-home end uses [refrigerator, washing machine, TV, microwave oven, air conditioner, personal computer (PC), induction cooker, and rice cooker]. It is observed that 27% of rural migrant workers owned a vehicle (car: 11%, electric bike: 9%, motorcycle: 7%), and the ownership rates of in-home end uses were 80% for rice cooker, 78% for TV, 70% for washing machine, 52% for induction cooker, 42% for PC, 32% for air conditioner, and 17% for microwave oven. The 27% value is a very low rate of vehicle ownership, and most of the rural migrant workers only owned in-home end uses (cookers and machine machines) for meeting fundamental needs in their lives.
- Residential behavior
  It includes residential location, residential duration, residence type, and housing size. Only 16% of rural migrant workers owned a house,

Table 10.1 Major questionnaire items used in this study.

Category	Description
Energy consumption	Electricity, gas (liquefied gas and coal gas), fuel (gasoline, kerosene, and diesel), water, and so on (RMB/month)
The built environment attributes (km)	Distances from residence to workplace and the nearest daily facilities, including transit station, bus stop, supermarket, department store, park, hospital, preschool, elementary school, secondary school, and high school
Residential behavior	Residential location (residence city, district, and street), residential type (e.g., own a house, rent a house, live in a company dormitory), housing size (m^2), residential duration (years)
Ownership of energy-consuming end uses	Vehicles (electric bicycle, motorcycle, and car) and in-home end uses (refrigerator, washing machine, TV, microwave oven, air conditioner, PC, induction cooker, rice cooker)
Commuting behavior and time use	Commuting mode, commuting time, usual departure time from residence, and daily time use (primary work, secondary work, sleeping, recreation and leisure, others)
Employment conditions	Job type (temporal or permanent), employment contract (yes or no, and how long), bonus per year, weekly holidays, extra working, number of jobs, job training experience, and skills
Insurances purchased	Pension, medical, unemployment, employment injury, maternity insurances, and housing provident fund
Current difficulties encountered in daily life	Lack of employment security, bad job rules, low income, lack of skill training, *hukou* discrimination, difficulty in finding a school for children, separate living from family, lack of social security, poor living environment, boring cultural life, difficulty in finding a doctor, and inconvenient transportation
Current barriers to job hunting	Lack of skills, low education level, lack of employ ment information, gender discrimination, *hukou* discrimination, and discrimination of birthplace
Future concerns	Taking care of parents, children's education, children's job, stability of job, spouse's job and life, housing problem, social security, medical care, relationship with neighbors, relationship with colleagues, personal health, family members' health, and personal social status

(*Continued*)

Table 10.1 (Continued)

Category	Description
Governmental supports	Familiarity with the current governmental policies supporting rural migrant workers, request on further governmental supports (to provide more employment opportunities, to improve social security system, to strengthen human rights protection, to abolish the *hukou* system, to provide affordable houses, and to secure children's education opportunities)
Factors affecting permanent urban residence	Economic difficulty, lack of skills, unstable job, housing problem, children education, *hukou* discrimination, and difficulty in finding a partner
Subjective evaluation	Satisfaction with the current income level, satisfaction with the current workplace, satisfaction with the current job position, satisfaction with the current life as a whole, happiness as a whole life, willingness to continue the current job, suitability to the current urban life, expectation on the level of children education, confidence about children's future development, perception about the position of the current household income at hometown, willingness to permanently reside in cities
Individual and household attributes	Gender, age, income, marital status, education level, number of children, education fee for children, cohabitants (living together with spouse/children/parent), food expenditure share, driving license, ownership of land/field/mountain assigned by government at hometown, etc.

PC, Personal computer; *TV*, television.

while 68% lived in a rental house, and 14% lived in dormitories prepared by workplaces. The average duration living in the current residence was 4.4 years, and the average housing size is about 50 m^2. It seems that the housing size is not small, but actually, many rural migrant workers lived in a shared house/apartment. This means that several workers lived together. Unfortunately, such a fact could not be properly captured during the survey. This is one of the limitations of this study.

- Commuting behavior and time use

  It is found that 54% of rural migrant workers walked to workplaces and 31% commuted by bus. On average, rural migrant workers spent 10.2 hours on work, 7.6 hours on sleep, and 2.8 hours on leisure.

Jiang et al. (2017) reported that the car ownership rate of residents with an urban *hukou* was about twice higher than that of rural migrant workers. Once rural migrant workers become a resident with an urban *hukou*, it is expected that the shares of walking and bus for commuting may decrease largely. Their time use patterns may also change considerably. Therefore it is important to understand how vehicle ownership and time use affect rural migrant workers' energy consumption.

- The built environment attributes
  Distances from residence to workplace and the nearest major daily facilities (bus stop, transit station, supermarket, department store, park, hospital, preschool, elementary school, secondary school, high school) were reported. The daily facilities nearby residence with the shortest distance were bus stops and supermarkets (both are 0.7 km), and the distances to high school and workplace are the longest (both are 4.0 km). Other daily facilities with longer access are park (3.6 km), secondary school (2.6 km), and hospital (2.5 km). The average distance to transit station was 1.6 km in the case with a station nearby residence, which is equal to the distance to the nearest preschool. The access to the nearest elementary school was a distance of 1.8 km, which is shorter than the distance to department store nearby residence (2.0 km).
- Employment conditions
  Rural migrant workers with a temporal job accounted for 62%, those receiving a job training were 42%, and those with a certain skill were 32%. Rural migrant workers with a formal contract occupied 56%, and the average contract period was just about 2 years.
- Insurances purchased
  Purchasing insurances is very important to survive in cities. Reflecting this, 64% of rural migrant workers purchased a medical insurance, 43% purchased a pension insurance, and 30% purchased an employment injury insurance. In addition, 21% purchased an unemployment insurance, 17% purchased a maternity insurance (covering medical treatment, health care, economic compensation for temporal leave from workplace during the pregnancy and childbirth period), and 16% purchased a housing provident fund (a long-term saving for house). However, 25% did not purchase any insurance. This is very a worrying fact.
- Current difficulties and barriers
  As for the current difficulties encountered by rural migrant workers, 53% mentioned about low salary, and 39% complained about no job security. Other difficulties were no social security (20%), separate living

from family (19%), difficulty in finding a school for children (18%), bad built environment (17%), and difficulty in finding a doctor (16%). Lack of cultural life was pointed out by 22% of rural migrant workers. *Hukou* discrimination was experienced by 10% of rural migrant workers. Major barriers in finding a job were low level of education (63%), no skill (50%), discriminations (22%: 8% due to without an urban *hukou*, 8% due to the area of hometown, and 6% due to gender), and no job information (20%).

- Future concerns
  What rural migrant workers mostly concerned about their future were taking care of their parents (48%), children education (42%), unstable job (41%), housing (37%), children job and family health (28% for each), social security and personal health (26% for each), and medical issue (24%). Furthermore, 15% worried about their personal social status, and 13% concerned about their spouses' employment. Related to future concerns, 60% of rural migrant workers had a land, field, and/or mountain at their rural hometowns, which is surely an issue if they decide to permanently reside in a city.
- Individual and household attributes
  More than 70% of rural migrant workers were male (73%), 64% were married, 55% lived together with their spouses, 44% lived with their children, and only 18% lived with their parents. As for education levels, 14% graduated from a high school, 59% from a secondary school or below, and 16% from a vocational school. There were 8% who graduated from a university.

10.4 Modeling analysis of rural migrant workers' energy consumption

10.4.1 Conceptual framework

Fig. 10.1 shows the assumptions made in this study, about the relationships between rural migrant workers' energy consumption and its influential factors. Each arrow indicates a hypothesis that should be tested by using empirical data. Because these hypotheses may work jointly in affecting energy consumption, they need to be tested together. To this end, a structural equation model (SEM) (Golob, 2003) is adopted in this study.

The above factors are grouped into seven categories: residential behavior, built environment, ownership of energy-consuming end uses, insurance purchase, employment conditions, current difficulties encountered in

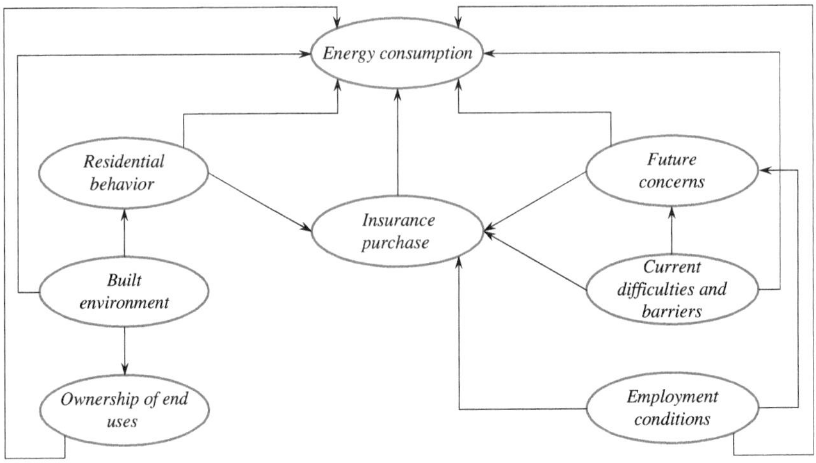

Figure 10.1 Assumptions about the relationships between rural migrant workers' energy consumption and its influential factors.

daily life and barriers to job hunting, and future concerns. These factors are selected from Table 10.1. Each category indicates a latent variable in the above SEM, where energy consumption is also a latent variable. Hereafter, these latent variables are shown in *italic*: *Residential behavior*, *Built environment*, *Ownership of end uses*, *Insurance purchase*, *Employment conditions*, *Current difficulties and barriers*, *Future concerns*, and *Energy consumption*.

As for the cause–effect relationships between these latent variables, it is first assumed that *Built environment* affects both *Residential behavior* and *Ownership of end uses*, as assumed in existing studies (e.g., Yu et al., 2012; Jiang et al., 2017). Because the purchased insurances include a housing provident fund, *Insurance purchase* is assumed to be affected by *Residential behavior*. As shown in Table 10.1, there are six types of insurances related to medical treatment, employment/unemployment, childbirth, retirement, and housing. As an expenditure decision, more expenditure spent on insurances indicates a decline of available household budget for other purposes and may consequently lead to less money spent on energy consumption. If a rural migrant worker worries about his/her unemployment because of a temporal job and a limited contract period, he/she may consider purchasing a relevant insurance. It is therefore natural to assume that *Employment conditions* affects *Insurance purchase*. Similarly, if a rural migrant worker often encounters various difficulties (e.g., lack of employment security, difficulty in finding a school for children, and lack of social security) and barriers to job hunting (e.g., lack of skills, and low education level), he/she is more likely to purchase relevant insurances. This argument motivates us to make an additional assumption: *Current difficulties and barriers* affects *Insurance purchase*. Considering that purchasing these insurances is for preparing for the occurrence of various unexpected/expected life events in the future, it is not illogical to further assume that *Future concerns* is influential to *Insurance purchase*. Thus *Insurance purchase* is a core concept to connect energy consumption and its other influential factors. Finally, it is assumed that *Current difficulties and barriers* is associated with *Future concerns*.

10.4.2 Modeling results

Estimation results about the SEM shown in Fig. 10.1 are summarized in Fig. 10.2 and Table 10.2 (all parameter values are standardized). Fig. 10.2 is about the relationships between the eight latent variables via structural equations (the relevant parameter is called a structural parameter), and Table 10.2 is about the results of the measurement equations. The modeling accuracy is sufficiently high, and all the measurement parameters are statistically significant.

- *Findings from structural equations*

  Looking at the standardized total effects, *Residential behavior* shows the largest influence on *Energy consumption* (total effects: 0.7683), followed by *Built environment* (total effects: 0.7176), and *Ownership of end uses*

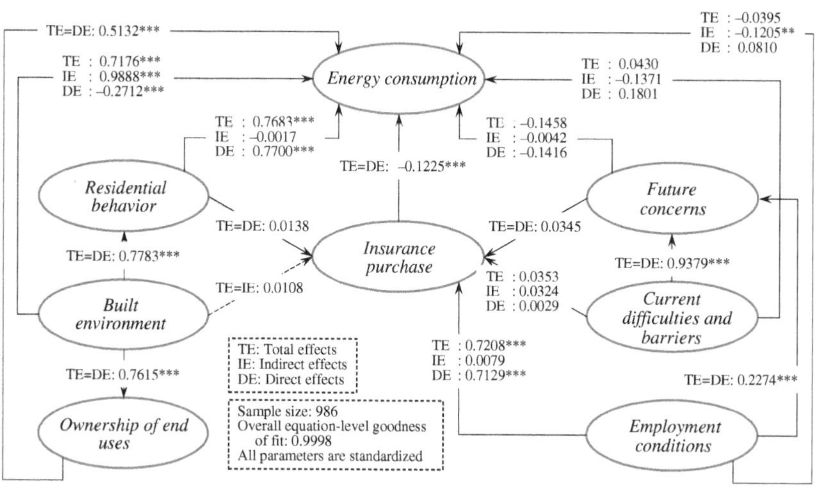

Figure 10.2 Rural migrant workers' energy consumption model: estimated relationships between latent variables (**significant at 5% level; ***significant at 1% level).

Table 10.2 Estimation results of standardized measurement parameters.

Energy consumption		
Electricity	0.8616	***
Gas	0.5021	***
Water	0.8480	–
Fuel	0.2964	***
The built environment		
dis_subway	0.6122	***
dis_bus	0.6303	***
dis_market	0.6462	***
dis_dept_store	0.6897	***
dis_park	0.7788	–
dis_hospital	0.7702	***
dis_preschool	0.7088	***
dis_ele_school	0.7992	***
dis_sec_school	0.7992	***
dis_highschool	0.8455	***
dis_workplace	0.6816	***
Residential behavior		
live_GuiYang	0.5880	***
live_ChongQing	0.4949	***
own_house	0.4461	***
rentl_house	0.7329	***
housing_size	0.8830	***
res_duration	0.6697	–
Ownership of end uses		
vehicle	0.5716	***
refrigerator	0.8887	***
wash_machine	0.9187	***
TV	0.9260	–
micro_oven	0.5310	***
air_cond	0.6251	***
PC	0.7301	***
induc_cooker	0.7493	***
rice_cooker	0.8948	***
Employment conditions		
temp_job	0.4682	***
contract	0.8166	***
with_skill	0.7410	***
job_training	0.8404	***
bonus	0.4747	***
rest_days	0.6446	–

(Continued)

Table 10.2 (Continued)

Energy consumption		
tm_job	−0.2901	***
tm_leisure	0.1562	***
income	0.2626	***
own_land/field/mt	−0.1609	***
Purchased insurances		
pension	0.7454	***
medical	0.6677	***
unemployment	0.9000	***
emp_injury	0.8215	—
maternity	0.8777	***
housing_fund	0.7566	***
Current difficulties and barriers		
no_job_security	0.7145	***
bad_job_rule	0.4453	—
low_salary	0.7334	***
no_skill_training	0.5419	***
hukou_discrim	0.4228	***
family_sepa	0.4595	***
no_social_security	0.5919	***
bad_living_env	0.5029	***
boring_cul_life	0.5556	***
difficult_medical	0.4617	***
barr_no_skill	0.7478	***
bar_low_edu	0.7472	***
barr_no_info	0.5429	***
barr_discrim	0.4077	***
Future concerns		
f_caringparent	0.6833	***
f_chd_edu	0.6595	***
f_chd_job	0.5712	***
f_unstablejob	0.6698	***
f_spousejob	0.4455	***
f_housing	0.6447	***
f_social_secu	0.6002	***
f_medical	0.5708	***
f_self_health	0.5717	—
f_fam_health	0.5697	***
f_self_status	0.4406	***

***Significant at 1% level; −, the corresponding parameter is fixed at 1 during the structural equation model estimation.

(total effects: 0.5132). More variables of life conditions were expected to have a larger influence on energy consumption; however, only *Insurance purchase* is influential, but its influence (total effects: −0.1225) is much smaller than those of the above three latent variables. *Employment conditions* also affects *Energy consumption* but in an indirect way. Unexpectedly, *Current difficulties and barriers* and *Future concerns* do not impose any significant influence on *Energy consumption*. Thus it seems that objective life conditions are associated with energy consumption; however, rural migrant workers' perceptions about their life conditions (as observed via *Current difficulties and barriers* and *Future concerns*) are not relevant to energy consumption.

- *Findings from measurement equations*
As for the latent variable *Energy consumption*, electricity and water show much larger standardized parameters (0.8616 and 0.8480, respectively) than gas and fuel. Thus this latent variable can be interpreted as a preference for electricity and water, which are most fundamental to rural migrant workers' lives. *Residential behavior* is mostly characterized by housing size (0.8830), living in a rental house (0.7329), and residential duration (0.6697). TV (0.9260), washing machine (0.9187), rice cooker (0.8948), and refrigerator (0.8887) are the most featured end uses for *Ownership of end uses*. All the distance parameters do not vary remarkably across facilities (between 0.6122 and 0.8455), indicating that rural migrant workers prefer a diverse set of the built environment for their urban lives. *Employment conditions* can be used to well explain job training (0.8404), contract (0.8166), and skills (0.7410). The measurement parameters of *Insurance purchase* range between 0.6677 and 0.900, where the top three predictive observed variables are unemployment insurance, maternity insurance, and employment injury insurance. *Current difficulties and barriers* can be used to well explain only the following four observed variables: that is, barriers due to no skill (0.7498), barriers due to low education level (0.7472), low salary (0.7334), and no job security (0.7145). As for *Future concerns*, the relevant measurement parameters range between 0.4455 and 0.6833, gaps of which are not that remarkable in comparison with those of *Current difficulties and barriers*. This means that all these future concerns are perceived to be troublesome for living in cities by rural migrant workers. Modeling results also revealed that various difficulties encountered in daily life and barriers to job hunting directly lead to the perception of a verity of future concerns.

10.5 Discussion

The top three influential factors, that is, *Residential behavior*, *Built environment*, and *Ownership of end uses*, have a positive parameter to *Energy consumption*. Looking at larger measurement parameters, more energy consumption is associated with residential behaviors of larger housing size, living in a rental house, and longer residential duration, the built environment of longer distances from residence to schools/park/hospital, and end uses of TV, washing machine, rice cooker, and refrigerator. Especially, the above factors result in more consumption of electricity and water. With the abolition of *hukou* system, more rural migrant workers may permanently reside in cities together with their families, which need a larger house and consequently consume more energy. With more rural migrant workers becoming permanent urban residents, ownership of end uses will increase, also resulting in more energy consumption; however, densities of various daily facilities will become higher and as a result distances from residence to facilities may become shorter, meaning a decrease in energy consumption. Thus an increase in urban residents transformed from rural migrant workers may change energy consumption in two ways at the same time: that is, increasing and reducing energy consumption, and which change direction is stronger is unknown, needing careful simulations. For this, the SEM approach will not work because of inability to make any prediction, and it is therefore necessary to develop better prediction models reflecting the derived complicated cause–effect relationships from this study.

As for the latent variable *Insurance purchase*, all its measurement parameters are positive, and its structural parameter to *Energy consumption* is negative (−0.1225). This means that purchasing any insurance is associated with a decline of energy consumption. In other words, rural migrant workers with more insurances are more likely to consume less energy. This observation is understandable, considering that households need to spend their limited income on various life purposes, which usually involves a variety of trade-offs. Then, who are more likely to purchase insurances? There is only one latent variable affecting *Insurance purchase*. That is *Employment conditions*. Modeling results further confirm that *Employment conditions* does not have a direct effect on *Energy consumption*, but an indirect effect is surely revealed. Aggregate analyses show that rural migrant workers, who received a job training, had a formal contract, and possessed certain skills are more likely to buy more insurances. Actually, these workers also have more income than other workers. Even though income (measurement parameter is 0.2626) is not the key variable to define *Employment conditions*, but its parameter is positive. This suggests that higher income is associated with less energy consumption. This result may be interpretable from an income perspective. If income is low, people may try to save money even though they have a higher income than those in the same low-income group, because they worry about their future and need to prepare for it. Rural migrant workers belong to such a low-income group.

10.6 Conclusion

This study has reconfirmed major contributions of residential behavior, ownership of energy-consuming end uses, and the built environment to energy consumption, and further identified the roles of various factors of living conditions, in the context of rural migrant workers. Especially, it is found that life conditions of rural migrant workers have limited effects on energy consumption. Even though the study is based on a small-scale questionnaire survey, the findings have various policy implications. Because a SEM was adopted, it has been an important research issue of how to incorporate behavioral decision-making mechanisms.

Because of the China's National New-type Urbanization Plan (2014–20), more and more rural migrant workers are expected to become a permanent urban resident. Meanwhile, becoming a permanent urban resident also means that they will no longer have any right to enjoy benefits from having a rural *hukou*, with which many rural migrant workers are assigned with a piece of land from the government (in China, land is a state-owned asset). Such a land ownership has various economic benefits. Such benefits from having a rural *hukou* together with various difficulties and barriers encountered in daily life in cities and the resulting future concerns discourage many rural migrant workers to make up their mind to give up a rural *hukou* and permanently reside in cities. Jiang et al. (2017) revealed that the share of rural migrant workers intending to become a permanent urban resident was just 39%. Different from the years of 2014 and 2015 when Jiang et al. (2017) conducted their study, nowadays, various supporting policies encouraging rural migrant workers to permanently reside in cities have been implemented in more and more cities (especially those cities beyond the coastal areas of China), and as a result, the above share will become larger. However, it will still take a longer time to eliminate most of the major difficulties and barriers encountered in rural migrant workers' daily lives. It is further unclear how China's economic slowdown will worsen employment opportunities for rural migrant workers. China is now trying to make economic transformations from quantity-driven development to quality-driven development. During this process, skill development will be a key for rural migrant workers to survive in severe competitions in cities. To make rural migrant workers themselves more competitive, vocational education and training will become more and more important for the current generation of rural migrant workers. Even though various social issues bothering rural migrant workers will be fully resolved, it will be still difficult for the current generation of rural migrant workers to have a truly fair competition with urban residents, who have already better educated. But the children of the current generation of rural migrant workers may enjoy a truly fair competition with urban residents, if they are given a better education opportunity than their parents. Over the past decades, *hukou* system did bring unfair life conditions and barriers to many Chinese residents, and social resources were not equally distributed among population groups. In the coming future, with the abolition of *hukou* system, rural migrant workers will "disappear," and more working mobilities across cities of the whole country will be expected, because they will not need to worry about the issues caused by the current *hukou* system. In the coming new era of working mobilities and Chinese economic transformation, the various needs of current rural migrant workers should be well addressed by policy makers. At the same time, various policies supporting rural development in China have also been implemented, which will provide rural people with much better development opportunities in rural areas. Even from a perspective of energy consumption, rural and urban policies are not independent of each other. Such push–pull policies will have various impacts on rural people's future rural-to-urban or urban-to-rural migration decisions. Thus to properly evaluate the effects of various policies supporting rural migrant workers on urban energy consumption, it is necessary to monitor the changes in rural migrant workers' and their children's lives over a longer period.

10.7 Future research

Considering their social status, research on rural migrant workers needs to pay more attention to the concept of capability (e.g., Chikaraishi et al., 2018; Perez Barbosa et al., 2018; Stanley, 2018), which is crucial to understand how rural migrant workers make various decisions on their lives in consideration of their abilities of surviving in cities. Some aspects of capability have already been captured in this study, as shown in Table 10.1; however, a theoretical framing of capability will be required in the future to overcome the issues caused by adopting statistically oriented analysis approaches, such as the SEM adopted in this study. Meanwhile, to understand rural migrant workers' various life decisions in a comprehensive and theoretical way, a life-oriented approach (Zhang, 2017) will be a powerful option, because it argues and also has revealed the existence of various interdependencies in life choices/activities, not only in a certain point of time but also over a longer period such as a life course (Zhang et al., 2014; Yu and Zhang, 2017). In this regard, Zong et al. (2019a,b) have shown the effectiveness of the life-oriented approach in terms of reflecting various life choices in the analysis of energy consumption. However, various causalities involved in the analyses by Zong et al. need to be better understood based on more sound behavioral decision models.

Rural-to-urban migration and transformation from rural migrant workers in cities to permanent urban residents will be a major power to push up China's urbanization. Such a progress of urbanization has various impacts on not only energy consumption but also social and economic developments. Needless to say, social and economic developments are further closely associated with energy consumption. In this regard, the improved population—economy—environment analytical model built by Yu et al. (2018) becomes relevant, which emphasizes the importance of heterogeneous human behaviors and consequently incorporates households' inter- and intralife-stage variations in time allocation and consumption. With such a model, both micro- and macrolevel factors can be simultaneously captured. In this regard, Zhang's (2019) integrated macroeconomic, transport, and climate change model is also relevant.

With the above integrated behavioral models and integrated system models, it is expected that transport and energy policies can be evaluated in a much better and more comprehensive way. Public policymaking is for bringing expected changes to our society. However, policy implementation will not automatically lead to expected changes. Such changes usually take time. Actual behavioral changes occur only if people's sociopsychological states (norm, attitude, perception, intention, etc.) change based on the theory of planned behavior (Ajzen, 1991). It is a great challenge to incorporate behavioral change processes into the preceding integrated behavioral and system models.

Acknowledgments

This research has been financially supported by a Grants-in-Aid for Scientific Research (B), Japan Society for the Promotion of Science (JSPS) (no. 26303003), and a Grants-in-Aid for Scientific Research (A), JSPS (no. 15H02271).

References

Ajzen, I., 1991. The theory of planned behavior. Organ. Behav. Hum. Decis. Processes. 50, 179–211.

Chan, K., Zhang, L., 1999. The hukou system and rural-urban migration in China: processes and changes. The China Quart 160, 818–855.

Chen, Q., Song, Z., 2014. Accounting for China's urbanization. China Econ. Rev. 30, 485–494.

Chikaraishi, M., Stanley, J., Fujiwara, A., Zhang, J., 2018. Social exclusion and transport in an aging and shrinking society. In: Zhang, J., Feng, C.M. (Eds.), Routledge Handbook of Transport in Asia. Taylor & Francis, Routledge, pp. 261–284.

Golob, T.F., 2003. Structural equation modeling for travel behavior research. Transp. Res. B 37 (1), 1–25.

Jiang, Y., Zhang, J., Jin, X., Ando, R., Chen, L., Shen, Z., et al., 2017. Rural migrant workers' intentions to permanently reside in cities and future energy consumption preference in the changing context of urban China. Transp. Res. D 52 (B), 600–618.

Jiang, Y., Zhang, L., Zhang, J., 2019. Energy consumption by rural migrant workers and urban residents with a hukou in China: quality-of-life-related factors and building environment. Nat. Hazards, (in press).

Perez Barbosa, D., Stanley, J., Zhang, J., 2018. Social exclusion and the place of urban passenger transport in developing Asian countries. In: Zhang, J., Feng, C.M. (Eds.), Routledge Handbook of Transport in Asia. Taylor & Francis, Routledge, pp. 243–260.

Stanley, J., 2018. Social exclusion and land passenger transport in Asia. In: Zhang, J., Feng, C.M. (Eds.), Routledge Handbook of Transport in Asia. Taylor & Francis, Routledge, pp. 229–242.

Yu, B., Zhang, J., 2017. Dynamic life course analysis on residential location choice. Transp. Res. A 104, 281–292.

Yu, B., Zhang, J., Fujiwara, A., 2012. Analysis of the residential location choice and household energy consumption behavior by incorporating multiple self-selection effects. Energy Policy 46, 319–334.

Yu, B., Wei, Y.M., Kei, G., Matsuoka, Y., 2018. Future scenarios for energy consumption and carbon emissions due to demographic transitions in Chinese households. Nat. Energy 3, 109–118.

Zhang, J., 2017. Life-Oriented Behavioral Research for Urban Policy. Springer.

Zhang, R., 2019. The role of the transport sector in energy transition and climate change mitigation: insights from an integrated assessment model. In: Zhang, J. (Ed.), Transport and Energy Research: A Behavioral Perspective. Elsevier.

Zhang, J., Yu, B., Chikaraishi, M., 2014. Interdependences between household residential and car ownership behavior: a life history analysis. J. Transp. Geogr. 34, 165–174.

Zong, W., Zhang, J., Jiang, Y., 2019a. Japanese young people's car ownership and usage: research from an expenditure perspective. In: Zhang, J. (Ed.), Transport and Energy Research: A Behavioral Perspective. Elsevier.

Zong, W., Zhang, J., Jiang, Y., 2019b. Life-oriented household energy consumption research. In: Zhang, J. (Ed.), Transport and Energy Research: A Behavioral Perspective. Elsevier.

CHAPTER 11

Japanese young people's car ownership and usage: research from an expenditure perspective

Weiyan Zong, Junyi Zhang* and Ying Jiang
Hiroshima University, Higashi Hiroshima, Japan

11.1 Introduction

A decreasing number of young people owning a car in Japan have been observed since the late 1990s (Zhang, 2017), as in some other developed countries (Metz, 2010; Davis et al., 2012; Goodwin and Van Dender, 2013; Kuhnimhof et al., 2012a, 2013). To date, many researchers have explored such a decline of car ownership and usage among young people based on psychological, statistical, and/or econometric approaches. Urbanization, increased singlehood and childlessness, delaying life stages, improved preferences for multimodal transport systems, the diminishing popularity of cars among young men, and the continuously strengthening climate change concerns (Oakil et al., 2016; Kuhnimhof et al., 2012b; Hopkins, 2016; Hopkins and Stephenson, 2016; Simons et al., 2017) have been documented to have contributions to weaker car dependence of young people. And a number of studies on Japanese young people's car ownership and usage also derived similar conclusions as in other developed countries (Nishimura, 2012; Fujioka et al., 2012; Yotsumoto, 2012).

Considering that owning and using a car spend a lot of money, choices and factors from other life domains may affect the decisions of young people on car-related expenditure behavior under the constraint of limited household budget. Choo et al. (2007) examined car-related expenditure and communication expenditure based on data from the US Consumer Expenditure Survey. Ferdous et al. (2010) examined possible changes of expenditures on 18 categories from the abovementioned US survey by building a multiple discrete–continuous nested extreme value

* Corresponding author.

Transport and Energy Research
DOI: https://doi.org/10.1016/B978-0-12-815965-1.00011-9

model with four nests; however, the interactions between car-related expenditure and other expenditures have not been revealed. However, different decision makers may have different nested structures for maximizing their own utility. If this is the case, the previous nested choice model is problematic. A later study used a data extracted from the Survey of Household Spending in Canada over 13 years and developed household budget allocation models with 13 types of nontransport-related expenditure and 8 types of transport-related expenditure by adopting a multiple discrete continuous extreme value (MDCEV), scaled MDCEV and mixed MDCEV models (Anowar et al., 2017). But these MDCEV models cannot address the interactions among expenditures.

Motivated by the abovementioned background, this study aims to reveal how decisions on various expenditures affect young people's car ownership and usage, where energy consumption expenditure is jointly incorporated into the analysis, together with other types of expenditures. For this purpose a new type of multiple discrete–continuous model for representing expenditure allocation decisions is built based on the concept of multilinear utility (MLU), where various interactions across expenditures are incorporated in a systematic way. Empirical analyses are conducted by focusing on not only car-related expenditure but also public transport expenditures, energy, residence, health, communication, recreation, foods, and others, which were extracted from a household-level national survey "National Survey of Family Income and Expenditure" in Japan, which was conducted in 2009.

The remanding part of this chapter is made up of five parts. Section 11.2 explains Japanese young people's car ownership and its usage using a longitudinal data from the National Survey of Family Income and Expenditure in Japan, for confirming the turning points of different groups of young people. Section 11.3 introduces the MLU-based multiple discrete–continuous model built in this study, and Section 11.4 explains and discusses modeling results for the year 2009. The final section summarizes main conclusions and future research issues.

11.2 Japanese young people's car ownership and usage: an expenditure analysis

Changes in Japanese young people's car ownership is first summarized in Table 11.1 using the aforementioned national data in the years 1984, 1989, 1994, 1999, 2004, and 2009. The car ownership ratios of all income

Table 11.1 Car ownership by age and income.

Income group	1984 (%)	1989 (%)	1994 (%)	1999 (%)	2004 (%)	2009 (%)
Low	60.2	82.5	74.5	77.3	76.2	76.0
Middle	79.2	86.3	89.0	91.1	89.5	89.3
High	88.9	88.2	91.8	95.1	91.1	88.9

groups decreased after 1999 (i.e., low income: from 77.3% to 76.0%; middle income: from 91.1% to 89.3%; high income: from 95.1% to 88.9%).

Observing changes in the expenditure patterns by life domain, as illustrated in Fig. 11.1, Japanese young people reduced their car-related expenditure (i.e., car ownership and usage) by 1.8% in low-income group and 1.8% in middle-income group, from 1994 to 2009, and 4.6% in high-income group from 2004 to 2009. As for public transport a decreasing trend was also observed, especially in low-income and middle-income groups.

The cumulative distributions of expenditures for different life domains in 2009 are displayed in Fig. 11.2. This is because the following modeling analysis only focuses on analyses of influential factors to car ownership and usage in 2009. Looking at the starting points of the curves, the shares of zero expenditure are higher in public transport and car ownership/usage domains than in other domains. And the curves of energy, health, public transport, and communication reach 90% before 30,000 JPY, indicating that these four domains are usually allocated less budgets. Moreover, food, residence, and others are three most costly domains as their curves rise later than other domains.

11.3 Modeling Japanese young people's car ownership and usage behavior

To represent the above-described expenditure allocation decisions an expenditure allocation model based on the concept of MLU (Zhang et al., 2002, 2005) is built, by referring to as Yu and Zhang (2015).

Here, household h needs to allocate its limited budget E_h to different life domains, the expenditure for life domain l is e_{hl}, and corresponding utility is u_{hl}. The objective of this allocation decision is to maximize the household utility U_h, which is composed of additive and interactive utilities [Eq. (11.1)], conditional on a limited budget constraint [Eq. (11.2)]. Expenditure e_{hl} is nonnegative.

$$\text{Maximize } U_h = \sum_l u_{hl} + \sum_l \sum_{t>l} \lambda_{lt} u_{hl} u_{ht} \tag{11.1}$$

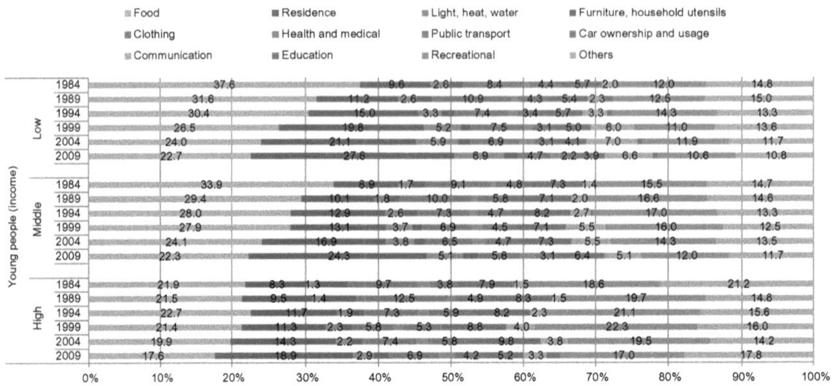

Figure 11.1 Changes of young people' expenditures over time by income.

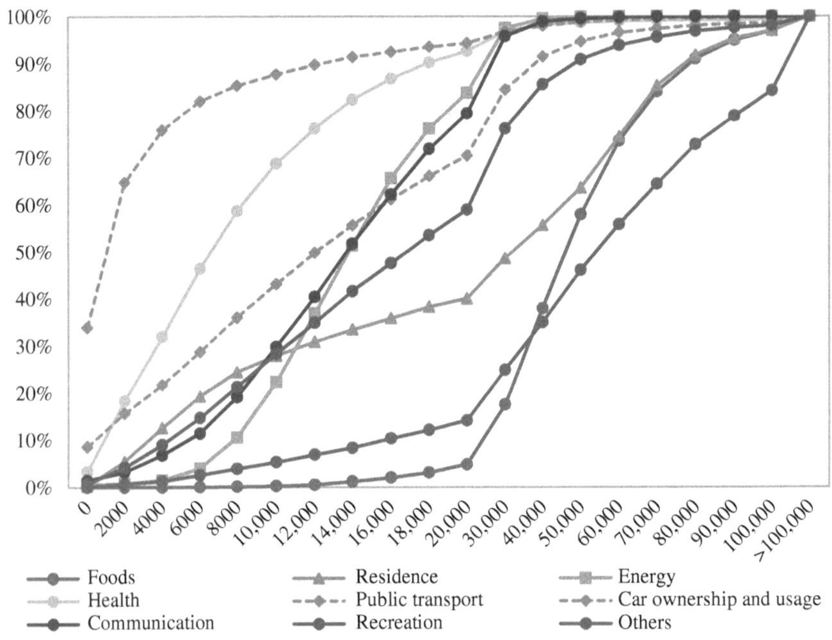

Figure 11.2 Cumulative distributions of expenditures by life domain in 2009.

subject to

$$\sum_{l} e_{hl} = E_h, \quad e_{hl} \geq 0 \tag{11.2}$$

$$u_{hl} = \exp(\beta_l x_{hl} + \varepsilon_l) \cdot \ln(e_{hl} + 1) \tag{11.3}$$

As shown in Eq. (11.3), utility u_{hl} is a logarithmic function of expenditure e_{hl}. This function is adopted by reflecting the diminishing feature of utility u_{hl}: that is, the household h derives positive utility from spending the amount of money e_{hl}, but the marginal utility ($\partial u/\partial e$) decreases with e_{hl}. This is because expenditure is a scare source. More spending on life domain l means less money left for other life domains. In other words the diminishing feature of expenditure utility results from competitions between life domains in spending limited household income. To reflect positivity of u_{hl} an exponential function is further introduced. Factors affecting expenditure e_{hl} are all introduced into the exponential function, including both the observed (x_{hl}: a vector of observed factors and β_l is a corresponding vector of unknown parameters) and unobserved factors (represented in the form of error term ε_l).

Here, the above utility-maximizing model comprising Eqs. (11.1)–(11.3) is called MLU-based expenditure allocation model in this study.

As illustrated by Yu and Zhang (2015), to calculate probabilities of multiple discrete–continuous choices, one life domain should have a nonzero consumption, called outside goods. In this study, food is chosen as the outside goods. Then, the probability of household h allocating its budget to F_h life domains in total L life domains is formulated, as shown in Eqs. (11.4) and (11.5), where $\varphi(\cdot)$ and $\Phi(\cdot)$ are a probability density function and a cumulative density function of standard normal distribution, respectively.

$$P\left(e_{h2}, e_{h3}, \ldots, e_{hF_h}, 0, 0, \ldots, 0\right) = \left[\prod_{l=2}^{F_h} \frac{1}{\sigma_l} \phi\left(\frac{\Omega_{hl}}{\sigma_l}\right)\right]\left[\prod_{l=F_h+1}^{L} \Phi\left(\frac{\Omega_{hl}}{\sigma_l}\right)\right] \tag{11.4}$$

$$\Omega_{hl} = \frac{1 + \sum_{m \neq 1} \lambda_{1m} u_{hm}}{\left(1 + \sum_{m \neq l} \lambda_{lm} u_{hm}\right) \cdot \exp(\beta_l x_{hl})} \cdot \frac{e_{hl} + 1}{e_{h1} + 1} \tag{11.5}$$

In Eq. (11.1) the additive utility is a sum of all expenditure utilities, and the interactive utility is a binary-type interaction function, which can include combinations of any pair of expenditures. Considering the research purpose of this study, the interactions are assumed to exist with respect to car-related expenditures, energy expenditures, and other expenditures: in total, eight types of interactions are introduced, as displayed in Fig. 11.3.

- Interaction A (IT-A, λ_a): Interaction between car-related expenditure (e_6) and residence expenditure (e_2);

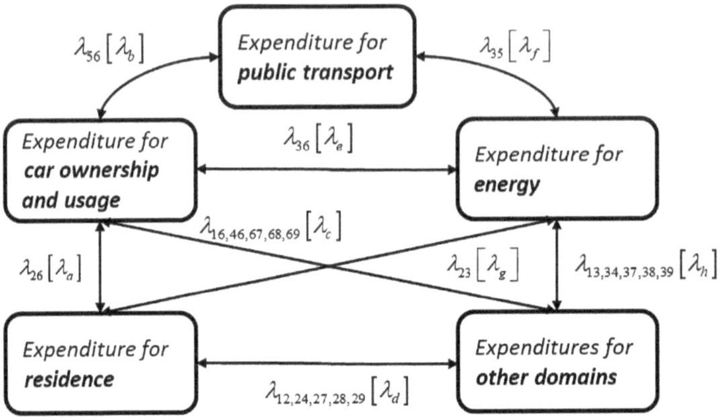

Figure 11.3 The settings of interaction terms.

- Interaction B (IT-B, λ_b): Interaction between car-related expenditure (e_6) and public transport expenditure (e_5);
- Interaction C (IT-C, λ_c): Interaction between car-related expenditure (e_6) and expenditures of food (e_1), health (e_4), communication (e_7), recreation (e_8), and others (e_9);
- Interaction D (IT-D, λ_d): Interaction between residence expenditure (e_2) and expenditures of food (e_1), health (e_4), communication (e_7), recreation (e_8), and others (e_9);
- Interaction E (IT-E, λ_e): Interaction between energy expenditure (e_3) and car-related expenditure (e_6);
- Interaction F (IT-F, λ_f): Interaction between energy expenditure (e_3) and public transport expenditure (e_5);
- Interaction G (IT-G, λ_g): Interaction between energy expenditure (e_3) and residence expenditure (e_2);
- Interaction H (IT-H, λ_h): Interaction between energy expenditure (e_3) and expenditures of food (e_1), health (e_4), communication (e_7), recreation (e_8), and others (e_9).

Thus this MLU-based expenditure allocation model is a multiple discrete–continuous choice model, similar to the popular MDCEV model (Bhat, 2005). Different from the MDCEV model, the MLU-based model incorporates interdependencies between choices based on observed utilities in an explicit way, rather than using the unknown satiation parameters. Yu and Zhang (2015) further empirically confirmed that the MLU-based model outperforms the MDCEV model, using a data of household energy consumption in Beijing, China. Furthermore, the MLU-based model does not suffer from the issue of heterogeneous-nested structures that are ignored in the nest-type MDCEV model, because the MLU-based model does not need to prespecify any nest for represented the interactions. Thus the MLU-based model is more flexible to accommodate various types of interactions without sacrificing the representation of true decision-making mechanisms.

11.4 Empirical results

The survey data collected in 2009 is adopted for model estimation, which contains valid data from 4835 young households. The dependent variables are expenditure for nine life domains listed next, including foods (outside goods, e_1), residence (e_2), energy (e_3), health (e_4), public transport (e_5), car ownership and usage (e_6), communication (e_7), recreation (e_8), and others (e_9).

- *Food*: Cereals, fish and shellfish, meat, dairy products and eggs, vegetables and seaweeds, fruits, oils, fats and condiments, cooked food, beverages, alcoholic beverages, eating out
- *Residence*: Rents for dwelling and land, repairs and maintenance, household durables, interior furnishings and decorations, bedding, domestic utensils, domestic nondurable goods, domestic services
- *Energy*: Electricity, gas, other fuel and light, water and sewerage charges
- *Health*: Medicines, health fortification, medical supplies and appliances, medical services
- *Public transport*: Railway fares, bus fares, taxi fares, airplane fares, highway fares, other public transport
- *Car related*: Purchase of cars, gasoline, automotive parts, articles related to private transport, automotive maintenance and repairs, yearly and monthly rent for parking, other rent for parking, charges for rental car and car sharing, other services related to private transportation, insurance premium
- *Communication*: Postage, telephone charges, mobile telephone charges, mobile telephone, other communication equipment, door-to-door trucking service, other carriage fees
- *Recreation*: Recreational durable goods, recreational goods and services, books, and other reading materials
- *Others*: Japanese clothing, shirts and sweaters, cloth and thread, other clothing, footwear, services related to clothing, school fees, school textbooks and reference books for study, tutorial fees, miscellaneous (personal care services, toilet articles, personal effects, tobacco, etc.), pocket money, social expenses, remittance

This study selected 29 explanatory variables that are divided into common (9) and domain-specific (20) variables.

- *Common variables* (nine variables): Due to changes in survey contents, the following UBA is excluded from the years 1984, 1989, and 1994 models, and N65 + is excluded from the 1984 model, respectively.
  - household size (*HHS*)
  - number of employed members (*NEM*)
  - number of members aged 65 + years old (*N65 +*)
  - number of members aged under 18 years old (aged under 15 years old for 1989 and 1984) (*N18 −*)
  - whether to belong to low-income group or not (*LIG*, 1: yes; 0: no)
  - whether to live in a megacity area or not (*MCA*, 1: yes; 0: no)
  - whether to live in urban area or not (*UBA*, 1: yes; 0: no)
  - whether to live in a cold area or not (*COA*, 1: yes; 0: no)
  - whether to live in a warm area or not (*WAA*, 1: yes; 0: no)
- Domain-specific variables (20 variables)
  - *Car-related expenditure (4 variables)*
    - number of cars owned by household (*NCO*)
    - number of motorbikes owned by household (*NMO*)
    - sum of all vehicle displacement (*SVD*)
    - years of usage of most frequently used car (*YOU*)
  - *Residence expenditure* (5 *variables*)
    - housing tenure (*HST*, 1: yes; 0: no)
    - wooden house (*WDH*, 1: yes; 0: no)
    - concrete house (*CCH*, 1: yes; 0: no)
    - house area (*HOA*, m^2)
    - detached house (*DEH*, 1: yes; 0: no)
  - *Energy expenditure* (11 *variables*)
    - house area (*HOA*, m^2)
    - detached house (*DEH*, 1: yes; 0: no)
    - (nine variables) ownership of water heater (*OWH*), electric toilet (*OET*), refrigerator (*ORF*), washing machine (*OWM*), IH cooking (*OIC*), air conditioner (*OAC*), television (*OTV*), personal computer (*OPC*), microwave (*OMW*): dummy variable (1: yes; 0: no)

The MLU-based household expenditure allocation model was estimated based on a maximum likelihood method, and the estimation results are summarized in Table 11.2. The model accuracy is sufficiently high as the adjusted McFadden's Rho-squared value is 0.5355. Concerning the

Table 11.2 Estimation results of young Japanese people's expenditure allocation model in the year 2009.

2009	Residence		Energy		Health		Public transport		Car related		Communication		Recreation		Others	
	Estimate	VR (%)	Estimate	VR (%)	Estimate	VR (%)	Estimate	VR (%)	Estimate	VR (%)	Estimate	VR (%)	Estimate	VR (%)	Estimate	VR (%)
Common variables																
HHS	0.1305***	4.38	− 0.0201	1.52	− 0.2048***	19.81	− 0.8668***	68.39	− 0.3940***	36.59	− 0.2135***	45.48	− 0.2784***	60.63	− 0.0157	1.51
NEM	− 0.0134	0.01	− 0.0712***	4.34	− 0.3923***	16.49	− 0.2222***	1.02	− 0.0808***	0.35	− 0.1000***	2.26	− 0.0408*	0.30	0.0178	0.44
N65 +	− 0.3394***	1.01	− 0.0062	0.00	0.2880***	1.34	0.6817***	1.44	− 0.2533***	0.52	− 0.0275	0.03	0.1323**	0.47	− 0.2136***	9.51
N18 −	− 0.1912***	6.32	0.0839***	17.83	0.2754***	24.04	0.5407***	17.86	0.3774***	22.54	0.1960***	25.73	0.1897***	18.90	0.0720***	21.26
LIG	− 0.0695***	0.16	− 0.2050***	21.02	− 0.4406***	12.15	− 0.7600***	6.97	− 0.5891***	10.84	− 0.0690***	0.63	− 0.4026***	16.81	− 0.2679***	58.12
MCA	0.0727***	0.22	− 0.2161***	28.36	− 0.3625***	9.99	0.2574***	0.97	− 0.3231***	3.96	− 0.2590***	10.78	0.0011	0.00	− 0.0896***	7.90
UBA	0.1251***	0.20	− 0.2649***	12.99	− 0.7108***	11.71	− 0.7620***	2.59	− 0.3273***	1.24	− 0.5195***	13.21	− 0.1648***	1.04	0.0640***	1.23
COA	− 0.0312	0.02	− 0.0027	0.00	− 0.2818***	3.47	− 0.2665***	0.60	− 0.1210***	0.32	− 0.1379***	1.76	− 0.0886**	0.57	0.0072	0.03
WAA	0.0244	0.01	− 0.0495**	0.78	− 0.1585***	1.00	− 0.1403*	0.15	− 0.1171**	0.27	− 0.0379	0.12	− 0.1401***	1.30	0.0053	0.01
Domain-specific variables																
HST	− 1.4663***	87.23														
WDH	0.0592	0.15														
CCH	0.0062	0.00														
HOA	0.0135	0.01	− 0.0221***	0.42												
DEH	− 0.0803**	0.27	0.0494***	1.49												
OWH			0.0255*	0.37												
OET			0.0461***	2.62												
ORF			− 0.0531***	1.00												
OWM			− 0.0117	0.03												
OIC			− 0.0968***	3.56												
OAC			0.0215***	2.02												
OTV			0.0141	0.39												
OPC			− 0.0249***	0.99												
OMW			− 0.0358*	0.27												
NCO									0.4736***	21.87						
NBO									0.1652***	0.77						
SVD									0.0052	0.01						

(Continued)

Table 11.2 (Continued)

2009	Residence		Energy		Health		Public transport		Car related		Communication		Recreation		Others	
	Estimate	VR (%)	Estimate	VR (%)	Estimate	VR (%)	Estimate	VR (%)	Estimate	VR (%)	Estimate	VR (%)	Estimate	VR (%)	Estimate	VR (%)
YOU									0.0152***	0.72						
Variances																
	0.5931***		0.2096***		0.8744***		1.9617***		1.3340***		0.5568***		0.6330***		0.5240***	
Interaction terms																
	IT-A (λ_a)		−0.0136***		*IT-B* (λ_b)		−0.0523***		*IT-C* (λ_c)		−0.0557***		*IT-D* (λ_d)		−0.0496***	
	IT-E (λ_e)		−0.0171***		*IT-F* (λ_f)		−0.1517***		*IT-G* (λ_g)		−0.0169***		*IT-H* (λ_h)		−0.0545***	
Sample size			4835		*Initial log-likelihood*				−99542.21		*McFadden's Rho-squared*				0.5366	
Number of parameters			108		*Converged log-likelihood*				−46125.46		*Adjusted Rho-squared*				0.5355	

Note: *significant at 10% level, **significant at 5% level, and ***significant at 1% level. *VR*, Variance Ratio.

statistical significance of parameters, 59 out of 72 common parameters, 14 out of 20 domain-specific parameters, all eight variances and all eight interaction terms are significant at least at 10% level, and most of them are actually significant at 1% level.

11.4.1 Influential factors to household expenditures

11.4.1.1 Car-related expenditure

Most explanatory variables show significant influences on young people's car-related expenditure. Among all explanatory variables, household size accounts for 36.59% of total variance, with a negative influence. The number of cars owned by household and the number of household members aged under 18 years old are positively related to car-related expenditure, accounting for 21.87% and 25.23% in total variance, respectively. Besides, young people in the low-income group, living in a megacity or an urban area, are more likely to allocate less budgets on car ownership and usage.

11.4.1.2 Public transport expenditure

Household size plays an important role in public transport expenditure with a variance ratio of 68.39%, meaning that the variations in public transport expenditure are caused by household size.

The second largest influencing factor is the number of members aged under 18 years old, which is negatively associated with public transport expenditure and accounts for 17.86% of total variance.

11.4.1.3 Energy expenditure

As for energy expenditure, most influential factors are whether to live in a megacity, whether to belong to the low-income group, whether to live in an urban area and the number of household members aged under 18 years old, where the first three factors show a negative influence, and the last one positively affects the energy expenditure. People living in a megacity or an urban area tend to allocate less budgets on energy, probably due to variations in energy prices across provinces and cities.

As for domain-specific variables, living in a detached house is related to high energy expenditure, while house area has a negative influence. Regarding the ownership of nine end uses, it explains 11.25% of total variance, while both the positive (for water heater: the share is 29.70%, electric toilet: 42.77%, and air conditioner: 77.52%) and negative parameters (for refrigerator: 94.48%, IH cooking: 16.84%, personal computer:

76.30%, and microwave oven: 94.02%) are estimated, among those being statistically significant. Interestingly, the shares of end uses with a negative parameter are mostly larger than those with a positive parameter. Considering such ownership shares, the parameter signs may be better explained by usage frequency and energy intensity per unit of consumption, because for a certain end use, it is obvious that the more frequent the usage is, the more the energy will be consumed, and even though usage frequency is low, more energy may be consumed if energy intensity per unit of consumption is high. Unfortunately, usage frequency and energy intensity are not available in the national survey adopted in this study.

11.4.1.4 Residence expenditure

The main influential factor is housing tenure, which occupies 87.23% in the total variance. As the rent for dwelling is included in residence expenditure, renting a house is truly a big expense for most people. Besides, household size and the number of household members aged under 18 years old account for 4.38% and 6.32% of total variance, respectively. The positive parameter sign of HHS indicates that the more the household members, the more the residence expenditure. The negative parameter sign of N18 − may suggest that household members aged under 18 years old spend less money related to residence than other members.

11.4.1.5 Expenditures for other domains

All common explanatory variables are of statistical significance for health expenditure, and only two of them have positive effects: the number of members aged 65 + years old and the number of members aged under 18 years old. This is understandable as elderly people and children have more needs for health services. As for communication expenditure, seven out of nine explanatory variables are significant, and main influential factors are household size, the number of household members aged under 18 years old, whether to live in an urban area, and whether to live in a megacity, which respective variance ratios are 45.48%, 25.73%, 13.21%, and 10.78%. Regarding recreation expenditure, three main influential factors are identified: that is, household size, the number of household members aged under 18 years old, and whether to belong to the low-income group.

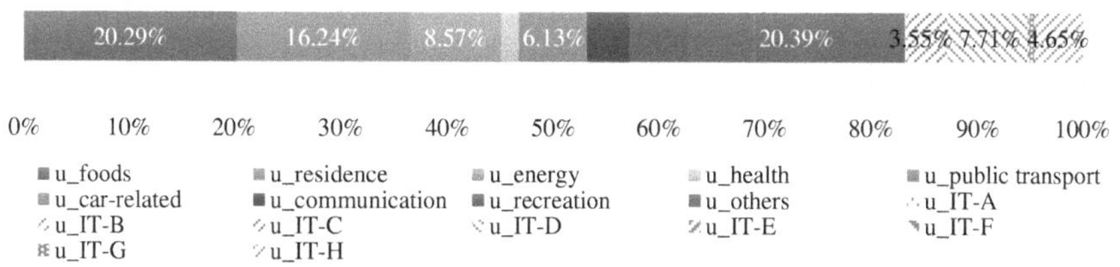

Figure 11.4 Shares of expenditure utility by type (both additive and interactive utility).

11.4.2 Interactions across expenditures for different life domains

The estimation results of all interaction terms are significantly negative. This indicates the presence of trade-offs between expenditures in different life domains. To clarify the composition of household utility and the roles of different life domains, the shares of additive and interactive utilities are illustrated by type in Fig. 11.4. The additive utility accounts for 83.18% of total household utility, and interactive utility does the remaining 16.82%. This suggests that the interactions of expenditure allocations across life domains are not ignorable, supporting the use of MLU concept in this study.

The top five life domains with higher utility shares are other expenditures (20.39%), foods expenditure (20.29%), residence expenditure (16.24%), energy expenditure (8.57%), and car-related expenditure (6.13%). The smaller utility shares are observed with respect to recreation expenditure (5.54%), communication expenditure (4.10%), health expenditure (1.65%), and public transport expenditure (0.28%). It is manifested that energy and car-related expenditures are an important part of young people's expenditure decisions. Looking at the shares of eight interactive utility parts, the interactions between residence expenditure and other domains (IT-D), between energy expenditure and other domains (IT-H), and between car-related expenditure and other domains (IT-C) are dominant, accounting for 7.71%, 4.65%, and 3.55% of the total household utility, respectively.

11.5 Conclusions and future research issues

This study has presented additional evidence on how young people in developed country decide car ownership and usage as a part of household expenditure decision and the first case study in Japan. A national expenditure data at the household level in Japan was used to estimate a MLU-based expenditure allocation model with car-related and energy-related interactions with other types of expenditures. Modeling results confirmed the modeling applicability and the importance of incorporating various interactions between expenditures. Various expenditure-specific influential factors have been revealed for the first time. This study has shown that the MLU model can service as a new behavior-oriented analysis approach to derive sound and scientifically sophisticated conclusions on factors affecting the trend of young people's car ownership and usage. One extension of the current study is to confirm the abovementioned findings by comparing all data from the years 1984, 1989, 1994, 1999, 2004, and 2009, especially by focusing on the turning points of different income groups of Japanese young people's car ownership and usage.

The significance of this study is that it presents a new way of understanding young people's car ownership and usage in terms of modeling various interactions between expenditures in a consistent and systematic manner and using a national household-level expenditure data. This is partially in line with the life-oriented approach. A joint analysis of transport and energy expenditures is also a unique feature of this study. The concept of MLU can also be applied to represent other types of multiple discrete–continuous choices, which can be observed in many life choices. Obviously, future studies should be conducted by incorporating more life choices in the analysis of car ownership and usage, where decision-making mechanisms of unexplored life choices should also be reflected.

To reduce energy consumption from the household transport domain, it seems important to encourage behavioral changes toward a less car-dependent lifestyle. However, such a lifestyle should be promoted only if the total household energy consumption will be reduced. In other words, if a less car-dependent lifestyle induces extraenergy consumption in other household domains, which is bigger than the saved energy from car, due to rebound effects, it is meaningless to reduce car-related energy consumption. In other words, cross-sectoral policymaking mechanisms should be developed for assisting households to realize a truly energy-saving lifestyle. Such mechanisms should include not only economic incentives and technological developments but also behavioral interventions. Due to lifestyles may change over time, it is important to capture temporal changes in household car ownership and usage as well as other life choices and to monitor the effects of various energy policies in a continuous manner.

References

Anowar, S., Eluru, N., Miranda-Moreno, L.F., 2017. How household transportation expenditures have evolved in Canada: a long term perspective. Transportation 105, 1–21.

Bhat, C.R., 2005. A multiple discrete–continuous extreme value model: formulation and application to discretionary time-use decisions. Transp. Res. B 39 (8), 679–707.

Choo, S., Lee, T., Mokhtarian, P.L., 2007. Do transportation and communications tend to be substitutes, complements, or neither? U.S. consumer expenditures perspective, 1984-2002. Transp. Res. Rec. 2010, 121–132.

Davis, B., Dutzik, T., Baxandall, P., 2012. Transportation and the New Generation: Why Young People Are Driving Less and What It Means for Transportation Policy. Frontier Group—US P.I.R.G. Education Fund.

Ferdous, N., Pinjari, A.R., Bhat, C.R., Pendyala, R.M., 2010. A comprehensive analysis of household transportation expenditures relative to other goods and services: an application to United States consumer expenditure data. Transportation 37 (3), 363–390.

Fujioka, K., Ishigami, T., Takahashi, K., 2012. Macro-analysis of travel behavior of youth in the Tokyo Metropolitan Area with a focus on the life stages of women. IATSS Res. 37 (2), 31–38.

Goodwin, P., Van Dender, K., 2013. 'Peak car'—themes and issues. Transp. Rev. 33, 243–254.

Hopkins, D., 2016. Can environmental awareness explain declining preference for car-based mobility amongst generation Y? A qualitative examination of learn to drive behaviours. Transp. Res. A 94, 149–163.

Hopkins, D., Stephenson, J., 2016. The replication and reduction of automobility: findings from Aotearoa New Zealand. J. Transp. Geogr. 56, 92–101.

Kuhnimhof, T., Armoogum, J., Buehler, R., Dargay, J., Denstadli, J.M., Yamamoto, T., 2012a. Men shape a downward trend in car use among young adults—evidence from six industrialized countries. Transp. Rev. 32, 761–779.

Kuhnimhof, T., Buehler, R., Wirtz, M., Kalinowska, D., 2012b. Travel trends among young adults in Germany: increasing multimodality and declining car use for men. J. Transp. Geogr. 24, 443–450.

Kuhnimhof, T., Zumkeller, D., Chlond, B., 2013. Who made peak car, and how? A breakdown of trends over four decades in four countries. Transp. Rev. 33, 325–342.

Metz, D., 2010. Saturation of demand for daily travel. Transp. Rev. 30, 659–674.

Nishimura, H., 2012. Sociological perspectives on youth and automobiles with special reference to contemporary Japan. IATSS Res. 37 (2), 21–30.

Oakil, A.T.M., Manting, D., Nijland, H., 2016. Determinants of car ownership among young households in the Netherlands: the role of urbanisation and demographic and economic characteristics. J. Transp. Geogr. 51, 229–235.

Simons, D., De Bourdeaudhuij, I., Clarys, P., De Geus, B., Vandelanotte, C., Van Cauwenberg, J., et al., 2017. Choice of transport mode in emerging adulthood: differences between secondary school students, studying young adults and working young adults and relations with gender, SES and living environment. Transp. Res. A 103, 172–184.

Yotsumoto, M., 2012. Loss of interest among youth toward automobiles: analysis of the present state and the possibility of a break through. IATSS Res. 37 (2), 39–47.

Yu, B., Zhang, J., 2015. Modeling household energy consumption behavior: a comparative analysis. Transp. Res. D 39, 126–140.

Zhang, J., 2017. Life-Oriented Behavioral Research for Urban Policy. Springer, Tokyo.

Zhang, J., Timmermans, H., Borgers, A., 2002. A utility-maximizing model of household time use for independent, shared and allocated activities incorporating group decision mechanisms. Transp. Res. Rec. 1807, 1–8.
Zhang, J., Timmermans, H.J.P., Borgers, A., 2005. A model of household task allocation and time use. Transp. Res. B 39 (1), 81–95.

CHAPTER 12

Behavioral research on transport and energy in the context of aviation

Tuqiang Zhou* and Junyi Zhang
Hiroshima University, Higashi Hiroshima, Japan

12.1 Introduction

Global carbon emissions will grow at a rate of 3%–4% annually, and one of the major contributors comes from the aviation industry (Hinnen et al., 2017), which emits greenhouse gases (GHGs) to the planet, exacerbating the process of global warming. In general, there are four kinds of gases emitted from the aviation industry, as shown in Fig. 12.1.

The environmental efficiency can be improved by modifying design of aircraft engines, optimizing air traffic management, and improving efficiency of routine operation; however, due to economic improvements, such as liberalization of the market, the growth of airline traffic is conspicuous and surpasses the savings from environmentally efficient operation. Further technological improvements are less likely to bring a significant reduction in energy consumption and emissions in the short run, and travelers' behavioral changes become more and more important (Davison et al., 2014). However, a distinct disharmony can be observed about how air travelers and policy makers consider air travel, in relation to energy-intensive activities. Aviation is thought to be the most difficult sector for formulating sustainable policies, due to the direct conflict between environmental impacts and economic development as well as lack of sustainable and substitutable alternatives for people to choose.

This chapter aims to provide a better understanding of the relationship between the aviation sector and its energy consumption from a behavioral perspective.

* Corresponding author.

Transport and Energy Research
DOI: https://doi.org/10.1016/B978-0-12-815965-1.00012-0

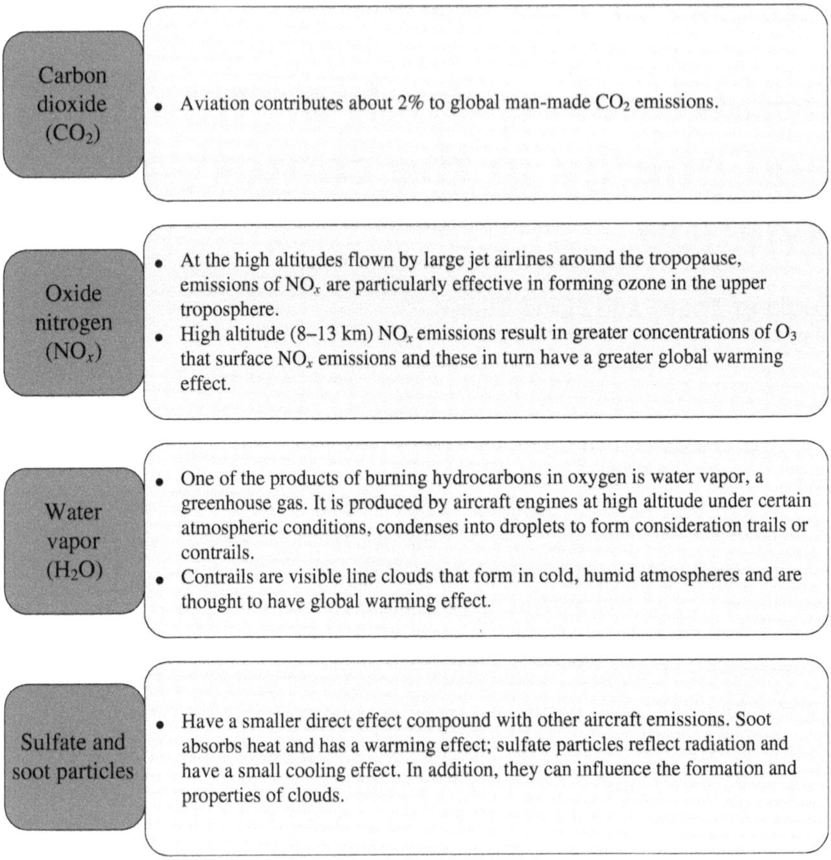

Figure 12.1 Environmental influence of the aviation industry on climate change (Maria, 2014).

12.2 From an air traveler's perspective

12.2.1 Attitudes toward flying and climate change

Although global climate is partially influenced by air travel, until recently there has been little sociocultural research focusing on whether travelers realize the fact that air travel has impacts on the climate system, and whether travelers' behaviors might be altered in response to the impacts.

Ipsos MORI (2007) divided air travelers into different behavioral styles ranging from the "ultragreens" to the "disengaged," based on the reaction intention or ability to respond to policy, such as government guidance, information provision, and increase in air travel cost (fares or taxes). Interestingly, results indicated that while the cluster of "ultragreens" indicated frequent flyers choosing budget aviation corporations for regular short-haul flights, the "disengaged" were not actually frequent travelers. Other research found that individuals' understanding or awareness about climate change was not often reflected in their reactions to transport, concluding that it was necessary to formulate measures that could support actual behavioral changes, instead of providing information (Ryley et al., 2013).

Cohen and Higham (2011) assessed British travelers' attitudes toward long-distance flight strategies to New Zealand in association with climate-change issues. Respondents' views about airline emissions from long-distance air travel in New Zealand were expressed as a range of perceptions and attitudes toward air travel and climate change. Three types of people were grouped, such as people who were not aware of the impacts caused by flights, people who knew the influence but would not alter their air-travel behaviors, and people who were aware of air travel's climate impacts but were beginning to show signs of consuming air travel with a conscience. It was concluded that an effective guidance on how to mitigate climate influences should be made by governments and diffused via mass media, if air travelers' behavioral changes were helpful to meet the governments' climate-change targets.

Another issue is about air travelers' attitudes toward green purchase, which are related to environmental responsibility of travelers who pay sufficient attention to the climate impacts of purchase, usage and recycling of different green goods, or receiving services. The concept of green products refers to additional contents added to the basis of air-travel services. Recently, airlines have offered green goods on the basis of the typical air-ticket, such as organic on-board food. When discussing green purchase behavior, several studies have found gaps between travelers' positive attitudes and real buying experiences (Hinnen et al., 2017). This means that travelers' favorable mind toward green goods may not always transfer into real action continuously. Therefore it's indispensable to investigate why air travelers' environmental attitudes result in weaker effects on their green purchase behaviors.

12.2.2 Willingness to pay for carbon offsets in air travel

Carbon offset, a very hot issue, often represents air travelers' purchase of emission reductions to cancel out carbon emissions caused by their activities. A series of condemnations about carbon offsets can be found in previous studies (Andy, 2015). The critiques of carbon offsetting schemes correlate to the low reliability, confusion and complexity of the schemes, and low levels of transparency. Other researches have shown that carbon offsetting may have no direct effect of emission reduction (Gössling et al., 2007). Nonetheless, it is currently the only available tool for both airlines and air travelers to mitigate the impacts of air-travel activities on climate change using voluntary carbon offsets, and they are expected to pay for it only if compulsory systems to decrease carbon emissions are taken into action.

Despite a majority of air travelers are aware of the fact that their travel partially contributes to climate change and expresses their willingness to pay (WTP) for a mandatory carbon tax, a very small portion (1%–2%) of the oversea travelers made extra payments for the voluntary offsets in reality, in comparison to about 1 in 10 people expressed their intention to the payments (Andy, 2015). Lu and Shon (2012) adopted a stated preference nonmarket valuation method to measure WTP for airline carbon offsets and reported their per capita estimation of \$14–\$30 per ton of carbon dioxide. At the same time, the carbon emission reduction value of WTP may be sensitive to changes in regulatory regimes. It is also observed that when potential payments change from mandatory cases (carbon taxes) to voluntary cases, the ratio of unwillingness to pay increases substantially (Stithou and Scarpa, 2012). One reason for the increase is that travelers are generally worried about how other people bear the liability. Meanwhile, the relationship between compulsory policy and voluntary action might depend on different factors, such as whether travelers support the policy, how much they share the environmental responsibility in concern, and whether they believe voluntary actions are effective.

It is expected that WTP for voluntary offsetting is affected by crowding-out effects, which indicate that reducing GHG emissions and mandatory payments may result in reduced voluntary donations. However, contrary to this expectation, it is also possible that the crowding-out effects may be limited or that new economic incentives may even motivate individuals' voluntary actions, namely, a "crowding-in" effect.

12.2.3 Behavioral models of air travelers

The behavioral models of air travelers mainly focus on the relations between knowledge, attitude, and behavior. Lu and Wang (2018) adopted the knowledge–attitude–behavior model to measure travelers' attitudes toward carbon offset, followed by a path model to analyze the various sources of knowledge on travelers' attitudes toward environmental impacts and their willingness to mitigate these impacts. The theory of planned behavior (TPB) has been widely applied to investigate consumers' behavioral changes, for understanding air travelers' attitudes, intentions, and real purchase behaviors in the case of green goods (Joshi and Rahman, 2015). Chen (2013) developed a model of goal-directed behavior (MGB) to explore the purpose and its key factors influencing air travelers' attitudes toward carbon offset schemes. The MGB broadens and deepens the traditional TPB by using new concepts of desires, active and passive expectations, and past action that address the importance of emotions, perceptions, and aims in decision process.

As for detailed models, choice models are the most widely used ones in the aviation field. Hinnen et al. (2017) developed an adaptive choice-based conjoint analysis to simulate decision-making process that is influential to actual choices. This conjoint analysis approach is especially applicable when decision makers employ a noncompensatory decision-making rule, assuming that people made complicated choices by establishing a consideration set (usually cut-off rules are used) and then chose an option within this set. Lu and Shon (2012) used a contingent valuation method with a double-bounded dichotomous choice format to explore airline travelers' WTP for carbon offsets. Such a double-bounded choice format is useful to assist respondents to report the true WTP. In order to have an economical estimation of voluntary carbon offsets, Choi et al. (2018) adopted a coupled method with forced choices and certainty responses to modify discrete choices with a "no-choice" option. The modified method with forced choices can better perform in reducing hypothetical biases and enhancing trade-offs across attribute levels.

12.2.4 Behavioral mechanisms of air travelers

The behavioral research involving air travelers' energy-related attitudes have given rise to the following concerns: what is the formation process of air passengers' environmental opinions? How does the traveler know the problems of climate change related to flying? As far as air travel is concerned, does environmental consciousness, including, but not limited to, consciousness of climate-change issues, have effects on actual passenger behavior?

Several researches indicated that if effective knowledge viewing CO_2 emissions and the environmental influence of air traffic was shown in an efficient way, it would help one to gain the individual's knowledge and thus lead to a different attitude and travel behaviors (Lu and Wang, 2018). However, most of the studies found a feeble connection between the conveyed favorable idea of travelers upon purchasing green goods and their real purchase action, generally known as the attitude–behavior gap (Joshi and Rahman, 2015).

There is no doubt that attitude is one of key factors affecting travel behavior. Based on the previous studies (Birgelen et al., 2011; Davison et al., 2014; Sonnenschein and Smedby, 2019), and in order to better understand how the attitude affects behavior, it is of vital importance to study other types of mechanisms and attitudes corresponding to rationality. For the issue of air travel the gap between awareness and understanding, and an individual's actions, is arguably reinforced by political decisions that prioritize the advantages provided through airport connectivity and national competitiveness over the need to reduce emissions from air travel.

Therefore the behavioral mechanisms of air travelers mainly involve two aspects. First, psychological impetus of behavioral change has been proved to be more effective than infrastructural changes in air travelers' environmental consciousness. Accordingly, it is indispensable to comprehend the significance of various impetuses of environment-friendly behaviors involving air travel. Second, regarding air travelers' WTP issues, information gap between the aviation industry and travelers regarding the environmental influence of air travel and the actions that can be actually applied to lighten these influences is the most important factor influencing air travelers' actual behavior.

12.3 From an airline perspective

When dealing with air fleet assignment, commercial airlines need to balance the environment impacts with operational cost, which are decided under uncertainty.

12.3.1 Airlines' fuel usage and carbon emissions

As more and more people pay attention to climate change, the issue of carbon emissions in the aviation industry has attracted a growing concern. For airline emissions, obtaining a greater comprehension of the relationships between individual airlines' emissions and characteristics of their fleets as well as flight operation is of great significance. Because the emissions are roughly in proportion to fuel usage, the abovementioned relationships can be explored by checking the connection between an individual airline's fuel consumption and relevant characteristics.

Several airline studies have recognized the significance of emissions reduction from a viewpoint of fuel consumption. But it is precisely due to the correlation of emissions and fuel usage; industrialists also realize that aimlessly focusing on environmental performance will bring some obstacles to the pursuit of economic interests (Brueckner and Abreu, 2017).

Existing studies have often paid attention to minimizing fuel consumption, because airline companies usually care about costs, rather than carbon emissions. When the task of emission reduction is determined, a majority of airline companies consider it consistent with reducing fuel usage, lowering fuel cost, and improving energy efficiency (Ma et al., 2018).

12.3.2 Airlines' choices of aircraft sizes

Capacity decisions are very important for airline companies. In general, with the increase of air traffic, airline services will be provided with larger aircrafts. To satisfy the increasing traffic demand, airline companies can increase aircraft sizes, service frequency, and/or the load factor. High load factors are considered as a precondition for profitable operation, as fuel consumption per traveler relies deeply on it. Conversely, fuel usage is probably the single most important item in operational fees. Accordingly, airline companies are bound to maximize the load factor all the time, where the current industry average is close to 75% (Givoni and Rictveld, 2009).

Givoni and Rietveld (2010) developed a representative model for comparing environmental influence of single-aisle and twin-aisle aircrafts plane. The Boeing B747-400 was selected as a typical wide-body (large) aircraft and the Airbus A320-200 as a representative small-capacity aircraft. It was found that the choice of aircraft directly affects operation costs. However, small-capacity aircrafts are used for short-distance trips, while large-capacity aircrafts are operated for long-distance trips. Because of this, it is difficult to empirically compare the difference between operation costs of these two aircraft types. It was further confirmed that choice of aircraft size had a direct effect on environmental emissions from aircrafts. Some studies have shown that environmental emissions from airline operation can be reduced by adopting large-capacity aircrafts (e.g., Peeters et al., 2005). In order to decrease the influence of airplane on the environment, particularly at large airports, larger airplanes designed for short-haul operation are recommended by Givoni and Rietveld (2009).

12.3.3 Airlines' environmental friendliness

In recent years the trend of airline companies' claiming their goods and services to be environmentally friendly has become more and more obvious. Initially, the focus is only on green food products; however, recently, more and more large airlines use the green concept or eco-labels to highlight their concern about the environment (Baumeister and Onkila, 2017). The main function of "green" or "eco" concept is to serve as a component of consumer choice and a reminder to take environmental issues into account. In addition, worldwide standards and regulations are becoming more and more important. The International Standard Organization (ISO) 14001 series and the EU Eco-Management and Audit Schemes are two main efforts for forcing airlines to pay more attention to environmental issues (Hagmann et al., 2015).

The low degree of transparency in the aspect of airlines is another critical issue. Certain people criticized that many airlines were likely to understate the effects of environmental pollution while exaggerate the latent profits, for example, new airplanes have higher fuel effectiveness (Gössling and Peeters, 2007). On the contrary, other industries have adopted criteria that promote the degree of transparency, for instance, on food products or logistics services (Jansen and Hamm, 2012; Connekt, 2013). Nevertheless, the International Air Transport Association has put environmental issues at first place of the developing process of airlines and settled several standards based on environment.

Researchers have studied Corporate Social Responsibility (CSR) initiatives released by top leaders of the airline industry, who are members from three main airline alliances (Hagmann et al., 2015). Popular CSR measurements are declined in CO_2 emissions and air pollutants, lower energy usage, and improvements of environmental protection programs. Although the airlines emphasized their great efforts, a lack of sufficient statistical data makes interairline comparison difficult. Some claimed that they had gained ISO 14001 authentication for environmental management systems; however, efforts to enhance air traveler's awareness of behaving in a green way are still too limited.

12.3.4 Energy efficiency of airlines

Improving the energy efficiency of airlines is a key to achieve the goals of energy saving and emissions reduction in the aviation industry. Energy efficiency in aviation is mainly known as fuel efficiency due to high fuel prices (Baharozou et al., 2017).

Existing studies evaluated energy efficiency using various measurements and models. Babikian et al. (2002) analyzed the fuel efficiency of different aircraft types with specific energy usage and energy intensity. Miyoshi and Merkert (2010) evaluated carbon and fuel efficiency of 14 European airlines to understand the relationship between fuel efficiency and fuel price, distance flown, and load factors. Cui and Li (2016) adopted a network slacks-based measure (SBM) model, which divided airline energy efficiency into two stages, such as operations and carbon abatement, to measure the overall efficiency and divisional efficiency of 22 international airlines. SBM-data envelopment analysis was used for exploring the influencing factors of energy efficiency, and results showed that the energy efficiency of airlines was affected by the fleet era and the modalities of proprietorship passively; numbers of passengers in one flight, and market share exerted great positive effects (Zhang et al., 2017).

However, as pointed by Cui and Li (2016), majority of existing studies treated energy efficiency measurements with a single production process, and CO_2 emissions were just treated as an undesirable output rather than a part of the production process of airline energy efficiency. Further research should focus on the multistage structure of airline efficiency.

12.3.5 Behavioral models for airlines

Regression models are the most frequently used tools for representing airline choice outputs associated with energy usage, such as airline fuel usage, carbon emissions, and aircraft size. Brueckner and Abreu (2017) built a regression model based on a theoretical framework that divided the aircraft operating costs into fuel cost and the leasing (or ownership) cost of aircraft, to reveal crucial factors affecting airline fuel usage and carbon emissions. Except for basic regression, two temporal effects of time trend and carrier-fixed effects were also considered. The chapter also identified another operational change that reduces fuel consumption and can be achieved through public policies: a reduction in flight delays. Regarding the choice of aircraft size, a regression analysis of more than 500 airways of the United States, Europe, and Asia showed that the choice of aircraft size was primarily influenced by features of airways, while almost not at all by airport characteristics (Givoni and Rietveld, 2009). Sibdari et al. (2018) used a generalized linear regression model to analyze three capacity decisions (flight frequency, aircraft size, and load factor) of seven airlines. The relationship between the decisions, and the level and fluctuations of three exogenous factors (fuel cost, total passenger demand, and unemployment rate) were discussed.

For overcoming the shortcomings of existing bilevel market models a new air transport market model was proposed by Takebayashi (2011) to deal with airline network design strategy, by introducing service frequency and aircraft size. The model adopted a stochastic user equilibrium situation, where passengers are assumed to choose the best routes by considering airfare, travel time, flight frequency, and reservation availability. And, it is further assumed that airlines control the flight frequency for profit maximization. As for fleet assignment problems, Ma et al. (2018) developed a multicriteria decision-making method, which further accommodates random demands, fare and fuel price, and identified the intrinsic relationship between profit and fuel cost of airlines.

12.4 From a perspective of aviation logistics

Airfreight is one of the main choices for cross-border logistics, which is faster than shipping and safer. It is especially suitable for fresh goods, valuables, and precise instruments. AirWay Bill (AWB) is a certificate of a receipt made by a shipper (or an agent) and a carrier. Data of freight, including freight region of origin and destination, transfer locations (if has), amount, variation tendency, transfer process, and transport costs, are recorded in the AWB. Different airlines vary with freight charge, total delivery time and distance, safety, and service. For a manufacture, it can take advantage of large amounts of aviation logistics data to analyze transport costs so that it can decide the most proper airline(s), transport mode (s) and route(s), and freight forwarder(s)/shipper(s) (Guan, 2018).

The aviation logistics industry is very important in today's global business because of its value of time and cost to support supply chains. The logistics industry is not only a key contributor to economic development but also plays a significant role in resolving environmental and social issues. The logistics industry consumes a considerable amount of energy resources worldwide and consequently emits voluminous GHGs and other emissions.

Nowadays, climate change has become a serious issue for both the international and domestic communities. To address this issue, it is necessary to effectively control the growth of GHG emissions from air transport by means of emission reduction market mechanisms (ERMM), among other measures. Under the ERMM the costs spent by companies to achieve emissions reduction targets have already constituted an important part of the cost of aviation logistics companies. Some countries have participated in the International Civil Aviation Organization intercontinental aviation carbon offset and emission-reduction plans in order to realize their commitments to reducing emissions and promote the low-carbon development of the aviation logistics industry. If emission reduction targets can be achieved, the participation and implementation of the international and domestic ERMM will have direct and indirect impacts on the carbon emission costs of aviation logistics companies (Luo, 2018).

Carriers have various ways to become more environment friendly via, such as technical investments, production reduction, route choices, and/or the permits purchase (e.g., a low-cost abatement carrier can sell carbon emissions right to a high-cost abatement carrier). The important issue for carriers is the carbon price. If the carbon price remains low, air carriers will have to purchase carbon allowances higher than investment for reducing emissions (Fukui and Miyoshi, 2017).

12.5 Discussion

12.5.1 Key factors influencing air travelers' energy consumption behavior

Air travelers would not be willing to pay for a carbon tax if they could not find evidence of the payment being spent on measures to reduce the environment impacts of air travel on climate change. In this regard, government plays a key role in encouraging or coercing both the airline industry and the public to take actions, either directly, or through an independent body. Its role is to ensure that the aviation industry aims to produce and utilize cleaner, greener airplanes, and maximize efficiency and offset their emissions.

Income and fuel price levels are also key factors identified by existing studies (Cohen and Higham, 2011). Income refers to the money available to spend on air-travel activities, and fuel price affects the benefits of energy efficiency of technologies used and of switching to alternative fuels. Moreover, the implementation of a carbon tax, which is the commonly used mitigation policy instrument, will impact fuel prices.

It is important to consider equity in developing economic instruments such as fuel tax and an emissions-trading system. For instance, how to charge small emitters (small fuel users) in a socially acceptable way? To make fuel tax a more effective policy instrument for reducing fuel consumption and CO_2 emissions, it is necessary to undertake further research on analysis and design of a tax scheme that could address the uneven effects of fuel taxation and help reduce fuel consumption and CO_2 emissions more effectively.

As far as green purchase behavior is concerned, environmental concerns and values are the prime factors that push an air traveler to think about green products. In the future, policy makers should nurture and develop this tendency through environmental education. Air travelers generally remain skeptical of environmental claims made by manufacturers and feel difficult to identify green products. Therefore environmental education should impart knowledge about how an air traveler can identify green products.

12.5.2 Implication for the aviation industry

The first implication is that airline companies should deepen the cooperation on control of emissions. According to several studies, better cooperation can help reduce most airlines' GHG abatement costs. For instance, airlines can cooperate in the form of navigation technologies, which can bring out a significant reduction in carbon dioxide emissions.

Another implication is that if nonalliance airlines want to select an alliance, they should select a reasonable alliance as different airline alliances have different help on reducing GHG abatement costs. A recent study indicated that Star Alliance and Skyteam could help their aligned airlines reduce GHG abatement costs, but the cooperation in Oneworld had a small effect. Probably, this will be a crucial reference point for the airlines which have not joined any alliance (Li and Cui, 2018).

With regard to manufacturers, they are bound not just to solve the "green" feature of their goods but ought to also focus on functional property. Airline companies need to introduce innovative products, which combine green and functional attributes. Producers and marketers should not only introduce products with eco-labels but also make efforts to develop consumer trust in the eco-labels.

12.6 Conclusion

To reduce energy consumption and emission from the aviation industry, behavioral changes of all stakeholders are required. Airline companies need to introduce more energy-efficient aircrafts and further enhance the load factor of flight. The former efforts need a strong support from the international society that pushes the aviation industry to commit more to mitigating the global warning and impacts of other environmental issues. The latter efforts need more collaboration of different airline companies to make collective decisions on more effective marketing strategies by making full use of aircraft capacities as much as possible. In addition, airport operators also play an important role by introducing more energy-saving business models to operation of airport facilities and services. Furthermore, local governments can make more efforts to make public transport more convenient for becoming the dominant access mode to the airport.

However, the success of all the abovementioned efforts/improvements heavily depends on air travelers. Efficiency analysis has been a focus in the aviation research; however, how air travelers' behaviors affect the realization of efficient operation of aviation business has remained unknown. A lack of knowledge about the impacts of their travel on the environment and a lack of information about the benefits of their pro-environmental behaviors have been identified to be an important reason of their low engagement in environment-friendly actions (Lu and Shon, 2012; Cliff, 2014; Lu and Wang, 2018). Even if air travelers are aware of the impacts and benefits and express their willingness to behave in an environment-friendly way, an attitude–behavior gap surely exists, which hinders travelers to take environment-friendly actions being consistent with their willingness, because of concerns about social fairness (travelers generally worry about how other people share the responsibility), policies (compulsive or voluntary), and effectiveness (whether travelers think voluntary actions are helpful or not). Therefore future behavioral research on energy consumption and emissions from air traffic should first focus on how to fill the above knowledge and information gaps and the attitude–behavior gap that encourage air travelers to behave voluntarily in an environment-friendly way. For this, moral decision-making theories (Chorus, 2015) may be helpful and it is worth applying them in the future.

Filling the aforementioned gaps needs more efforts from the whole aviation industry, airline companies, and governments as well as other international and domestic stakeholders. Then, it becomes crucial to figure out how air travelers' decisions and behaviors are affected by efforts from different stakeholders and the ways of communicating the efforts to air travelers. Reducing the attitude–behavior gap further needs to make use of sociopsychological instruments that can encourage air travelers to interact with each other toward a socially desirable collective action. It is therefore necessary to develop choice models with social interactions and sociopsychological factors motivating behavioral changes, for addressing individual choice issues, and to develop effective public involvement theories that allow all stakeholders to work together at various decision-making levels toward a globally sustainable aviation society.

References

Andy, S.C., 2015. An experimental study to explore WTP for aviation carbon offsets: the impact of a carbon tax on the voluntary action. J. Environ. Plann. Manage. 58 (9), 1617–1634.

Babikian, R., Lukachko, S.P., Waitz, I.A., 2002. The historical fuel efficiency characteristics of regional aircraft from technological, operational, and cost perspectives. J. Air Transp. Manage. 8 (6), 389–400.

Baharozou, E., Soykan, G., Ozerdem, M.B., 2017. Future aircraft concept in terms of energy efficiency and environmental factors. Energy 140, 1368–1377.

Baumeister, S., Onkila, T., 2017. An eco-label for the airline industry? J. Cleaner Prod. 142, 1368–1376.

Birgelen, M., Semeijn, J., Behren, P., 2011. Explaining pro-environment consumer behavior in air travel. J. Air Transp. Manage. 17, 125–128.

Brueckner, J.K., Abreu, C., 2017. Airline fuel usage and carbon emissions: determining factors. J. Air Transp. Manage. 62, 10–17.

Chen, F.Y., 2013. The intention and determining factors for airline passengers' participation in carbon offset schemes. J. Air Transp. Manage. 29, 17–22.

Choi, A.S., Gössling, S., Ritchie, B.W., 2018. Flying with climate liability? Economic valuation of voluntary carbon offsets using forced choices. Transp. Res. D 62, 225–235.

Chorus, C.G., 2015. Models of moral decision making: literature review and research agenda for discrete choice analysis. J. Choice Modell. 16, 69–85.

Cliff, A.D., 2014. An Investigation of UK Passenger Attitudes Towards the Carbon Offsetting of Both Flight and Airport Emissions (Unpublished Master's thesis). University of Chester, United Kingdom. Available at: <http://www.hdl.handle.net/10034/346175>.

Cohen, S.A., Higham, J.E.S., 2011. Eyes wide shut? UK consumer perceptions on aviation climate impacts and travel decisions to New Zealand. Curr. Issues Tourism 14 (4), 323–335.

Connekt, 2013. Logistics award step-by-step plan. <http://lean-green.nl/en-GB/logistics/logistics-award-step-by-step-plan>.

Cui, Q., Li, Y., 2016. Airline energy efficiency measures considering carbon abatement: a new strategic framework. Transp. Res. D 49, 246–258.

Davison, L., Littleford, C., Ryley, T., 2014. Air travel attitudes and behaviors: the development of environment-based segments. J. Air Transp. Manage. 36, 13–22.

Fukui, H., Miyoshi, C., 2017. The impact of aviation fuel tax on fuel consumption and carbon emissions: the case of the US airline industry. Transp. Res. D 50, 234–253.

Gössling, S., Peeters, P., 2007. 'It does not harm the environment!' an analysis of industry discourses on tourism, air travel and the environment. J. Sustain. Tourism 15 (4), 402–416.

Gössling, S., Broderick, J., Upham, P., Ceron, J.P., Dubois, G., Peeters, P., et al., 2007. Voluntary carbon offsetting schemes for aviation: efficiency, credibility and sustainable tourism. J. Sustain. Tourism 15 (3), 223–248.

Givoni, M., Rietveld, P., 2009. Airline's choice of aircraft size—explanations and implications. Transp. Res. A 43, 500–510.

Givoni, M., Rietveld, P., 2010. The environmental implications of airlines' choice of aircraft size. J. Air Transp. Manage. 16, 159–167.

Guan, T., 2018. Big Data in Aviation. Publishing House of Electronics Industry, Beijing (in Chinese).

Hagmann, C., Semeijn, J., Vellenga, D.B., 2015. Exploring the green image of airlines: passenger perceptions and airline choice. J. Air Transp. Manage. 43, 37–45.

Hinnen, G., Hille, S.L., Wittmer, A., 2017. Willingness to pay for green products in air travel: ready for take-off? Bus. Strategy Environ. 26, 197–208.

Ipsos MORI, 2007. Attitudes to aviation and climate change. In: Final Report, Research Study Conducted for the Commission for Integrated Transport.

Jansen, M., Hamm, U., 2012. Product labelling in the market for organic food: consumer preferences and willingness-to-pay for different organic certification logos. Food Qual. Preference 25 (1), 9–22.

Joshi, Y., Rahman, Z., 2015. Factors affecting green purchase behaviour and future research directions. Int. Strategic Manage. Rev. 3 (1–2), 128–143.

Li, Y., Cui, Q., 2018. Investigating the role of cooperation in the GHG abatement costs of airlines under CNG2020 strategy via a DEA cross PAC model. Energy 161, 725–736.

Lu, J.L., Shon, Z.Y., 2012. Exploring airline passengers' willingness to pay for carbon offsets. Transp. Res. D 17 (2), 124–128.

Lu, J.L., Wang, C.Y., 2018. Investigating the impacts of air travellers' environmental knowledge on attitudes toward carbon offsetting and willingness to mitigate the environmental impacts of aviation. Transp. Res. D 59, 96–107.

Luo, R., 2018. Research on Carbon Emissions Reduction Costs of China's Aviation Logistics under the Background of Emissions Reduction Market Mechanism. Civil Aviation University of China, Beijing (in Chinese).

Ma, Q., Song, H., Zhu, W., 2018. Low-carbon airline fleet assignment: a compromise approach. J. Air Transp. Manage. 68, 86–102.

Maria, M., 2014. Sustainable development: the key for green aviation. INCAS Bull. 6 (1), 109–122.

Miyoshi, C., Merkert, R., 2010. Changes in carbon efficiency, unit cost of firms over time and the impacts of the fuel price-an empirical analysis of major European airlines. In: Proceedings of the 14th Air Transport Research Society (ATRS) World Conference. Porto, Portugal.

Peeters, P., Rietveld, P., Schipper, Y., 2005. Deregulation in Europe aviation and the evolution of hub-and-spoke networks: impacts on the environment. In: Atzema, O., Rietveld, P., Shefer, D. (Eds.), Regions, Land Consumption and Sustainable Growth. Edward Elgar, Cheltenham, pp. 137–153.

Ryley, T., Burchell, J., Davison, L., 2013. Valuing air transportation and sustainability from a public perspective: evidence from the United Kingdom and the United States. Res. Transp. Bus. Manage. 7, 114–119.

Sibdari, S., Mohammadian, I., Pyke, D.F., 2018. On the impact of jet fuel cost on airlines' capacity choice: evidence from the U.S. domestic markets. Transp. Res. E 111, 1–17.

Sonnenschein, J., Smedby, N., 2019. Designing air ticket taxes for climate change mitigation: insights from a Swedish valuation study. Clim. Policy 19 (5), 651–663.

Stithou, M., Scarpa, R., 2012. Collective versus voluntary payment in contingent valuation for the conservation of marine biodiversity: an exploratory study from Zakynthos, Greece. Ocean Coast. Manage. 56, 1–9.

Takebayashi, M., 2011. The runway capacity constraint and airlines' behavior: choice of aircraft size and network design. Transp. Res. E 47, 390–400.

Zhang, J., Fang, H., Wang, H., Jia, M., Wu, J., Fang, S., 2017. Energy efficiency of airlines and its influencing factors: a comparison between China and the United States. Resour., Conserv. Recycl. 125, 1–8.

CHAPTER 13

A systematic review on tourism energy consumption, sustainable tourism, and destination development: a behavioral perspective

Linghan Zhang* **and Junyi Zhang**
Hiroshima University, Higashi Hiroshima, Japan

13.1 Introduction

13.1.1 Research background

Tourism has become an important industry in the past few decades, and it has brought enormous economic benefits to the visiting destinations. According to World Travel & Tourism Council (WTTC) (2018), travel and tourism grew faster than any other industry sector in 2017, generating US$8.3 trillion in GDP and supporting 313 million jobs. On the other hand, tourism is also claimed to bring different kinds of negative impacts, especially environmental emissions (Tang and Abosedra, 2014). Heavy tourist traffic, air pollution, and waste of resources cause a significant damage to local environment and resources due to tourism development (Katircioglu, 2014; Zhang and Gao, 2016). Tourism brings many long-distance travels between destinations, which are considered as the major contributor to energy consumption and carbon emissions in the tourism industry (Dubois and Ceron, 2006; Liu et al., 2011). Meanwhile, with the increase of tourists' diversified activities, energy consumption will be increased to meet various needs of tourism experiences. Understanding tourism energy consumption is becoming an indispensable way for destinations to improve environmental degradation and promote tourism development in an environmentally sustainable way. Nevertheless, the

*Corresponding author

Transport and Energy Research
DOI: https://doi.org/10.1016/B978-0-12-815965-1.00013-2

relationships between energy consumption and tourism development have remained unclear in the literature.

With the increasing concerns about environmental protection and climate change, many scholars stated that tourism and environment should be managed in an integrated and interdisciplinary way (Beccali et al., 2009; Janusz and Bajdor, 2013). Sustainable tourism and destination management and development require effective energy and environment management policies (Erdogan and Baris, 2007). Measuring energy consumption is useful to understand tourism and environment in a quantitative and objective way (Becken and Simmons, 2002). If tourism energy consumption can be quantified, it is easier to control the negative environmental impacts under the critical thresholds; thus tourism and destinations can be sustainably developed (Gössling et al., 2005). However, a relatively small number of existing studies have investigated tourism energy consumption, and its relationship between sustainable tourism and destination development has been poorly discussed in a systematic way (Gössling, 2000; Day and Cai, 2012). In this respect, an investigation of the relationships between tourism energy consumption, sustainable tourism, and destination development will be of interest to both industry and policy makers.

To fill such research gaps, this chapter aims to propose a comprehensive framework to discuss energy consumption and related environmental problems within sustainable goals. Tourism is a dynamic and complex process that includes mobilities between destinations, the involvement of multiple stakeholders, and destinations management (Mill and Morrison, 2002; Bornhorst et al., 2010). Energy consumption should be examined from various angles at different levels. At each level of the system, there are studies related to energy consumption and sustainability (Day and Cai, 2012). Studies on tourism energy consumption have been done in a less systematic way from a behavioral perspective, suggesting the necessity of more future research.

13.1.2 Conceptual framework

The conceptual framework of this study is built based on the framework of tourism (Leiper, 1979) and stakeholder theory (Saito and Ruhanen, 2017) (see Fig. 13.1). Three main features of this framework are emphasized. First and foremost, the tourism industry can be viewed as a linked process, stretching across tourist generating regions, transit routes, and

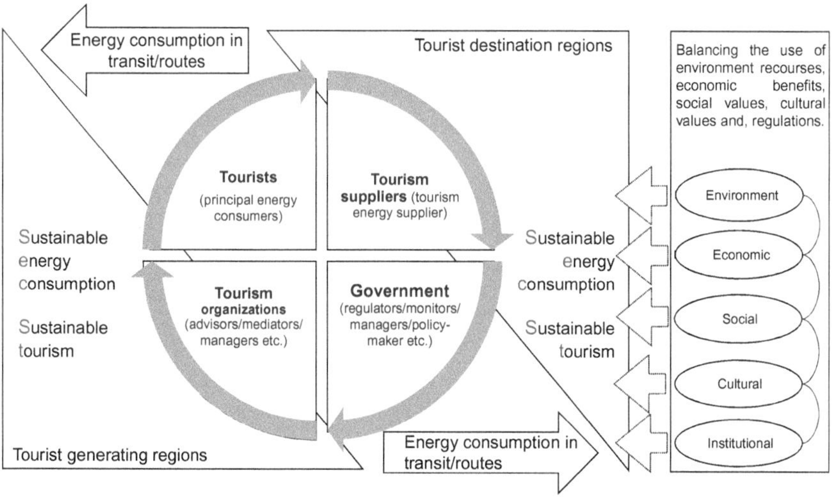

Figure 13.1 The conceptual framework of tourism energy consumption in destination.

destination regions (Leiper, 1979). The links in the process are diverse, which include informational and business cooperation between tourists and firms and facilities, and the connections through wholesaling deals. Transportation is one of the most important sectors in the tourism industry, which connects one place to the other place, including origin, destination, and transit place(s) (Palhares, 2003; Van Truong and Shimizu, 2017). A larger percent of energy is consumed on the tourism routes, and tourists' travel mode choices and activity participation are directly linked with energy consumption.

The second feature of this framework is that it emphasizes that tourism energy consumption is not solely attributable to tourists. Rather, it is an outcome generated by actions of various stakeholders. Tourism energy consumption is not only a source of energy demand but also the energy audit of many direct business activities (Becken and Simmons, 2002). Different stakeholders play different roles in tourism energy consumption, but they are all responsible for sustainable tourism and destination management. Four main stakeholders are chosen in the framework: tourists, tourism suppliers, organizations, and government. Tourists are the principal consumers in transit routes and visiting destinations. They participate in various activities and seek for tourism services and experiences, which cause energy consumption not only at destinations but also at other places/routes along all traveling courses starting from a departure place.

It is claimed that the tourist energy consumption behaviors and decisions might have different impacts on reducing energy consumptions (Kelly et al., 2007). Tourism suppliers are also tightly connected to local energy use at destinations. From a behavioral perspective, this study will summarize and discuss existing research on tourism entrepreneurs' behaviors in transportation and energy use. Tourism organizations and government have different but overlapped roles in tourism energy consumption. Although they are not the direct users or providers of tourism energy, their roles in environmental regulation, resources management, and destination policymaking determine the trend of local transportation planning and energy use. However, the behaviors of different stakeholders have not been explored in the tourism energy research.

The third feature of this framework is that it tries to explain tourism energy use based on the goal of sustainable tourism (Høyer, 2000). Sustainability of local tourism and destination development has already been acknowledged as an essential factor to define a successful tourism (Hunter, 1997; Cernat and Gourdon, 2012). The way of tourism transportation planning and energy use should match with the aims of sustainable tourism development, which are balancing the use of environment resources, economic benefits, social values, cultural values, and regulations (Janusz and Bajdor, 2013). Tourism energy consumption could not be discussed without considering sustainable tourism.

Based on this framework, in this chapter, we first summarize existing studies of tourism energy from different stakeholders' perspectives. Then, we discuss how tourism energy consumption is linked with sustainable tourism and destination development. Policy implications related to sustainable energy consumption and future directions on energy consumption are given in the last part. This conceptual framework provides an integrated picture to combine energy consumption and sustainable tourism development together.

13.2 Stakeholder behavior in transportation and energy consumption

13.2.1 Tourist behavior

Needless to say, tourists are the most important stakeholders in tourism. They are the participants of tourism activities and the direct users of tourism energy consumption. Tourist behaviors and choices have significant impacts on energy use, and proper tourist decisions are important to enhance energy efficiency and realize sustainable tourism development. Tourists make joint choices of transport, accommodation, attractions, and activities, and the resulting energy demand differs across joint choices (Becken et al., 2003a).

The first and foremost choice in tourism is about transportation modes and routes connecting origin and visiting destinations and travel within visiting destination areas. Despite transportation takes the largest share of tourism energy consumption (Gössling, 2000), there has been limited research about the influence of transportation choices on energy consumption, and what behavioral change will lead to sustainable tourism. Between tourist's home and destination, air transport is increasingly dominating the market (Gössling et al., 2007). According to Stohl (2008), 90% of the tourism-related carbon emissions arise from aviation. Tourists have increased the energy consumption and resultant greenhouse gas (GHG) emissions via air transportation (Nepal, 2008; Liu et al., 2011). However, little research is about how to manage the energy behavior patterns of tourists and then reduce the energy use. As air travel is responsible for the major tourism energy use, research and programs need to have in-depth understanding of tourist behavior, especially for the long-haul tourists who depend on air travel, aiming to increase energy efficiency. Another important transportation decision is about route choice. Tourist preference for environment-friendly travel routes contributes to sustainable tourism development. The recent emerging forms of tourism, such as ecotourism and rural tourism, are the representatives of low-carbon tourism form (Lane, 1994; Rinzin et al., 2007). The routes of trekking mount trail, cycling, and hiking are important for ecotourism and rural tourism. Tourists are encouraged to participate in the low-carbon tourism route choice to cope with the environment problems (Luo and Zhang, 2011). Transportation choices also take a paramount role in the energy consumption of visiting destination. Tourists are encouraged to use more energy-efficient transportation modes for their travel (Høyer, 2000; Holding, 2001). However, very little research has examined the impacts of these preferred travel modes on energy consumption or GHG emissions. To address such gap, Kelly et al. (2007) used a discrete-choice experiment to examine tourist-destination travel mode choices and forecasted the resulting environmental impacts. It was found that innovative transportation management strategies could encourage tourists to use public transport modes rather than private or rental vehicles to access tourism destinations and further reduced energy consumption and GHG. Furthermore,

Reilly et al. (2010) measured tourist reactions to a range of transportation strategies designed to shift skiers from private to public transport modes and found that long-haul tourists were more willing to accept changes. Studies on the impacts of transportation travel modes on energy consumption are still a potential area for future research.

There have been some studies on tourist preferences for different types of hotels and attractions; however, the impact of tourist choices on energy consumption is a largely neglected area. For example, it can be surely beneficial to energy saving, if tourists choose to stay at a low energy-use lodging. However, in reality, low energy use in accommodation often means fewer options of comfortable facilities and services. It is therefore unclear whether tourists are willing to choose to stay while sacrificing their comfort. If more tourists would not like to choose such an energy-saving lodging, the lodging owner would not choose to run such a business. From such a consideration, encouraging tourists to make more energy-efficient choices would be a better strategy. There is an increasing trend of studies on measuring tourist preference for energy-saving installation (ESI) or renewable energy sources (RES). Kostakis and Sardianou (2012) examined the determinants of tourists' willingness to pay (WTP) for renewable energy in hotel and revealed that middle-aged, male, and highly environmentally conscious tourists were willing to pay more than other tourists. Meanwhile, Tsagarakis et al. (2011) showed that tourists from countries with higher energy awareness were more willing to choose hotels with ESI and RES. Do Valle et al. (2012) studied tourist attitude toward an accommodation tax earmarked for environmental protection and found that a very high percentage of tourists would prefer hotels invested with ESI and RES. Similarly, Chou and Chen (2014) measured the WTP of tourists for green hotel attributes by multinomial and latent class models and found that tourists with high levels of the green consumption trait were more likely to choose hotels with environment-friendly attributes. It is not difficult to find that, in the field of tourist choice for accommodation and attraction, the stated preference approach and discrete choice models are mostly applied in measuring the impacts of tourist choices on energy-related problems. Measuring the intention and willingness of tourists for ESI and RES helps to predict the potentially relevant market segments for developing sustainable tourism. However, there is barely research discussing the impacts of revealed preference or actual choice behaviors on tourist energy consumption.

There are several main defects of current tourist behavior research. First, the impacts of tourist choices on energy consumption are examined in different sectors but not focusing on the whole decision-making and behavior process of a tourist. Tourists not only make decisions in transportation or accommodation but also in multiple subsectors of tourism. For example, a tourist may choose a bus to travel from a hotel to a destination (the two places are not the same administrative area) and rent a jeep-type vehicle for traveling within the destination after arrival. For the tourist, riding the bus is for saving money for renting the jeep. If this is the case, the so-called indirect rebound effect occurs. In such a case, a policy supporting bus services in the hotel area is not a better energy-saving option for the whole region covering the two areas. Related to the above situation, Becken et al. (2003b) examined the total "energy bill" of tourist choices within transport, accommodation, attraction, and other subsectors of tourism process and proved that the energy consumption of transportation dominated the bill. Unfortunately, such rebound effects are fully ignored in their analysis. Second, existing studies cannot well explain how an individual tourist behavior is linked with energy consumption, and its impact on sustainable tourism and destination development is missing. In addition, there are not enough empirical studies on different types of tourism energy behavioral patterns. For instance, international tourists traveling by air and cruise would be much more different in energy consumption behavior than other types of tourists (Brida and Zapata-Aguirre, 2009). Tourist segmentation is important to understanding different energy consumption behavior patterns. The case studies of both developed and developing countries are also needed. Tourist behaviors of energy consumption differ across regions. However, many of the above aspects have not been explored in existing research.

13.2.2 Tourism entrepreneurs' behaviors

Multiple industry sectors are involved in tourism and destination development. The range and scale of tourism suppliers are wide, which include hotels, restaurants, tourism attractions, travel agents, and tourism retailers. Managing the behaviors of tourism suppliers in energy consumption is crucial to local community and destination management. Among all of the suppliers, hotel is one of the hottest topics in tourism energy conservation and carbon reduction (Ali et al., 2008; Beccali et al., 2009;

Tsai et al., 2014). Previous studies confirmed that the energy consumption in hotels was 10–20 times higher than that of residential buildings.[1] It is believed that a significant amount of energy and resources in hotel was wasted (Sheng et al., 2018). To reduce waste and have more efficient energy use, many hotel chains have engaged in the green hotel programs and adopted energy conservation policies (Teng et al., 2012). There are different strategies to promote energy conservation activities. The most common way is to install energy-efficient equipment, such as lights, showering, bathing, laundry, and swimming pool (Sloan et al., 2009; Chan and Lam, 2003). The second one is to apply new technologies of ESI and RES (Zografakis et al., 2011; Zhao et al., 2012). The third one is through the management and involvement of staff. Employees can be trained for proper energy consumption and management behaviors (Erdogan and Tosun, 2009). The fourth way is to increase consumers' recognition and cooperation in hotel management. Hotel guests who are encouraged to participate in green hotel programs would be helpful in energy saving (Kasim, 2007). The hotel industry is facing a lot of challenges in reducing energy consumption. The biggest dilemma of the hotel industry is that very few tourist accommodations, particularly those operated by small and medium enterprises in developing countries, have the knowledge and skills required to improve their energy conservation and carbon reduction performance (Teng et al., 2012). It is not sure whether the hotel operators are willing to make energy-efficient strategies when facing profit allure. Dalton et al. (2007) measured the tourist operators' attitudes toward RES in accommodation. The majority of the tourist operators have interests in RES while having doubts in its reliability and economic viability. The hotel industry has already become the core sub-sector of tourist supply side. Dalton et al. (2009) further compared different types of RES options among small to medium–sized tourist accommodations and valued the feasibility. Although not every hotel can adopt renewable energy supply, the growing awareness of environmental factors in hotel design and management may lead to the sustainable energy consumption and tourism development (Karagiorgas et al., 2006).

Another important part of tourism supplier is related to tourist attractions and activities. Visiting local tourism spots is almost in every tourist's travel schedule. However, compared to the research on hotels, very

[1] Building Energy Saving Research Center of Tsinghua University, 2012. Annual Report of China's Building Energy Efficiency. Building Industry Press, Beijing.

limited studies can be found about tourism attractions and activities. Becken and Simmons (2002) categorized tourist attractions and activities in New Zealand and made an initial attempt to analyze the energy consumption of tourist attractions and activities under each categorization. It was found that cultural visitation in museums consumed less energy consumption than activities involving motorized travel did. It was further claimed that to improve the energy efficiency in tourism, energy benchmarks should be established, and energy use patterns for visiting tourist attractions and participating in activities should be explored. In the past decade, research on energy use caused by tourist attraction visitation and activity participation increased in a slow pace. For example, Cheng et al. (2013) integrated a Delphi method and an analytic hierarchy process approach to develop an evaluation index system for low-carbon tourist attractions and verified its applicability in a wetland of China. Shi et al. (2013) argued that wastes from tourist attractions could be a source of renewable energy, and they had a great potential to reduce tourism waste and improve the way of energy use.

The above review suggests that there lacks in-depth research and comprehensive scientific evidence on the patterns of tourism energy use (Shi and Peng, 2011). First, the existing detailed patterns of energy consumption among different tourist markets and categorizations of the tourism industry have not been well investigated, especially in the developing countries. Second, there are even no standards or commonly used benchmarks for tourism energy use or conservation for tourism suppliers. Third, from a behavioral perspective, tourism suppliers' behaviors, awareness, and attitudes in energy consumption and conservation have not been understood in a satisfactory way. It is expected that all the tourism industry sectors could adapt their behaviors toward a higher energy-efficient industry; however, it is unknown how tourism suppliers make trade-offs between economic efficiency and environmental benefits.

13.2.3 Governmental and organizational behaviors

Governments and organizations are usually involved in various forms and play multifaceted roles in tourism development. Governmental behavior can be observed at different levels and stages of the whole tourism development process. At earlier stages of tourism development, governments provide infrastructure and fundamental services to tourists, which directly affect the scale of local tourism. With the growth of tourism, governments usually set up legislations/regulations to manage tourism development and assist the tourism industry to achieve its economic and sustainable goals. Governments also play an important role in balancing between energy demands of tourist and local residents. While local energy supply should meet both types of demands, energy efficiency and saving cannot be neglected either. Meanwhile, multiple tourism organizations (e.g., UNWTO, national tourism organizations, destination marketing organizations, event organizers, nonprofit organizations) help to implement sustainable development strategies for planning, marketing, advising, and operation (Day and Cai, 2012). Both governmental and organizational behaviors would affect tourism energy consumption.

However, behavioral decisions by tourism organizations and governments have been largely neglected, and the impacts of policies on tourism energy consumption have remained unclear in the literature. There are almost no guidelines for many aspects of tourism consumption.

13.3 The relationship between energy consumption, sustainable tourism, and destination management

Unlike the above discussions of energy consumption from the behavior perspective of individual stakeholders, destinations, where tourism is mostly performed, represent a macrolevel perspective for energy-related studies. Tourism energy consumption should be researched within the goals of sustainable tourism development and successful destination management.

Existing studies on tourism energy consumption and sustainability are mostly related to two environmental themes. The first theme is that tourism energy consumption is responsible for sustainable tourism development, and thus relevant policies should be made to reduce GHG, pollution, and other negative impacts on the environment. For example, Gössling et al. (2002) and Hunter and Shaw (2007) adopted an ecological footprint analysis to assess tourism sustainability. Becken and Patterson (2006) used both bottom-up and top-down analyses to estimate the degree to which tourism contributes to national carbon dioxide emissions. The second theme is that how energy conservation and renewable energy can contribute to sustainable tourism (e.g., Bohdanowicz et al., 2001; Michalena et al., 2009; Michalena and Tripanagnostopoulos, 2010).

The research on tourism energy consumption and destination management is tightly connected to sustainable tourism. Energy consumption is rapidly increasing with a fast growth in tourist arrival and changes in tourism destination. Tourism-generated energy and its impact on destination should be evaluated (Tabatchnaia-Tamirisa et al., 1997). First, energy use is often related to the environmental problems in destination, such as carbon dioxide emissions. Wu et al. (2015) estimated the energy consumption and carbon dioxide emissions in the tourism sector of China and compared the total emissions among major provinces. Katircioglu (2014) and Katircioglu et al. (2014) investigated the long-run tourism energy consumption patterns and their impacts on environmental degradation in Turkey and Cyprus, respectively. Understanding the energy consumption in destination could promote energy conservation and decrease GHG emissions through making regulations and guidelines. Second, the impacts of energy consumption on destination development have been evaluated. The impacts of energy consumption on destination development could be both positive and negative. Becken and Simmons (2005) examined fossil fuel consumption and its negative impacts on the global climate. Tang and Abosedra (2014) revealed that tourism and energy consumption had positive impacts on local GDP growth but reacted negatively to political instability. Zaman et al. (2016) investigated the relationship between carbon dioxide emission, economic growth, and tourism development and explored the complicated causal relationships. The above research is helpful to resolve energy and environmental problems via effective solutions.

The relationships between tourism energy consumption, sustainable tourism, and destination management are complicated. A successful tourism destination could not exist without sustainable tourism development (Ritchie and Crouch, 2003). Tourism energy consumption is a key to achieving tourism and destination sustainability. More research on energy consumption should be promoted.

13.4 Policy implications to tourism energy consumption

Here, the strategies and policy implications mentioned in the literature are summarized with respect to every type of stakeholders (see Table 13.1).

From the table, we can find that most of the policies and strategies are set up at industry or government level and with the aim of sustainable tourism and destination development. It is, however, difficult to force tourists to accept the strategies of energy saving, pay extra for new technologies, or contribute to sustainable development. Instead, marketing segmentation seems to be a more feasible strategy. Tourist marketing

Table 13.1 The strategies and policy implications toward tourism energy consumption.

Policies and strategies	Tourism marketing	Tourism supplier	Organization and government
Transport	Marketing segments, attracting tourists who like environmentally-friendly travel modes	Provide multiple travel combinations that minimize travel distances and encourage walking and cycling (Kelly and Williams, 2007) Promote short-distance tourism destination Increase the efficiency of transport modes (Becken et al., 2003a,b)	Organize networks of pedestrian and bike paths (Lumsdon, 2000); no-vehicle zone, limit speed of motorized vehicles in certain areas (Holding, 2001)
Accommodation	Marketing segments, attracting tourists who prefer green hotel attributes	Join green hotel programs/ project ESI technology (Zhao et al., 2012)	Set up energy conservation and carbon reduction criteria (Teng et al., 2012)
Tourist attraction	Self-management in tourist attraction	Improve environmental protection and quality of tourism attraction and develop low-carbon tourism (Chan and Lam, 2003) Categorize tourist attractions and establish energy benchmarks (Becken and Simmons, 2002)	Ecologization and rational distribution of tourist attraction (Cheng et al., 2013)

(*Continued*)

Table 13.1 (Continued)

Policies and strategies	Tourism marketing	Tourism supplier	Organization and government
RES	Marketing segments, attracting tourists who are willing to use RES	Reuse of green waste resources (Shi et al., 2013) Increase the use of RES	Apply the resources evaluation system Provide technological help with RES Encourage tourism industry to use green waste and apply RES (Tsagarakis et al., 2011)

ESI, Energy-saving installation; *RES*, renewable energy resources.

segmentation requires a grounded understanding of tourists through their behaviors, individual characteristics, attitudes, preferences, and even habits and lifestyles. The main policies on tourism suppliers are to encourage them to provide multiple choices of tourism products/routes and adopt ESI and RES in operation. The similar strategies of applying new technologies are also useful for tourism organizations and governments. In addition, more policies using infrastructure, facilities, and regulations to achieve energy-saving and sustainable development goals are advocated among organizations and governments.

13.5 Future directions toward tourism energy-related problems

Although the total number of studies related to tourism energy consumption is not small and research topics are broad, the behaviors of various stakeholders have not been analyzed/compared in a consistent way in association with tourism energy consumption. To generate a relatively general picture of various behavioral research gaps, the conceptual framework in Fig. 13.1 was proposed. Eventually, four main future research directions related to tourism energy consumption can be summarized as follows:

13.5.1 Research on individual energy-related behavior and benchmark

Literature review suggests that individual energy-related behavior has been underresearched, which should be given more attention in the future. For tourist behavior research, both revealed and stated preferences as well as socio-psychological factors motivating behavioral changes toward an energy-saving tourism lifestyle should be further examined. Studies based on stated preference approaches should be deepened, while actual tourist behavior and the hidden preference behind should not be neglected either, because they are important to identify important factors encouraging and discouraging behavioral changes for reducing energy use. Behavioral interventions should be given a similar importance as technologies have been given, especially considering the existence of various rebound effects and the importance of achieving a long-lasting sustainable lifestyle. In this regard, tourist segmentation will become very important, not only for successful marketing but also for achieving environmentally sustainable tourism development. Such a segmentation should be conducted by looking at not only tourism visiting behaviors, but also energy-related consumption behaviors as well as objective and subjective factors affecting these behaviors. Furthermore, the influences of tourist decisions and choices on energy use should be investigated by associating with different tourism subsectors in a consistent and systematic way.

For tourism suppliers, two aspects should be emphasized in future research. First, it is necessary to investigate not only the hotel sector and tourist attractions, but also other various suppliers, such as small-/medium-sized retailers, especially in developing countries. Small-/medium-sized retailers play important roles not only in bringing tourism benefits to local residents and communities, for example, by employing local residents in tourism business operation, but also in creating local-specific and culture-oriented tourism services, which are crucial to socially sustainable tourism and regional development. Second, it is necessary to establish energy consumption benchmarks or evaluation systems. Energy consumption is a good parameter to quantify environmental problems. But if there is no standardized benchmark of energy consumption, it is difficult to compare destinations and propose effective policy strategies. Aimless policy goals will not be helpful to any stakeholder. Such benchmarks or evaluation systems should pay careful attention to specific features of tourism subsectors.

Related to organizations and government, measuring the impacts of various tourism policies on energy use and destination development will still be an important research focus in the future, considering the importance of monitoring in the whole policymaking process. However, it has remained unclear how organizational and governmental behaviors affect tourists' and other stakeholders' energy consumption. For this, it has been difficult to collect relevant data from organizations and government, and it is also a challenge to develop sound theories reflecting true decision-making mechanisms by relaxing various unrealistic assumptions made in existing studies.

13.5.2 Energy tourism: an emerging field

Tourism has come to the forefront in promoting innovative clean energy solutions, such as applying ESI in accommodation facilities and use of RES (Michalena and Tripanagnostopoulos, 2010). Energy facilities and landscape can also become tourism attractions. A new niche of industrial tourism, called energy tourism, has been just raised (Frantál and Urbánková, 2017). A specific segment of tourists are attracted to such sites. Many aspects of energy tourism remain unknown. For instance, what is the position of energy tourism in the whole tourism market? What are the motivations and characteristics of tourists visiting such energy sites? How about the future perspectives of such a new energy tourism market?

13.5.3 Building systematic and multidisciplinary approaches for tourism energy research

Tourism energy consumption occurs at all stages of the whole tourism participation process, which is supported by various firms, organizations, local communities, and government from both the tourism sector and other relevant sectors. To address the various issues of tourism energy consumption, systematic and multidisciplinary approaches are required, where behaviors of various stakeholders and dynamic interactions between stakeholders as a multimarket demand—supply system at both micro and macrolevels should be reflected, closely linking with major tourist activities at different locations. For this, Fig. 13.1 presents just a rough sketch for starting to draw the whole picture. Furthermore, behavioral research is surely important, but it is just a part of the whole tourism and energy system. Systematic approaches should also be built by considering mechanisms of the whole system by making full use of multidisciplinary wisdoms.

References

Ali, Y., Mustafa, M., Al-Mashaqbah, S., Mashal, K., Mohsen, M., 2008. Potential of energy savings in the hotel sector in Jordan. Energy Convers. Manage. 49 (11), 3391–3397.

Beccali, M., La Gennusa, M., Coco, L.L., Rizzo, G., 2009. An empirical approach for ranking environmental and energy saving measures in the hotel sector. Renew. Energy 34 (1), 82–90.

Becken, S., Patterson, M., 2006. Measuring national carbon dioxide emissions from tourism as a key step towards achieving sustainable tourism. J. Sustain. Tourism 14 (4), 323–338.

Becken, S., Simmons, D.G., 2002. Understanding energy consumption patterns of tourist attractions and activities in New Zealand. Tourism Manage. 23 (4), 343–354.

Becken, S., Simmons, D.G., 2005. Tourism, fossil fuel consumption and the impact on the global climate. Tourism Recreation Clim. Change 192–206.

Becken, S., Simmons, D., Frampton, C., 2003a. Segmenting tourists by their travel pattern for insights into achieving energy efficiency. J. Travel Res. 42 (1), 48–56.

Becken, S., Simmons, D.G., Frampton, C., 2003b. Energy use associated with different travel choices. Tourism Manage. 24 (3), 267–277.

Bohdanowicz, P., Churie-Kallhauge, A., Martinac, I., Rezachek, D., 2001. Energy-efficiency and conservation in hotels—towards sustainable tourism. In: The 4th International Symposium on Asia Pacific Architecture, Hawai'I, April 5–7.

Bornhorst, T., Ritchie, J.B., Sheehan, L., 2010. Determinants of tourism success for DMOs & destinations: an empirical examination of stakeholders' perspectives. Tourism Manage. 31 (5), 572–589.

Brida, J.G., Zapata-Aguirre, S., 2009. Cruise tourism: economic, socio-cultural and environmental impacts. Int. J. Leisure Tourism Mark. 1 (3), 205–226.

Cernat, L., Gourdon, J., 2012. Paths to success: benchmarking cross-country sustainable tourism. Tourism Manage. 33 (5), 1044–1056.

Chan, W.W., Lam, J.C., 2003. Energy-saving supporting tourism sustainability: a case study of hotel swimming pool heat pump. J. Sustain. Tourism 11 (1), 74–83.

Cheng, Q., Su, B., Tan, J., 2013. Developing an evaluation index system for low-carbon tourist attractions in China—a case study examining the Xixi wetland. Tourism Manage. 36, 314–320.

Chou, C.J., Chen, P.C., 2014. Preferences and willingness to pay for green hotel attributes in tourist choice behavior: the case of Taiwan. J. Travel Tourism Mark. 31 (8), 937–957.

Dalton, G.J., Lockington, D.A., Baldock, T.E., 2007. A survey of tourist operator attitudes to renewable energy supply in Queensland, Australia. Renew. Energy 32 (4), 567–586.

Dalton, G.J., Lockington, D.A., Baldock, T.E., 2009. Case study feasibility analysis of renewable energy supply options for small to medium-sized tourist accommodations. Renew. Energy 34 (4), 1134–1144.

Day, J., Cai, L., 2012. Environmental and energy-related challenges to sustainable tourism in the United States and China. Int. J. Sustain. Dev. World Ecol. 19 (5), 379–388.

Do Valle, P.O., Pintassilgo, P., Matias, A., André, F., 2012. Tourist attitudes towards an accommodation tax earmarked for environmental protection: a survey in the Algarve. Tourism Manage. 33 (6), 1408–1416.

Dubois, G., Ceron, J.P., 2006. Tourism and climate change: proposals for a research agenda. J. Sustain. Tourism 14 (4), 399–415.
Erdogan, N., Baris, E., 2002. Environmental protection programs and conservation practices of hotels in Ankara, Turkey. Tourism Manage 28 (2), 604–614.
Erdogan, N., Tosun, C., 2009. Environmental performance of tourism accommodations in the protected areas: case of Goreme Historical National Park. Int. J. Hosp. Manage. 28 (3), 406–414.
Frantál, B., Urbánková, R., 2017. Energy tourism: an emerging field of study. Curr. Issues Tourism 20 (13), 1395–1412.
Gössling, S., 2000. Sustainable tourism development in developing countries: some aspects of energy use. J. Sustain. Tourism 8 (5), 410–425.
Gössling, S., Hansson, C.B., Hörstmeier, O., Saggel, S., 2002. Ecological footprint analysis as a tool to assess tourism sustainability. Ecol. Econ. 43 (2–3), 199–211.
Gössling, S., Peeters, P., Ceron, J.P., Dubois, G., Patterson, T., Richardson, R.B., 2005. The eco-efficiency of tourism. Ecol. Econ. 54 (4), 417–434.
Gössling, S., Broderick, J., Upham, P., Ceron, J.P., Dubois, G., Peeters, P., et al., 2007. Voluntary carbon offsetting schemes for aviation: efficiency, credibility and sustainable tourism. J. Sustain. Tourism 15 (3), 223–248.
Holding, D.M., 2001. The Sanfte Mobilitaet project: achieving reduced car-dependence in European resort areas. Tourism Manage. 22 (4), 411–417.
Hunter, C., 1997. Sustainable tourism as an adaptive paradigm. Ann. Tourism Res. 24 (4), 850–867.
Hunter, C., Shaw, J., 2007. The ecological footprint as a key indicator of sustainable tourism. Tourism Manage. 28 (1), 46–57.
Høyer, K.G., 2000. Sustainable tourism or sustainable mobility? The Norwegian case. J. Sustain. Tourism 8 (2), 147–160.
Janusz, G.K., Bajdor, P., 2013. Towards to sustainable tourism–framework, activities and dimensions. Procedia Econ. Finance 6, 523–529.
Karagiorgas, M., Tsoutsos, T., Drosou, V., Pouffary, S., Pagano, T., Lara, G.L., et al., 2006. HOTRES: renewable energies in the hotels. An extensive technical tool for the hotel industry. Renew. Sustain. Energy Rev. 10 (3), 198–224.
Kasim, A., 2007. Towards a wider adoption of environmental responsibility in the hotel sector. Int. J. Hosp. Tourism Adm. 8 (2), 25–49.
Katircioglu, S.T., 2014. International tourism, energy consumption, and environmental pollution: the case of Turkey. Renew. Sustain. Energy Rev. 36, 180–187.
Katircioglu, S.T., Feridun, M., Kilinc, C., 2014. Estimating tourism-induced energy consumption and CO_2 emissions: the case of Cyprus. Renew. Sustain. Energy Rev. 29, 634–640.
Kelly, J., Williams, P.W., 2007. Modelling tourism destination energy consumption and greenhouse gas emissions: Whistler, British Columbia, Canada. J. Sustain. Tourism 15 (1), 67–90.
Kelly, J., Haider, W., Williams, P.W., 2007. A behavioral assessment of tourism transportation options for reducing energy consumption and greenhouse gases. J. Travel Res. 45 (3), 297–309.
Kostakis, I., Sardianou, E., 2012. Which factors affect the willingness of tourists to pay for renewable energy? Renew. Energy 38 (1), 169–172.
Lane, B., 1994. Sustainable rural tourism strategies: a tool for development and conservation. J. Sustain. Tourism 2 (1–2), 102–111.
Leiper, N., 1979. The framework of tourism: towards a definition of tourism, tourist, and the tourist industry. Ann. Tourism Res. 6 (4), 390–407.
Liu, J., Feng, T., Yang, X., 2011. The energy requirements and carbon dioxide emissions of tourism industry of Western China: a case of Chengdu city. Renew. Sustain. Energy Rev. 15 (6), 2887–2894.

Lumsdon, L., 2000. Transport and tourism: cycle tourism—a model for sustainable development? J. Sustain. Tourism 8 (5), 361–377.
Luo, J., Zhang, M., 2011. Route Choice of low-carbon industry for global climate change: an issue of China tourism reform. Energy Procedia 5, 2283–2288.
Michalena, E., Tripanagnostopoulos, Y., 2010. Contribution of the solar energy in the sustainable tourism development of the Mediterranean islands. Renew. Energy 35 (3), 667–673.
Michalena, E., Hills, J., Amat, J.P., 2009. Developing sustainable tourism, using a multicriteria analysis on renewable energy in Mediterranean Islands. Energy Sustain. Dev. 13 (2), 129–136.
Mill R.C. and Morrison A.M., The Tourism System, 2002, Kendall Hunt.
Nepal, S.K., 2008. Tourism-induced rural energy consumption in the Annapurna region of Nepal. Tourism Manage. 29 (1), 89–100.
Palhares, G.L., 2003. The role of transport in tourism development: nodal functions and management practices. Int. J. Tourism Res. 5 (5), 403–407.
Reilly, J., Williams, P., Haider, W., 2010. Moving towards more eco-efficient tourist transportation to a resort destination: the case of Whistler, British Columbia. Res. Transp. Econ. 26 (1), 66–73.
Rinzin, C., Vermeulen, W.J., Glasbergen, P., 2007. Ecotourism as a mechanism for sustainable development: the case of Bhutan. Environ. Sci. 4 (2), 109–125.
Ritchie, J.B., Crouch, G.I., 2003. The Competitive Destination: A Sustainable Tourism Perspective. Cabi.
Saito, H., Ruhanen, L., 2017. Power in tourism stakeholder collaborations: power types and power holders. J. Hosp. Tourism Manage. 31, 189–196.
Sheng, Y., Miao, Z., Zhang, J., Lin, X., Ma, H., 2018. Energy consumption model and energy benchmarks of five-star hotels in China. Energy Build. 165, 286–292.
Shi, C., Peng, J., 2011. Construction of low-carbon tourist attractions based on low-carbon economy. Energy Procedia 5, 759–762.
Shi, Y., Du, Y., Yang, G., Tang, Y., Fan, L., Zhang, J., et al., 2013. The use of green waste from tourist attractions for renewable energy production: the potential and policy implications. Energy Policy 62, 410–418.
Sloan, P., Legrand, W., Tooman, H., Fendt, J., 2009. Best practices in sustainability: German and Estonian hotels. Advances in Hospitality and Leisure. Emerald Group Publishing Limited, pp. 89–107.
Stohl, A., 2008. The travel-related carbon dioxide emissions of atmospheric researchers. Atmos. Chem. Phys. 8 (21), 6499–6504.
Tabatchnaia-Tamirisa, N., Loke, M.K., Leung, P., Tucker, K.A., 1997. Energy and tourism in Hawaii. Ann. Tourism Res. 24 (2), 390–401.
Tang, C.F., Abosedra, S., 2014. The impacts of tourism, energy consumption and political instability on economic growth in the MENA countries. Energy Policy 68, 458–464.
Teng, C.C., Horng, J.S., Hu, M.L.M., Chien, L.H., Shen, Y.C., 2012. Developing energy conservation and carbon reduction indicators for the hotel industry in Taiwan. Int. J. Hosp. Manage. 31 (1), 199–208.
Tsagarakis, K.P., Bounialetou, F., Gillas, K., Profylienou, M., Pollaki, A., Zografakis, N., 2011. Tourists' attitudes for selecting accommodation with investments in renewable energy and energy saving systems. Renew. Sustain. Energy Rev. 15 (2), 1335–1342.
Tsai, K.T., Lin, T.P., Hwang, R.L., Huang, Y.J., 2014. Carbon dioxide emissions generated by energy consumption of hotels and homestay facilities in Taiwan. Tourism Manage. 42, 13–21.
Van Truong, N., Shimizu, T., 2017. The effect of transportation on tourism promotion: literature review on application of the Computable General Equilibrium (CGE) Model. Transp. Res. Procedia 25, 3096–3115.

World Travel & Tourism Council (WTTC), 2018. <https://www.wttc.org/priorities/sustainable-growth/> (accessed 01.05.18).
Wu, P., Han, Y., Tian, M., 2015. The measurement and comparative study of carbon dioxide emissions from tourism in typical provinces in China. Acta Ecol. Sin. 35 (6), 184–190.
Zaman, K., Shahbaz, M., Loganathan, N., Raza, S.A., 2016. Tourism development, energy consumption and Environmental Kuznets Curve: trivariate analysis in the panel of developed and developing countries. Tourism Manage. 54, 275–283.
Zhang, L., Gao, J., 2016. Exploring the effects of international tourism on China's economic growth, energy consumption and environmental pollution: evidence from a regional panel analysis. Renew. Sustain. Energy Rev. 53, 225–234.
Zhao, X., Ma, C., Gu, P., 2012. Energy saving methods and results analysis in the hotel. Energy Procedia 14, 1523–1527.
Zografakis, N., Gillas, K., Pollaki, A., Profylienou, M., Bounialetou, F., Tsagarakis, K.P., 2011. Assessment of practices and technologies of energy saving and renewable energy sources in hotels in Crete. Renew. Energy 36 (5), 1323–1328.

CHAPTER 14

Behavioral interventions for sustainable transportation: an overview of programs and guide for practitioners

Reuven Sussman[1,*], Li Qin Tan[2] and Christine E. Kormos[3]
[1]Behavior and Human Dimensions Program, American Council for an Energy-Efficient Economy, Washington, DC, United States
[2]University of Victoria, Victoria, BC, Canada
[3]Sustainable Transportation Action Research Team, Simon Fraser University, Burnaby, BC, Canada

14.1 Introduction

The global transportation sector is a major contributor to worldwide greenhouse gas emissions, accounting for 28% of global final energy demand and 23% of global energy-related CO_2 emissions in 2014 (Masson-Delmotte et al., 2018). Although physical efficiency upgrades and structural infrastructure changes can reduce the impact of vehicle emissions, human behavior is the underlying root cause of the problem and, as such, is an important part of the solution. Behavior encompasses adoption of technology, driving habits, travel mode decisions, and more. Ultimately, even if technology can automate human transportation, humans design and demand those technologies. Thus, transportation efficiency is as much a human behavior problem as it is an engineering problem.

Social scientists in fields such as environmental psychology and behavioral economics have examined human behavior in the context of environmental sustainability for over four decades. Researchers in these fields have applied established theories of behavior change to conduct field experiments that encourage people to change their transportation behaviors. In recent years, this theoretical work has taken on a more central role in the design of real-world large-scale interventions. For example, behavior change research was highlighted as a key strategy for energy

*Corresponding author

Transport and Energy Research
DOI: https://doi.org/10.1016/B978-0-12-815965-1.00014-4

reduction by the International partnership for energy efficiency cooperation (IPEEC) which advises the G20.[1]

This chapter begins with a description of potential target behaviors that can be changed with the help of behavioral science. In the second section, we explain how theories of behavior change can inform effective interventions, with field research that supports each theory. The third section describes the examples of real-world programs and strategies that have been used to change behavior in several transportation behavior domains. The concluding section is a practical guide for practitioners, which describes "DRIVE"—the procedure for designing, implementing, and evaluating transportation behavior change programs.

14.2 Behaviors to target

An effective transportation behavior change program should start by specifying the target behavior to change. Clear and specific definitions allow the program implementer to more easily measure success and increase the likelihood of being successful.

Choosing a target behavior can be hard because it requires balancing the difficulty of changing the behavior, with the impact of doing so, and the feasibility of implementing an intervention to do it. Transportation-related behaviors, overall, tend to have high impacts on greenhouse gas emissions, relative to other behaviors such as daily energy use actions, solid waste disposal and recycling, or water use. However, the behaviors are also notoriously difficult to change because they are fundamental to peoples' lifestyles and can be costly (in terms of money, time, and effort). For some transportation behaviors the opportunity to intervene comes extremely infrequently, and for others, the behaviors are deeply ingrained habits. Generally, the most effective transportation behaviors to target are the so-called low-hanging fruit; actions that are easiest to change with the highest relative impact. Nevertheless, even hard-to-change behaviors can be targeted, depending on the population and goal of the intervention.

Each type of target behavior must be approached differently, and behavioral science can provide some insights into which programs may be more or less likely to work with each behavior. Nevertheless, interventions should be pilot tested and revised before large-scale implementation. No single intervention will work successfully for all behaviors among all

[1] https://ipeec.org/cms/28-events-.html.

target populations. In this section, we describe the types of behaviors that may be targeted by transportation behavior change programs. In the later sections, we describe some programs that may be used to change these behaviors.

14.2.1 Two transportation behavior domains

Transportation-related greenhouse gasses are emitted by two types of actors: (1) personal and (2) professional.

Personal transportation refers to short- or long-distance human travel, which includes personal travel from home to work, school, or any other local trip, as well as long-haul personal trips. Road vehicle transportation accounts for 83% of transportation greenhouse gas emissions in the United States (Environmental Protection Agency [EPA], 2018). People engaging in this type of travel can use a variety of short-distance modes of transportation, including, driving, walking, busing, biking, or carpooling. They may also use long-distance modes such as airplanes, cruise ships, boats, or trains.

Professional transportation refers to the use of vehicles by operators who transport goods or people for their job or career. This transportation category includes pilots, bus drivers, truck drivers, taxi drivers, boat captains, and others with similar duties. The behavior of this group is sometimes similar to personal transportation travelers, however, the opportunities and options for changing behavior are distinctly different. Individuals in this category may operate vehicles directly or may be responsible for a fleet of vehicles. Although relatively few individuals are professional "transporters," their behavior can have a disproportionately high impact on energy use and greenhouse gas emissions. Freight transport emissions are expected to rise from 42% in 2010 to 60% by 2050 (OECD/ITF, 2015). International freight transport is responsible for 33% of worldwide trade-related emissions and more than 75% of emissions from major manufacturing categories (Cristea et al., 2013).

14.2.2 Three types of behaviors

Within each type of transportation domain described earlier, policymakers may wish to target any of the following three types of behaviors: (1) efficient use of chosen transportation options, (2) purchase of efficient vehicles or upgrades, and (3) transportation mode switching (or overall transportation reduction).

Efficient use of chosen transportation options refers to encouraging commuters to travel as efficiently as possible without changing their mode of travel.[2] Behavior change interventions targeting this type of behavior may encourage drivers to accelerate and brake more smoothly, or remove roof racks that create drag. Often these behaviors are habitual or somewhat frequent and, as such, may allow for many opportunities for change. The goal may not be to change the behavior every time it occurs, but rather more often than it currently occurs. However, these behaviors may also have a smaller greenhouse gas impact than changing transportation modes or vehicle-purchasing decisions (Gardner and Stern, 2008). The additive individual impacts can have an important overall effect only if a large group of commuters engages in the behaviors frequently.

Purchase of efficient vehicles or upgrades refers to encouraging people to choose the most efficient option that is available. It can also refer to encouraging vehicle owners to purchase upgrades that increase their vehicle's efficiency or to discourage vehicle purchase entirely. A behavior change intervention targeting this type of behavior might encourage new car buyers to purchase electric vehicles, a behavior that reduces greenhouse gas emissions in most regions of the world (Anair and Mahmassani, 2012) or existing freight truck drivers to purchase a bolt-on vehicle upgrade that increases efficiency. It may also refer to interventions that convince young people to refrain from purchasing vehicles. Vehicle-purchasing decisions generally occur infrequently and are difficult to change, but they can have high impacts because vehicles may remain in use for years after their initial purchase.[3] Vehicle upgrades may have a smaller impact and may be easier to change than initial purchases, but they also occur relatively infrequently.

Transportation mode switching and overall reduction of travel refer to encouraging travelers or freight moving companies to choose a more efficient mode of transportation than they are currently using or choosing simply to travel less often or shorter distances. Behavior change interventions targeting this type of behavior may encourage drivers to use a bus or bike to get to work. They may encourage courier companies to hire cyclists for short trips rather than cars. They often have either the goal of

[2] We refer to "efficient use" as solely from a driver or operator perspective. The term could also be used to refer to efficient use of existing infrastructure (e.g., maximizing use of streets by cars and trucks), but this is not how we use the term in this case.

[3] The average age of light cars and trucks in the United States is about 11.6 years (U.S. Department of Transportation, 2017).

reducing one type of behavior (e.g., single-occupancy car trips), increasing another type of behavior (e.g., walking or cycling), or both. These behaviors can occur frequently (e.g., commuting to work) or infrequently (e.g., flying long distances) and vary in impact.

14.2.3 A matrix of potential target behaviors

Table 14.1 shows the range of potential behaviors that may be targeted by behavior change programs. It is not a comprehensive list of all possible interventions, but rather a sample of programs that have demonstrated potential and good research support. By far, the most commonly targeted transportation behaviors are personal travel. Within this domain, encouraging change of transportation modes or efficient use of vehicles is most common. However, given the potentially high impact of changing long-haul personal transportation behaviors and professional transportation behaviors, these may be excellent domains for behavior researchers to target in the future. An eco-driving program for truck drivers may only need to be implemented with a small number of total drivers to have a relatively large effect on greenhouse gas emissions. Similarly, other understudied areas can have potentially large impacts. In the section 14.4, we discuss several (of the large number of potential) examples of successful programs.

14.2.4 Bigger picture behaviors: activities and lifestyle

Although this chapter focuses on fuel-efficient transportation choices that people can use to maintain their lifestyles, some researchers focus on the bigger picture: changing lifestyle itself. Large-scale lifestyle changes can have high impacts but are challenging to implement and, in some cases, take a long time to achieve. For example, choosing to live close to one's workplace or choosing to live a simpler life with fewer trips, activities, and travels may have strong positive impacts on greenhouse gas emissions, but the opportunities for these changes are infrequent, and the trade-offs required to make them are often unacceptable by travelers. In this chapter, we touch briefly on some initiatives that governments or businesses can take to encourage these major lifestyle shifts (e.g., offering remote work options or planning cities to encourage living near the workplace), but many of these lifestyle changes are voluntary. To the best of our knowledge, no structured programs promoting simplified lifestyle choices have been studied. Preliminary research into the "voluntary simplicity"

Table 14.1 Potential transportation behaviors to target, with sample programs.

	Efficient use of vehicles	Purchase behavior	Mode switching or reducing travel
Personal transportation	Eco-driver training Eco-driver feedback devices/apps Pay-as-you-speed (variant of pay-as-you-drive) Idling laws Environmental design of cities and streets Carpooling programs	Monetary incentives HOV lane access for electric vehicles Vehicle labeling Free trials	Guaranteed ride home Parking cash out Congestion pricing to reduce SOV travel Street layout Free bus passes Bicycle infrastructure Bicycle facilities at work Alternative work options Carpooling programs
Professional transportation	Eco-driver/Pilot feedback Eco-driver training Eco-driver incentives or rewards	Resolve split-incentive (potential intervention) Monetary incentives	Encourage mode switching to bicycle or tricycle for "last mile" distribution (potential intervention) Encourage online shoppers to accept longer shipping times (potential intervention)

HOV, High-occupancy vehicle; *SOV*, single-occupancy vehicle.

movement suggests that, in some cases, living simply may be associated with higher life satisfaction (Boujbel and d'Astous, 2012). This promising work may provide an impetus for future research to tackle lifestyle simplification as a method for changing travel behavior.

14.3 Why behavior and behavioral science matters

Models of behavior change are a useful guide in the design of effective interventions. Researchers have applied various established models in their attempt to change behaviors. In this section, we introduce three established models of behavior change, as well as the concept of habits, and illustrate how they have guided applied research pertaining to sustainable transportation behaviors. Notably, many programs reviewed later in this chapter were not developed with these models in mind, or even for the purpose of reducing greenhouse gas emissions. Nevertheless, they could be deployed as part of a theory-based approach to emissions reduction by creative program designers seeking to address transportation behavior issues.

14.3.1 Transtheoretical model of change

14.3.1.1 About the model

The transtheoretical model (TTM) of change outlines a series of stages that people go through when they intentionally try to change their behavior (Prochaska and DiClemente, 1982). The TTM assumes that behavioral change involves a process, which gradually occurs when people are ready. The TTM's various stages of change were initially used to guide clinical therapy but are now widely applied to other types of behavioral change. There are six stages of change: (1) precontemplation, (2) contemplation, (3) preparation, (4) action, (5) maintenance, and (6) termination (Prochaska and Velicer, 1997). People go through these stages of change over time, progressing or even regressing from one stage to another. The further along one progress on the stages, the higher the likelihood of behavioral change occurring.

Precontemplation refers to the stage before change is considered. At this stage, people are not ready to make any changes to their behavior for at least 6 months. They could be unaware or ambivalent about changing their behavior.

Contemplation refers to the stage where people are getting ready to change their behavior in the next 6 months. Contemplators are aware of the pros and cons of changing their behavior and are weighing them to decide whether to act on their behavior. Contemplators are, true to the name of this stage, contemplating change and are, therefore, unable to commit to change immediately.

Those in the preparation stage are ready to change their behavior in the near future (i.e., next month). Typically, these people would have taken some step toward the direction of change and have a plan to implement changes to their behavior.

People in the action stage are those who have made observable behavioral changes in the last 6 months and are intending to continue with the modifications to their behavior. Those in the maintenance stage have sustained their behavioral changes for at least 6 months and are actively maintaining the changes to prevent regression to the earlier stages. Finally, those in the termination stage have successfully changed their behavior and have little chance of regressing to earlier stages. However, the termination stage is rarely reached, with most staying in the maintenance stage.

14.3.1.2 Application of the model

The TTM has been applied to transportation behavior in promoting sustainable behavioral change. Although the TTM is not a theory per se and does not predict outcomes, it is useful for identifying the level of readiness for change. In turn, practitioners can apply suitable interventions to increase participants' readiness for change.

To illustrate a simple intervention using the TTM was used to increase the readiness of participants in the precontemplation and contemplation stages to work on their transportation behavior (Mundorf et al., 2018). Utilizing a short video intervention targeting the precontemplation and contemplation stages, participants' attitudes toward using sustainable transportation were improved, and they displayed increased readiness to change their transportation behavior.

The provision of personalized feedback can help move people from one stage of change to the next. One study showed that 30% of participants who received personalized feedback regarding their travel behavior moved from the precontemplation stage to the contemplation stage, whereas those from the control group remained in the precontemplation stage (De Nazelle, 2016).

In a study examining the decision-making process of choosing cycling rather than driving as the preferred travel mode, differences were found in the perceived consequences of behavioral change between the TTM stages (Forward, 2014). Although participants largely agreed that cycling is better for the environment and their health as compared to driving, their perception of the benefits of cycling became more positive the further along they were on the TTM stages.

A recent review of transportation research using the TTM revealed that proper reporting was not done in a number of studies, hence limiting the ability to evaluate intervention effectiveness across different studies (Friman et al., 2017). Despite these limitations, the review demonstrated that some positive changes in transportation behavior occurred as a result of TTM interventions.

14.3.2 Diffusion of innovation theory

14.3.2.1 About the theory

The diffusion of innovation (DOI) theory explains how a new behavior or product spreads through a social system, such that people in that social system start to adopt and accept this new behavior or product (Rogers, 1982). The prerequisite to diffusion is that the new behavior or product has to be perceived as something innovative. The DOI outlines five adopter categories, named according to the timing in which people start to adopt the innovation: (1) innovators, (2) early adopters, (3) early majority, (4) late majority, and (5) laggards. Similar to the TTM, these categories are associated with certain characteristics, but the DOI and TTM are different in that the DOI's categories are distinct, whereas people cycle through the TTM stages.

Innovators are the earliest adopters of innovation, characterized by people who are interested and open to trying something new. Not much is required to get the innovators on board with an innovation. Early adopters are those who understand that change is necessary and are willing to embrace innovations. As this group of people is already motivated to change, they only require instructions on the implementation portion (i.e., a how-to guide) to get started on trying out the new behavior or product. The early majority are willing to adopt an innovation only when convinced that the innovation is effective in achieving a desired outcome. Hence, appealing to this group requires empirical evidence to prove that adopting the innovation is indeed advantageous. The late majority have their doubts about innovation, preferring to wait and observe the previous adopters of innovation before evaluating whether to jump onto the bandwagon. This group requires the success stories of many prior adopters to accept the innovation. The laggards are the last group to adopt innovation. This group is resistant to change and is difficult to persuade.

Strategies to appeal to this group would include normative influences, fear, and evidence of effectiveness.

14.3.2.2 Application of theory

The DOI has been used to examine the adoption of new travel behavior, in particular the use of public bikeshare programs. Public bikeshare programs are increasing in popularity, with more cities implementing them each year. The success of these programs is dependent on the uptake rate; thus cities keen to launch such programs would be interested to understand how to facilitate the uptake rate.

A study conducted in Vancouver, Canada, used the DOI as a guide to learn the public's opinion before the implementation of a public bikeshare program in the city (Therrien et al., 2014). Through a telephone survey, approximately 70% of respondents were supportive of the launch of the public bikeshare program, though only around 27% of respondents stated that they intended to use the program. This information, along with the DOI, was used to develop suggestions to increase the program's uptake rate, such as ensuring that the barriers to using the program are low. Many studies of electric vehicle purchasing also use this model to guide their efforts.

14.3.3 Theory of planned behavior

14.3.3.1 About the theory

The theory of planned behavior (TPB) states that behavior is predicted by behavioral intention, which is in turn predicted by three variables—attitudes, subjective norms, and perceived behavioral control (Ajzen, 1991). Attitudes refer to one's perception of the target behavior, which can be positive or negative. Subjective norms refer to the perception of the target behavior of important others, such as close family and friends. Perceived behavioral control refers to one's perceived ability to perform the target behavior. In general, the more positive the attitude toward the behavior, the stronger the subjective norms, and the higher the perceived control over the behavior, the stronger the intention to change one's behavior. This, in turn, predicts the likelihood of behavior change occurring.

14.3.3.2 Application of theory

The TPB has been successfully applied to a range of settings and travel behaviors, in some instances explaining up to nearly half of the variance of intention to use certain travel modes (e.g., Eriksson and Forward, 2011;

Zailani et al., 2016). In three major cities in China the TPB was applied in an attempt to reduce car use (Liu et al., 2017). Using a survey methodology involving 600 participants, the TPB variables—attitude toward car-transport reduction, perceived norm of car-transport reduction, and perceived behavioral control over car-transport reduction—were found to have significantly increased participants' intention to reduce car use. In Canada the TPB variables successfully predicted the use of public transport following the introduction of a university pass (i.e., a bus pass) for students (Heath and Gifford, 2002). Through a pre–post longitudinal study design a significant increase in the use of buses was observed, accompanied by a significant decline in the use of cars and bicycles.

The TPB can be further used to explain why some choose to take public transport rather than drive. An interesting study was done with participants who own cars and were intending to make an unfamiliar journey (Farag and Lyons, 2010). This particular study found that out of the three TPB variables, attitudes toward the usefulness of looking up public transport information largely accounted for the intention to take public transport instead of driving for an unfamiliar journey. Specifically, the more useful participants perceived public transport information to be, the stronger their intention to use public transport.

Apart from being used to examine travel mode choice, the TPB was also used to guide research on the purchase of energy-efficient vehicles. All three TPB variables significantly predicted the intention to adopt hybrid electric vehicles (HEVs) in a study conducted in China, with attitudes being mediated by environmental concern (Wang et al., 2016).

Note that studies typically measure behavioral intention but not actual behavior, despite known inconsistencies between the two (Kormos and Gifford, 2014). Given that the link between attitudes (i.e., behavioral intention) and behavior is not always clear; results of studies that measure solely behavioral intention without actual behavior should be interpreted with caution.

14.3.4 Habits and transportation behavior

The TTM and the TPB are models that generally assume that human actions are deliberate. Thus, if any change in behavior was to occur, careful consideration would have preceded the behavior change. In a way, habits are on the opposite end of such deliberate change in behavior, as habits refer to actions that occur without much thought given to them

(Verplanken and Aarts, 1999). Habits are closely related to transportation behavior, especially when transportation behavior occurs in a familiar environment (Thogersen and Moller, 2008). In a familiar environment, our commute from place to place forms part of a daily, habitual routine. Hence, once we settle into a routine mode of commuting, we are unlikely to think much about our transportation behavior.

Habits have a negative effect on the intention to switch travel mode. A study conducted in Taiwan found that those who habitually use private transport (motorcyclists and car drivers) were resistant to changing their travel mode to public transport (Chen and Chao, 2011). In addition, car drivers were more resistant to switching to public transport compared to motorcyclists, suggesting something unique about the perceived advantages of car driving.

As such, timing is crucial when attempting to encourage behavior change. As suggested by the habit discontinuity hypothesis (Verplanken and Roy, 2016), intervening at an appropriate juncture could increase its success rate. For instance, participants who received an intervention containing useful information on the public transport system shortly after relocating to a city in Germany did significantly increase their use of public transport, while a significant decline in private car use was observed (Bamberg, 2006).

Although habits, such as mental shortcuts, are useful in reducing our cognitive load, they inadvertently create a resistance to the unfamiliar. This resistance is especially pronounced in the adoption of new technologies, such as electric vehicles. A study examining the barriers to adoption of electric vehicles found that technology-savvy consumers are most likely to be early adopters of electric vehicles, but only if electric vehicles are perceived to outperform conventional vehicles (Egbue and Long, 2012). Perceptions may not necessarily reflect actual differences in performance. Previously formed transportation habits may have contributed to this perception and resistance to electric vehicles.

14.4 Sample intervention programs for different behaviors

14.4.1 Personal travel

This is the most commonly studied transportation behavioral domain. Programs and strategies in this domain generally aim to change the travel behavior for frequent commutes, such as those for work or school. Efficient-use and mode-switching programs are sometimes referred to as transportation demand management (TDM) programs. Using a behavioral science lens, we discuss a selection of popular TDM programs next, but we recommend that interested readers also consult the online TDM Encyclopedia, which includes dozens of excellent behavior change programs.[4]

14.4.1.1 *Efficient use*

Eco-driving refers to the constellation of postpurchase behaviors that can influence vehicle fuel consumption. Huang et al. (2018) reviewed the available literature on eco-driving programs and produced a detailed description of their effectiveness. They describe eco-driving as including acceleration, route choice, deceleration, driving speed, idling, and other factors. Each of these behaviors can influence fuel consumption of personal vehicles and does not involve vehicle purchase or changing of travel mode. Most eco-driving programs that have been the subject of peer-reviewed research target personal car drivers. They primarily rely on training, onboard technology, incentives, laws, or some combination of those strategies (Huang et al., 2018).

14.4.1.1.1 Training

The effectiveness of eco-driving training varies between approximately 2% and 15% fuel savings, based on evaluations using driving simulators, data loggers, and numerical modeling for car and bus drivers (Huang et al., 2018).

One example of a widespread personal-vehicle training program is ECOWILL. The ECOWILL project was piloted in 13 European countries with the goal of training 10 million new drivers.[5] In 2013, the program was integrated into the standard drivers' license exam for those countries.[6] The program teaches "five golden rules" and "ten silver rules" of eco-driving using both in-class and online methods. Recent Swedish graduates from this program expressed a much more concrete and well-defined understanding of eco-driving than experienced Swedish drivers who did not undergo training (Strömberg et al., 2015). Using a train-the-trainer model, this program is being spread to over 500 instructors across Europe, with the goal of standardizing driver training and

[4] http://www.vtpi.org/tdm/index.php.

[5] https://ec.europa.eu/energy/intelligent/projects/en/projects/ecowill.

[6] https://ec.europa.eu/energy/intelligent/projects/sites/iee-projects/files/projects/documents/ecowill_brochure_for_driving_schools_en.pdf.

improving overall fuel economy.[7] Similar driver training programs have been integrated into new driver courses in China (Wu et al., 2017).

ECOWILL and other training programs also teach eco-driving to experienced drivers of personal vehicles. In general, although training programs are usually found to reduce fuel consumption, they work differently under different conditions. Training appeared to be more effective for city than highway driving (Barla et al., 2017) and for manual than automatic transmission vehicles (Larue et al., 2014; Barla et al., 2017). In two studies, participants did not improve or performed worse after training (Beusen et al., 2009). Courses that offered intensive training (e.g., in-car training or half-day workshops) did not seem to necessarily improve eco-driving more than courses including only explanations of eco-driving (Andrieu et al., 2012; Jeffreys et al., 2018). Furthermore, short-term studies, lasting only a few days (Larue et al., 2014; Rolim et al., 2014; Ho et al., 2015; Rutty et al., 2013), tended to have larger effects than longer term studies that lasted weeks or months (Barla et al., 2017; Degraeuwe and Beusen, 2013). This suggests that the effects of eco-driver training may fade over time, and thus drivers could benefit from occasional "refresher" courses.

14.4.1.1.2 Technology

Devices installed in personal vehicles can advise drivers how to save gas in real time during trips. Evaluations of the effectiveness of these devices vary greatly, with some data modeling projections of real-time visual displays as high as a 30%–37% maximum savings (e.g., Barth and Boriboonsomsin, 2009; Orfila et al., 2015) and as low as 5% actual savings in extended field trials (e.g., Stillwater et al., 2017).

Numerous smartphone apps can be downloaded to provide onboard driver assistance with eco-driving, often using visual displays and verbal suggestions. One such example, evaluated in 2016 (Magaña and Muñoz-Organero, 2016), was called Artemisa. The app combined GPS, vehicle telemetry, and internet information about weather and road conditions to create custom real-time suggestions for drivers. It would offer optimal speeds for certain areas of roads and traffic conditions, places to improve on habitual journeys, and how to avoid traffic that would adversely affect efficiency. It also gamified the experience by presenting results as dollar and CO_2 savings, along with a "driving score" and potential "achievements" to unlock. The effectiveness of the app for saving fuel was

[7] https://ec.europa.eu/energy/intelligent/projects/en/projects/ecowill.

demonstrated in a series of controlled tests. In one comparison of 25 drivers doing 30 laps each (in cities, not a closed track), those with the app, including the gamification feature, reduced their fuel consumption by 11% (4% for those without gamification). Two drivers were asked to drive the same five-lap route for 4 months (five times per week at 8:30 a.m. each day) and both conserved the same amount of gas for the first 3 weeks using the app. However, the driver without the gamification option enabled, began diverging significantly (using more gas) from week four to the end of the simulation.

As with training programs, feedback systems range in effectiveness with different conditions. Aggressive drivers benefit more than mild or normal drivers (Rios-Torres et al., 2015; Kim et al., 2011), and savings are larger in urban congested traffic than on highways (Van der Voort et al., 2001; Barth and Boriboonsomsin, 2009; Kim et al., 2011; Rommerskirchen et al., 2014). Short studies of only 1–2 days tend to have higher savings (about 10%) than longer studies lasting weeks or months (less than 8%, Rolim et al., 2014; Vagg et al., 2013; Caulfield et al., 2014; Stillwater et al., 2017). Real-time feedback is likely to be more effective than delayed feedback (Rolim et al., 2014; Zhao et al., 2015; Dijksterhuis et al., 2015). Many feedback programs are combined with training.

The concern with onboard eco-driving technology is that it could reduce safety if drivers pay too much attention to the eco-driving information rather than the road. However, most studies suggest that eco-driving feedback systems do not actually pose a significant safety hazard (see Huang et al., 2018, for a review). One evaluation suggests that eco-driving feedback may be less distracting than changing CDs or using navigation assistants (Rouzikhah et al., 2013). Continuous visual feedback is likely to be the most effective for changing behavior, but it also requires the most attention. Haptic feedback, delivered through the accelerator pedal, is less distracting but also less effective (Jamson et al., 2015). Haptic feedback also has the disadvantage of usually only being provided through the gas pedal.

14.4.1.1.3 Incentives and punishments

Offering incentives or implementing laws (punishments) to encourage eco-driving can be challenging, but some examples exist. Laws can be used to deter noneco-driving behavior using fines, taxes, or other punishments. For example, Hong Kong implemented an antiidling law in 2011 that prohibits more than 3 minutes of idling in a 60-minute period, punishable by a HK$320 fine.[8] Many cities across the United States have implemented similar laws.[9] Laws against dangerous driving, including behaviors, such as speeding, are ubiquitous. Safe driving and eco-driving share many basic tenets and, as such, these laws help promote eco-driving (to the degree that they are effective). Differences, in how these laws are enforced (e.g., with traffic cameras or on-site police officers) and how frequently they are enforced, can change the degree to which they are effective. In most cases, laws may be effective in deterring a proportion of dangerous driving, but violations often still occur (e.g., speeding), and therefore other approaches may combine with laws in order to maximize the likelihood of safe, eco-friendly, driving.

One interesting market-based approach to incentivizing eco-driving is pay-as-you-drive (PAYD) insurance. With advances in technology and data availability, insurance companies can use new information about drivers, not previously available, to calculate more accurate insurance rates. Rather than base insurance rate information on general trends of whole groups and past accidents of the specific driver being insured, companies may now use GPS and telemetry-based devices, mounted voluntarily in customers' cars to get specific driving behavior information. That information can be collected and analyzed to award lower rates to safe drivers and higher rates to risky drivers (people who drive more are at higher risk for accidents, Litman 2004). Although the primary interest of insurance companies is to properly assess risk of accidents for each individual driver, the indirect bonus is that this monitoring, feedback, and incentivization can lead to reduced driving (up to 8%–12%, Harvey and Deakin, 1998; Adkins, 2004; Balcombe et al., 2004; Litman, 2004; Litman, 2005; Bordoff and Noel, 2008), and safer driving (Dijksterhuis et al., 2015). Given that both of these factors reduce fuel consumption, PAYD could be a promising intervention for reducing fuel consumption of already purchased vehicles.

Given that safe driving and eco-driving share many similarities, we present here one implementation of PAYD that specifically encouraged safer driving, the introduction of PAYD with pay-as-you-speed by five Dutch auto insurers (PAYD could also reduce overall mileage and, as such

[8] https://www.epd.gov.hk/epd/english/environmentinhk/air/prob_solutions/idling_prohibition.html.

[9] https://cleancities.energy.gov/technical-assistance/idlebox/.

could also be a strategy for reducing travel). In this intervention, 141 young drivers agreed to have their vehicles equipped with data loggers in exchange for up to 200 euros off of their annual insurance premiums (delivered as a payment by check, Bolderdijk et al., 2011). Control group participants received the incentive at the end of the year, regardless of their behavior. Intervention participants received up to 30 euros/month for not speeding, 15 euros/month for driving less overall, and 5 euros/month for driving less on weekend nights. The participants in the intervention group reduced the distance they drove above the speed limit by 14%.[10] A later study in a driving simulator suggested that the intervention could be improved by offering real-time onboard feedback (as opposed to website-based delayed feedback) and by monitoring and rewarding other driver actions, such as acceleration, breaking, and harsh cornering (Dijksterhuis et al., 2015). These additional behaviors also contribute to both safety and eco-driving.

14.4.1.1.4 Public campaigns

The US Environmental Protection Agency (EPA) Clean Cities Coalition Network has compiled a set of resources, available online, that facilitate organizers in creating local campaigns to reduce idling.[11] The resources, called *IdleBox Tookit*, use a number of behavioral insights to encourage drivers stop idling. Pledge forms, stickers, and signs use the principles of public commitment and social norms to increase the likelihood that drivers will take action and change the perception of normal behavior. Prewritten press releases and outreach letters also help spread the word about campaigns and fact sheets along with other technical reports arm campaign organizers with the knowledge they need to be persuasive. The resources separately target drivers of personal vehicles, fleet vehicles, and emergency vehicles. Although no formal evaluation of the program exists, the website contains quotes with strong anecdotal evidence of success from several Clean Cities Coalition coordinators. Since its launch in 2013, people viewed the website 22,283 times (Personal communication, Patricia Weikersheimer, January 8, 2019).

[10] No significant difference was found in overall driving. One reason could be that the largest monthly discount was for reducing speeding (30 euros) and a smaller discount was provided for reducing over mileage and weekend night driving (20 euros total). Another explanation is that reducing overall travel requires a different skill than avoiding speeding (planning in advance) and may be more difficult to learn and employ.

[11] https://cleancities.energy.gov/technical-assistance/idlebox/.

14.4.1.1.5 Physical environment

Efficient driving can be encouraged through physical changes in roadways and city design that discourage idling, speeding, acceleration, or deceleration. A variety of strategies can be used to affect these behaviors, but only a few have strong evaluations (Bigazzi and Rouleau, 2017; Luoma and Sivak, 2012).

In a general sense, designing cities to be less spread out, with higher population density, and higher density of pedestrian-scale retail stores will help reduce transportation-related emissions (Luoma and Sivak, 2012; Wang et al., 2014). Creating pedestrian-only urban areas may be among the most promising strategies for reducing emissions (Elvik et al., 2009). However, this may not be an option for cities that are heavily dependent on personal vehicle infrastructure.

For cities not willing or able to create pedestrian-only areas, changes in specific streets and intersections can reduce emissions. For example, replacing intersections controlled by traffic lights with intersections managed by roundabouts reduce CO_2 emissions from acceleration, deceleration, and idling (Várhelyi, 2002). Installing arterial roads in urban areas has some air pollution benefit (Klæboe et al., 2000). In one computer modeling study of a section of Antwerp, researchers found that when an urban area had an arterial road with synchronized traffic lights (allowing traffic to move with fewer stops), carbon emissions reduced by 10% (Madireddy et al., 2011). Unfortunately, like many transportation-related physical measures, installing physical measures that reduce congestion may eventually lead to a rebound effect and more traffic (Elvik et al., 2009).

Sometimes, traffic safety and eco-driving are at odds. Reducing overall traffic speed can reduce emissions as long as speed reduction measures do not increase deceleration and acceleration. So-called traffic calming measures, such as speed bumps, may increase emissions despite reducing average speed, especially among aggressive drivers (Ahn and Rakha, 2009; Ghafghazi and Hatzopoulou, 2014; Ghafghazi and Hatzopoulou, 2015). Traffic calming measures require drivers to drastically reduce their speeds, below a normal commuting speed, and then accelerate. One simulation of an urban Montreal neighborhood estimated that average CO_2 emissions would increase in traffic-calmed areas by an average of 1.5%, and up to as much as 83% in some places (Ghafghazi and Hatzopoulou, 2014). In some cases, these measures would also have the knock-on effect of diverting traffic to other streets that would then also become congested and experience additional emissions (Ghafghazi and Hatzopoulou, 2014).

Speed bumps, which must be traversed at only 5–10 km/h, tend to reduce efficiency and increase emissions more than speed humps that are lower and wider and can be traversed at 25–30 km/h (Ghafghazi and Hatzopoulou, 2014; Molan and Kordani, 2014).

One traffic calming measure that does appear to be effective for both reducing speed and emissions is speed limits. Some evidence suggests that reducing speed limits on both highways (e.g., from 120 to 100 km/h or 100–80 km/h) and on residential streets (50–30 km/h) can reduce carbon emissions (Den Tonkelaar, 1991; Baldasano et al., 2010; Madireddy et al., 2011). These reduced speed limits encourage drivers to slow down smoothly, avoiding decelerations and accelerations, while also preventing traffic jams that require idling. Dynamically changing speed limits using a computer algorithm is an innovative way to use speed limits to reduce emissions and congestion, but again, this strategy may lead to a rebound effect and an overall increase in traffic, mitigating the per-vehicle savings (Wang et al., 2011).

14.4.1.1.6 Carpooling

Encouraging commuters to share rides with others traveling in the same direction can be considered an efficiency behavior or a mode-switching behavior. Although not always the case, under the right conditions, carpooling can reduce emissions by removing cars from the road (Shewmake, 2012). It is an effective strategy for reducing emissions when a suitable public transportation option is not available, but it can be detrimental if commuters choose to switch from public transit or cycling to carpooling (Vanoutrive et al., 2012; Toth, 2015; Menczer, 2007). Strategies for encouraging carpooling usually involve either installing high-occupancy vehicle (HOV) lanes on busy streets, providing workplace-based programs, or using technology to facilitate connections between carpoolers.

HOV lanes encourage carpooling by offering drivers a faster lane if they have more than one occupant in their vehicle. Despite initial some initial negative reactions to HOV lanes (Van Vugt et al., 1996), they are now generally accepted. Recently, some regions have chosen to replace HOV lanes with high-occupancy toll (HOT) lanes. HOT lanes not only allow drivers who have two (or in some cases three) occupants to drive in the lane for free but also allow other drivers to pay a toll to use the lane. In some cases the toll is dynamically priced based on congestion levels. Early evidence suggests that HOT lanes may help move traffic through the area faster but may also be associated with fewer carpoolers and increased emissions (Xu et al., 2017; Burris et al., 2014). One might surmise that this is because drivers have less incentive to carpool (they can use the HOV lane without carpooling), but further research is needed.

As explained in the "Practitioner's Guide" section of this chapter, program designers should consider their target audience before deciding on which behavior change strategy to deploy. Programs encouraging carpooling should target commuters who are most likely to engage in the behavior. Households with lower incomes and education (two related factors), without children, or with multiple working adults are slightly more likely than others to carpool (Buliung et al., 2010; Canning et al., 2010; Ferguson, 1997; Teal, 1987). The ideal candidate for a carpool program is a female with full-time employment and a vehicle (Neoh et al., 2017). Carpool programs should address financial motives (rather than environmental concerns) as well as the barriers of fear (of strangers), lack of privacy, insurance concerns, and lack of solitary control over personal space in the vehicle (Gardner and Abraham, 2007; Canning et al., 2010).

Workplace programs can encourage carpooling in several ways. Given that subjective norm and corporate culture influence travel behavior (e.g., Bonham and Koth, 2010; Heinen et al., 2011; Van Acker et al., 2011), these programs will become more successful as they become more popular. Workplaces can encourage carpooling (or willingness to carpool) by providing preferential parking or free parking to carpoolers, employee shuttles, financial incentives for using vanpools, or providing flextime.[12] The benefits of each of these strategies should be verified with additional empirical validation, but they are some of the methods workplaces can use to be designated by the US Department of Transportation as "best workplace for commuting" (BWC) (described in section 14.4.1.3. "Mode switching and overall travel reduction"). Given that carpooling is more likely in larger organizations, nearby businesses that band together to form "superpool" opportunities for its employees may be particularly successful in encouraging carpooling (Neoh et al., 2017).

Flextime (the ability to work on a flexible schedule) reduces congestion and, therefore, emissions from frequent deceleration and acceleration (Van der Loop et al., 2019), but it may negatively affect the likelihood of carpooling. In a study of self-reported behavior in 1986, work commuters in a suburban region of Northern California were less likely to carpool if

[12] www.bestworkplaces.org.

they had flexible working hours (Cervero and Griesenbeck, 1988), but in a more recent stated choice experiment, researchers found that flexible hours could lead to an increase in carpooling (Habib et al., 2011). A key hypothesis for this difference is that the participants in the second study were only more likely to carpool if they already perceived carpooling as a viable commuting option.

Web-based platforms can encourage carpooling by linking multiple interested parties who are traveling in the same direction. One web-based platform in New Zealand called "Let's Carpool" increased self-reported carpooling to work significantly among its users from 12% to 27% (Abrahamse and Keall, 2012). Those who continued to drive solo did so because of beliefs about cost, comfort, and convenience. As of the writing of this chapter, over a dozen carpool apps are available on both the Android and iPhone platforms.

14.4.1.2 Purchasing vehicles

Vehicle-purchasing behavior is both impactful (Gardner and Stern, 2008) and difficult to change. Behaviors with high barriers (which may include purchasing fuel-efficient vehicles) require strong behavioral interventions, such as incentives (Schultz, 2014). Indeed, incentives are by far the most commonly studied intervention for encouraging the purchase of efficient vehicles, such as HEV, plug-in hybrid electric vehicle (PEV), or battery electric vehicle (BEV). However, other psychology-informed strategies, such as vehicle labeling and free trials, may also influence this behavior. Little research to date has examined interventions to discourage vehicle purchase entirely, but this could be a promising avenue for future work.

14.4.1.2.1 Monetary incentives

In 2005 the US federal government imposed the Energy Policy Act that supported financial incentives for HEVs. The act provided rebates for hybrid purchases until the manufacturer sold 60,000 units, after which the rebate would be cut back and eventually phased it out entirely. The initial incentive was designed to reduce the cost of the vehicles until the level of adoption was high enough that the market could sustain itself without assistance. This act successfully increased hybrid sales by 3%–20%, depending on the model (Jenn et al., 2013). When the rebate was at least \$1000, sales of hybrids increased by 0.0046% per dollar of incentive (Jenn et al., 2013). State-level analysis from 2010 to 2015 similarly revealed that for every \$1000 in model-specific electric vehicle incentives, new registrations of that model would increase by 5%–11% (Wee et al., 2018). This helped the US PEV sales triple between 2011 and 2015 (Zhou et al., 2015).

A recent systematic review of BEV and PEV financial incentives suggests that they work best under specific conditions (Hardman et al., 2017). Given that humans often do not act as rational economic beings, financial incentive can be made more or less effective, depending on how they are applied. Incentives should be applied at the point of purchase, when the customer is buying the PEV, as opposed to afterward (Hardman et al., 2017). This is possibly because of hyperbolic discounting, the phenomenon that consumers value rewards delivered sooner more than rewards delivered later. Rebates are generally more effective than federal tax credits for stimulating vehicle purchase, possibly because rebate money is delivered to the consumer earlier. However, VAT and purchase tax exemptions applied at the point of purchase for PEVs are most effective. A small incentive at the point of purchase can be as effective as a large incentive later on.

Education and awareness campaigns should promote incentives to consumers, in order to maximize their potential benefit. In terms of the vehicles that are chosen for incentives, policymakers would likely get the most benefit from incentivizing BEVs and PEVs with high electric ranges but not "high-end" BEVs (Hardman et al., 2017). A stated choice experiment also revealed that customers who are in a later stage of purchasing an electric vehicle, as described by the TTM, are less influenced by incentives than those who are in earlier stages (Langbroek et al., 2016). Importantly, policymakers should consider the longevity of incentives, as premature removal of incentives could damage the PEV market (Hardman et al., 2017). In Georgia (United States), for example, electric vehicle sales dropped from 1400 to 250 after a $5000 tax credit was removed and a $200 registration fee was instated (Walton, 2017).

Drivers in various US states can benefit from a wide variety of financial incentives (Hartman and Dowd, 2015). Depending on the state, these may include reduced electricity rates, tax credits, emissions inspection exemptions, energy credits, vehicle replacement vouchers, loans and grants, special financing rates, reduced or eliminated registration fees, pollution control equipment exemptions, or reduced insurance rates. However, many states also offer nonfinancial (or less direct financial) incentives, such as HOV lane exemptions, free parking, or vehicle toll exemptions. We now turn to these alternative incentives.

14.4.1.2.2 Nonmonetary incentives

Comparatively less peer-reviewed research is available on nonmonetary incentives than direct monetary incentives. One stated choice experiment demonstrated that, alongside financial considerations, consumers might be more likely to purchase electric vehicles if they received access to free parking or bus lanes (Langbroek et al., 2016). In Norway, a leading country in BEV adoption and a region offering a variety of incentives, the VAT and purchase tax exemptions are the most critical reasons that most drivers buy BEVs, but bus lane access and free road tolls are nevertheless decisive factors for a subsample of BEV drivers (Bjerkan et al., 2016).

California, a strong promoter of electric vehicles, incentivized hybrid sales using a program that allowed HEV drivers to access HOV lanes without the requisite number of passengers. Evidence for the effectiveness of this program is mixed. A phone survey of Californians in 2004 found that an access to HOV lanes when driving alone was an important motivator of interest in HEVs (Sangkapichai and Saphores, 2009). Indeed, a geographical analysis of electric vehicle registrations in California found that at least one quarter of new registrations between 2010 and 2013 was a result of HOV lane policy (Chen, 2018).[13] That said, the true cost of allowing HOV access is not zero. In Los Angeles, HEV access to carpool lanes slows down overall traffic and may have net negative welfare impacts (Bento et al., 2013). Although HOV lane access may stimulate sales of HEVs, one analysis suggests that HOV lane access is undervalued by the California program (Shewmake and Jarvis, 2014). These researchers suggest that HOV lane access is quite valuable, and simply selling it to anyone (regardless of vehicle type) at its true value could earn sufficient money for directly subsidizing HEV sales. They argue that this direct subsidy could be a better method of stimulating HEV adoption than providing HOV lane access (although they do not directly test this alternative hypothesis).

14.4.1.2.3 Vehicle labeling

One reason that car buyers do not buy efficient vehicles, even when they claim to care about fuel economy or environmental sustainability, is that they simply do not know which vehicles are most (or least) efficient. Labels on products, such as cars, act as sign posts that allow consumers to

[13] A similar analysis in Virginia found that HOV lane access promoted new HEV registrations in some regions, but not others (Diamond, 2008).

learn if (and to what degree) the product expresses their values and attitudes (Ungemach et al., 2017). Hence, survey respondents who care about the environment are more likely to select "green" vehicle options when vehicle labels include greenhouse emissions, rather than only fuel costs (Ungemach et al., 2017). Similarly, a randomized-stated choice experiment shows that the presence of fuel economy information could affect vehicle choices and that labels that include a variety of information (multiple types of sign posts) are most likely to facilitate selection of efficient vehicles (Kormos and Sussman, 2018). Vehicle fuel economy labels are most effective if they take into account insights about human behavior learned from psychology research (e.g., Camilleri and Larrick, 2014). Therefore the American EPA-mandated fuel-economy label underwent an extensive revision process. This new label was developed with the help of multiple experts in various social science disciplines and now includes a variety of information (EPA and DOT, 2010).

A good example of a vehicle labeling program comes from the European Union (EU). The original EU labeling directive was established in 1999 and evaluated in 2016.[14] The evaluation found that awareness of fuel efficiency increased steadily since the directive was instated (more than 75% awareness in 2016), and that labels that were consistent with other the EU efficiency labels were most preferred.[15] The cost of labeling vehicles is relatively low and, therefore, most likely worth the investment. Nevertheless, establishing the effectiveness of the label to independently affect vehicle sales and CO_2 emissions is difficult. One reason is that the EU countries implement the labeling program and also simultaneously implement other programs to encourage purchase of efficient vehicles. In France, evaluators believe that the label itself may have effectively reduced emissions, but in other countries, they believe the effect is lower. The evidence for effectiveness is thin and, as such, at least one group of researchers suggests potential improvements to the design of the label (Waechter et al., 2015).

14.4.1.2.4 Free or discounted trials

Mounting evidence shows that human decision-making is more commonly emotional than rational (Lerner et al., 2015). Personal direct

[14] https://ec.europa.eu/clima/sites/clima/files/transport/vehicles/labelling/docs/car_labelling_final_report20160728_en.pd.

[15] The standard efficiency label has a scale from D to A+ + +, with the product's specific rating placed along the continuum.

experience triggers more emotional responses than secondhand information (Tyng et al., 2017) and, as such, a positive direct driving experience may be more influential for encouraging vehicle purchases than impersonal statistics or facts. This is particularly noteworthy for BEVs, in which the car driving and fueling experience is qualitatively different from other cars. For example, refueling at home is quite different than refueling at a gas station, and this is not perceived as a salient benefit until consumers drive the vehicles for a few weeks or months (Bühler et al., 2014).

Car buyers state that one of the primary barriers to purchasing BEVs is "range anxiety," the concern that BEVs do not drive far enough and that the driver might become stranded (Bühler et al., 2014). Statistics show that range anxiety should not be an issue for most households, but this statistical information is not understood at a deep level, unless it is experienced. The average American household travels less than the range of almost any BEV currently on the market (only 39.5 miles per day, with 93% traveling less than 100 miles per day (National Household Travel Survey, 2009). However, education of this fact, alone, does not provide the deep emotional understanding that can affect decision-making. Thus, one strategy to encourage purchase of BEVs is to provide a trial to potential buyers to allow them direct experience and a deeper (emotional) understanding.

German researchers have conducted several innovative studies showing that direct experience with BEVs can reduce range anxiety and increase intentions to purchase. In one program, 75 Germans were provided an opportunity to lease a new BMW BEV for 3 months in exchange for completing surveys at four time points (before the trial, 1 week into the trial, 6 weeks into the trial, and upon return of the vehicle) and installing a data logger in the vehicle to measure distance and location (Franke et al., 2014). This nonrandom group of participants may have been motivated to provide positive survey responses, but they nevertheless showed signs of increased satisfaction with range over time (Franke et al., 2017). This range satisfaction translated to general BEV acceptance, perceived usefulness of BEVs, the intention to recommend BEVs to others, and a change in preferences of range and BEV purchase intentions. This study did find, however, that drivers experienced an initial drop in range satisfaction in the first week, before the steady increase to above-baseline levels. This may explain why a 24-hour German test drive did not elicit the same sort of significant increase in range satisfaction, despite an increase in positive perception of acceleration, fun, low noise, and other traits (Schmalfuß et al., 2017).

BEV studies in Germany also tested the in situ effects of actually experiencing a scenario in which they would run out of electricity (or come close) before reaching a destination. Novice drivers instructed to drive a route that was very close to the maximum range of their BEV felt some range stress at the beginning of their trip but subsequently felt reduced range anxiety about BEVs in general (Rauh et al., 2017). The anxiety reduction effect was particularly pronounced if they were given instructions on how to maximize their range before they started their journey. Similarly, another test found that experienced BEV drivers felt significantly less stress when driving close to their maximum range than novice BEV drivers (Rauh et al., 2015). Personality traits, coping skills, route familiarity, trust and understanding of the feedback system, and an overall ability to understand technology also affect the experience of temporary range stress (Franke et al., 2012, 2016).

14.4.1.3 Mode switching and overall travel reduction

Encouraging travelers to switch their mode of transportation or to travel less (in distance or frequency) can be challenging. In this section, we discuss two types of personal transportation. The first, and far more studied, is frequent or habitual travel, usually by commuters going to work. Policymakers tackling these behaviors usually aim to reduce the frequency or duration of single-occupancy vehicle (SOV) trips, often by discouraging driving, encouraging use of alternative transportation (e.g., public transit, biking, or walking), or discouraging travel in general (e.g., providing work-from-home options instead).

The second type of personal transportation behavior is almost never targeted but could be impactful from a greenhouse gas emissions perspective. This behavior includes infrequent but long trips, usually by airplane. Programs targeting these behaviors could seek to encourage travelers to switch to less carbon-intensive modes of transportation (e.g., from airplane to train) or to do fewer long-distance trips.

TDM research offers a large number of solutions for mode switching and reducing travel. Below, we present several effective examples, focusing (when possible) on those that apply social science research.

14.4.1.3.1 Generally encouraging alternatives to single-occupancy vehicle travel

Programs encouraging mode switching are often initiated by workplaces. In 2001 the US EPA and department of transportation (DOT) worked together to create an initiative that rewarded workplaces for these programs. This private–public partnership designated employers as "BWCs"[16] if they implemented several practices that discouraged driving and offered benefits for alternative transportation options. The benefits successfully reduced employee vehicle miles traveled, fuel consumption, and SOV trips by 7% if they included no financial incentives, or by 15% if they also included some financial benefits (Herzog et al., 2006). We review two of the practices that workplaces could chose to implement as part of this program (guaranteed ride home, and parking cash out), as well as two strategies that cities can implement (congestion pricing and physical street layout).[17]

The guaranteed ride home program is a requirement for designation as one of the BWCs. This program is a benefit that workplaces offer to employees who commute by non-SOV travel most of the time. These employees can occasionally use a taxi or ride-hailing app, paid by the employer, to commute home if they experience an emergency. The program successfully addresses one barrier for commuting by non-SOV travel and is relatively inexpensive (Menczer, 2007). Although it has not been evaluated using an experimental design, several survey-based studies find that commuters self-report it as an important reason for their choosing not to drive (Nelson/Nygaard Consulting Associates 2005; Vantage Marketing Research 2003; Menczer, 2007).

An optional strategy for workplaces to offer employees, as part of the BWC, is "parking cash out." This strategy involves workplaces offering employees at least $30 per month for them to relinquish their parking privileges at work. Employees with free parking at work commute by SOV 62%–72% of the time, whereas those who pay for parking do so 43%–50% of the time (Wilson, 1992; Miller, 1991; Shoup 1995; Hess, 2001). The parking cash out could incentivize workers to use alternative transportation options and simultaneously help them pay for those options. In places where parking cash out programs were implemented, SOV trips decreased by 7%–65%, from before to after the implementation (Shoup, 1995; Weinberger et al., 2010). This is one of many possible

[16] https://www.bestworkplaces.org/.

[17] Pay-as-you-drive is also an effective strategy that insurance companies can implement to reduce overall mileage (Buxbaum, 2006). It is described in the "efficient use" section of program examples.

parking policies that have the power to affect car ownership and driving (e.g., Weinberger et al., 2010).

Physical street layout can influence the likelihood that city dwellers will commute by car. Driving is encouraged by offering many intersections with arterial roads, more lanes per road and more curvilinear roads (Marshall and Garrick, 2012). Driving alternatives are encouraged by including more intersections, more dead-end streets, more bike lanes, sidewalks and street parking, and more land use density and mix (Marshall and Garrick, 2012).

Cities may also choose to reduce car traffic by implementing variable, congestion-dependent charges, which includes tolls for city drivers moving through certain areas at certain times. In London, congestion charging reduced CO_2 emissions by 19.5% (Beevers and Carslaw, 2005), and in the Netherlands, economic models predicted a 6%–15% reduction (Ubbels and Verhoef, 2006). An Irish study of actual drivers before and after implementing congestion pricing in Dublin suggests that these models may be accurate. Motorists who allowed the researchers to track their driving reduced their total number of trips by 6% and total mileage by 12%, when compared before and after congestion pricing (O'Mahony et al., 2000).

14.4.1.3.2 Encouraging public transit

Some mode-switching programs are specifically designed to encourage public transit use. The online TDM Encyclopedia describes at least 13 types of initiatives that have been successfully implemented to encourage public transit commuting.[18] Most interventions are either structural (roadways, traffic regulations, or facilities that serve riders) or incentives (Isaacs, 1980). The interventions change behavior by focusing on cost, convenience, or time (Isaacs, 1980). A few such programs were tested experimentally and are described next.

Beginning in the 1970s and 1980s, researchers experimented with creative incentive and commitment strategies to encourage bus ridership. One consistent finding was that providing free bus passes increased ridership (e.g., Katzev and Bachman, 1982; Bachman and Katzev, 1982), and this continues to be the case in newer studies (Fujii and Kitamura, 2003; Edwards et al., 2013). One study found that providing a reduced fare for frequent commuting was also an effective alternative to free bus passes

[18] http://www.vtpi.org/tdm/tdm112.htm.

(Katzev and Bachman, 1982). However, there was mixed evidence regarding persistence of the behavior change after the passes expired. In most cases, behavior persisted at follow-ups as much as 3 months later (e.g., Bachman and Katzev, 1982; Fujii and Kitamura, 2003). However, in at least one case, ridership returned to baseline levels after the intervention (Katzev and Bachman, 1982). This suggests a potential interplay between theories of habit formation and of operant conditioning. Theories of operant conditioning suggest that behavior should extinguish when rewards are removed, but theories of habit formation suggest that once a habit is established, it is easier to continue the behavior than to stop.

In a sign of the times, one interesting study that was conducted in 1977, when behaviorism and operant conditioning were more popular, used the principle of reward schedules to test an intervention to encourage bus riding (Deslauriers and Everett, 1977). For this study, drivers in an intervention condition rewarded bus passengers with a 10-cent token that could be spent on food or future bus trips. Later, researchers changed the reward so the token was only given to every third passenger. Buses giving tokens had significantly higher ridership than before the intervention, and there was no significant difference between tokens provided to every third rider and every rider. Although this demonstrates that full free passes may not be necessary to encourage bus ridership (smaller rewards at lower frequency may also work), this intervention, in its exact form, is unlikely to work for the most bus routes today; given that the choice to ride a bus might not be as flexible, day-to-day, as it was on this university campus in the 1970s.

14.4.1.3.3 Encouraging bicycle commuting

Bicycle commuting could also be a mode-switching target behavior. Effective solutions are to encourage bike riding both recruit new riders and encourage regular commuters to continue riding. Bike commuting is a day-to-day choice that is influenced by requirements of the day, as well as weather (Heinen et al., 2011). Regular commuters are deterred by practical limitations, such as the requirement to be at multiple locations in a day, or being unable to ride due to strong wind. Occasional riders, however, are unlikely to ride unless the weather is good. They are most motivated by the positive effect of good weather (Heinen et al., 2011). That said, structural changes and workplace initiatives could also increase bike-commuting rates.

The most common structural change for promoting cycle commuting is creating bike lanes on streets. People bike more in cities that have bike lanes (Buehler and Pucher, 2012), and bike lanes that offer connectivity and directness are most effective (Schoner and Levinson, 2014). Numerous studies show that cities with cycling infrastructure have more cyclists and that even veteran cyclists appreciate bike lanes (e.g., Dill, 2009), but to date, no study to our knowledge has experimentally tested whether the presence of bicycle infrastructure causes increases in cycling.

An analysis of work commute data in Washington, DC, revealed that the presence of bicycle parking spaces and showers at work are associated with increased rates of bike commuting (Buehler, 2012). Conversely, the presence of free car parking at work reduces the odds of bike ridership by 70% (Buehler, 2012). The same study finds that transit benefits do not affect bike-commuting rates. A program in several New Zealand cities, *Bike Now*, recruited cycling champions at workplaces to implement a series of bicycle initiatives (O'Fallon, 2010). These included bike mentoring/buddying, bike buses, cycle skills training, secure bike parking, and a "cycle fleet" at the workplace for travel during working hours. This program was not empirically tested and, as such, the effectiveness of the program, and the relative value of each initiative was difficult to gauge. Nevertheless, surveys showed that the program likely increased awareness of cycling and possibly influenced some employees to begin cycle commuting for the first time.

Some regions may implement *Bike-to-Work* events to encourage commuters to try cycling for the first time. Like test drives, events that encourage a single day of bicycle commuting may work through several behavioral science channels. We hypothesize that they (1) create an emotional experience, (2) employ the foot in the door technique, (3) reduce uncertainty, and (4) unfreeze habits (allowing new habits to form). In a multiyear survey-based study of *Ride-to-Work* day in Victoria, Australia, the researchers found promising signs that the event may, in fact, encourage commuters to try cycling to work for the first time and continue cycling 6 months later (Rose and Marfurt, 2007). Across years, 20% of cyclists registered for the event were cycling to work for the first time, and 27% of those new riders were still cycling to work in a follow-up survey 5 months later. Over half of the first-time cycle commuters (57%) claimed that the event influenced their decision to cycle to work 5 months later.

14.4.1.3.4 Alternative work options

Rather than discouraging SOV travel or encouraging alternative transportation options, policies can be implemented to reduce work travel altogether. The most common strategies of this type are tele-working and flexible schedules. Tele-working, or working from home, tends to remove drivers from the roads and reduce traffic. Although the previous researchers have theorized that tele-work could increase traffic and car use (e.g., Mokhtarian, 2009) and some studies have linked communications technology use with increased travel (Mokhtarian, 2002), recent analyses show that tele-working opportunities may actually reduce car travel (Van der Loop et al., 2019). Flexible schedules allow workers to commute during less busy times of the day and, as such, flatten the effect of peak congestion periods (Van der Loop et al., 2019). Less congestion means less greenhouse gas emissions per commute. Flexible work options are solutions that the Southern California Association of Governments accepts for workplaces to meet their mandatory vehicle reduction targets.[19] In the Netherlands the increasing prevalence of these two alternative work options between 2000 and 2016 led to significant decreases in traffic congestion and commuting times (Van der Loop et al., 2019). Modeling simulations show that car kilometers on working days would have increased by 2.6% without this increase in alternative work options, and commuters would have 18% slower commutes.

14.4.1.3.5 Long-haul travel

Research on long-haul travel is dwarfed by work on short commuter travel behavior, in both quantity and depth. To our knowledge, no evaluated program has been implemented to reduce long-haul travel or encourage mode switching (e.g., from air travel to rail or bus). However, the per-person and per-trip impacts of long-distance travel, specifically by air, are high. For Germans in 2008, for example, air travel contributed approximately 45% of all transportation-based carbon emissions, with the relatively small number of trips over 100 km contributing over half of this effect (Aamaas et al., 2013). Intercontinental flights and cruise ship vacations are two of the most carbon-intensive tourist options available (Chenoweth, 2009). With a relatively small number of trips, and specific target population (mostly high-income travelers), a small change in behavior could have an important impact on emissions (Aamaas et al., 2013).

[19] http://scagrtpscs.net/Documents/2016/final/f2016RTPSCS.pdf.

This underexplored behavior could be a key target for future behavior change initiatives.

Market segmentation studies show who travels by air and who is most likely to change their travel behavior and, thus, may be good targets for a behavioral intervention. In Germany, travelers with the highest income have a carbon impact 250% higher than those with the lowest incomes (Aamaas et al., 2013). Long-haul flyers in the country are the smallest market segment but contribute 80% of emissions, traveling more frequently, traveling longer distances, and using airplanes for over 60% of their trips (Böhler et al., 2006).

Unfortunately, most air travelers are aware of their negative impact on the environment but are unable or unwilling to change their behavior. In Hong Kong the regular international tourists who travel by air most are aware of environmental harm associated with flying, but are least willing to change, whereas infrequent flyers are most willing to curtail their air travel (McKercher et al., 2010). The "flyers' dilemma" (flying despite believing it contributes to harmful greenhouse gas emissions) is apparent in Norway, the United Kingdom, and Germany as well (Higham et al., 2014). When frequent air travelers (so-called binge flyers) experience cognitive dissonance from the awareness that their behavior contributes to climate change, they do not change their behavior but instead respond with guilt suppression, denial, and a reframing of what constitutes "excessive" holiday flying (Cohen et al., 2011). This may partly explain why people exposed to information about the adverse effect of aviation on the environment want to fly more than those not exposed to this information (Gilbey et al., 2011).

Tourism is an important part of long-haul travel and, as such, potentially important lever for behavior change programs. One type of program that could benefit from systematic testing is a campaign to promote "staycations," or local tourism. Local tourism can reduce carbon emissions by eliminating the requirement of long-haul travel. Several researchers have used a lifestyle segmentation approach to learn preferences of potential local tourists to develop potential tourism packages for them (e.g., James et al., 2017). These have not been converted into behavioral interventions but hold promise as strategies for informing potential programs.

Purchasing carbon-offset credits is one way to mitigate the carbon impacts of flying. It may not discourage travel or encourage mode switching, but it nevertheless relates to carbon emissions from long-haul travel. One experiment tested a behavioral approach for encouraging purchase of travel carbon offsets (Araña et al., 2013). The experimenters worked with a conference center in the Canary Islands to offer conference attendees the opportunity to purchase offsets when they registered. They varied the cost of the offsets as well as whether the offer was framed as opt-in (check this box to add $X to your registration costs for credits) or opt-out ($X in carbon offsets was included in your registration; check this box to cancel this purchase). They found that the opt-out framing caused significantly more carbon-offset purchases than the opt-in framing, although the effect was smaller at the highest price range.

Conferences and events contribute to long-haul travel by attracting large groups of people from disparate locations. An analysis of The University of British Columbia sporting events shows that although only 4% of fans travel by plane to attend games, they contributed 52% of the total emissions of the events (Dolf and Teehan, 2015). In 2008 the average carbon emissions from academics presenting papers at international conferences was 849 kg (Spinellis and Louridas, 2013). The average student in a 4-year Ph.D. program thus emits an estimated 21.5 t of carbon dioxide over the course of his or her studies; a total that could be cut in almost half if they attended virtual meetings and conferences instead of traveling in person (Achten et al., 2013).

The downside of virtual conferences is that they do not facilitate the same level of networking and interaction. As such, hybrid models of virtual conferences, with a hub (or multiple hubs) that host the event and broadcast it to one or more nodes (satellite locations hosting group viewings and discussions), may offer an effective compromise solution (Fraser et al., 2017). The effectiveness of this strategy as part of a behavior intervention has yet to be tested.

14.4.2 Professional operators

The vast bulk of research on transportation behavior change involves personal transportation choices for short commutes. Although this travel segment is important to study because it is responsible for a substantial proportion of overall emissions (e.g., roughly 27% of household emissions in the United Kingdom; Preston et al., 2013), professional vehicle operators have a higher per capita influence (M.J. Bradley & Associates, 2014). A single truck driver, bus driver, boat captain, or airplane pilot produces more emissions through his or her professional transportation work than individual commuters do. Professional operators frequently travel long routes with high-emissions vehicles. As such, changing the behavior of a single operator for some trips can have a disproportionally large effect on overall emissions and, therefore, this might be a promising sector to target with behavior change intervention programs.

A small number of studies have examined this group of travelers' potentials for behavior change. As with personal transportation, the behavioral domains are divided into three overarching classes: efficient use, vehicle purchase, and mode switching or overall travel reduction. The available research is reviewed in the following subsections.

14.4.2.1 Targeting professional operators

Before discussing examples of successful behavior change programs for professional operators, it is important to note differences between personal and professional operators. Special considerations should be made for programs targeting professional operators.

Professional operators rarely own and operate their vehicles (with the notable exceptions of some taxi drivers and a few others). This means that although behavior change programs may target operators, they must usually be implemented by fleet managers or higher level managers within organizations. This is particularly true for programs encouraging efficient purchases or mode switching. The decision-making process within organizations can be different than for individuals, and this may be part of the reason that behavioral science rarely addresses professional transportation behaviors. Behavior change programs targeting organizations should consider business decision-making motivations and key intragroup processes.

Organizations comprises individuals and, as such, their decisions are somewhat influenced by the same decision processes and biases that affect individuals. For example, managers' personal values can drive corporate sustainability initiatives (Hemingway and Maclagan, 2004), and hyperbolic discounting of future benefits prevents shipping companies from investing in transportation efficiency measures (Guerrero, 2014), just as it prevents individuals from acting on climate change (McDonald et al., 2015).

However, individuals acting on behalf of businesses may also make more financial-based decisions than when acting for themselves. Energy efficiency auditors have told us, for example, that commercial businesses prioritize efficiency upgrade decisions based almost entirely on price, whereas residential customers base their decisions on price, comfort, health, and other factors. Therefore behavior change interventions directed at actors within organizations may sometimes appeal to personal values of social responsibility but may more often be successful focusing on financial benefits.

Group decision-making biases may also play a role in organizations' transportation choices. When appealing to individuals within organizations, program implementers should be aware that individuals within groups make different decisions than they might on their own. People within organizations tend to conform to group norms (Ehrhart and Naumann, 2004), make more extreme choices (Isenberg, 1986), and avoid damaging group harmony by speaking out against popular choices (Esser, 1998). In addition, others' advice may play a critical role for decision-makers within organizations (Bonaccio and Dalal, 2006).

In this section, we focus on programs targeting professional operators who own their own vehicles (e.g., some taxi and truck drivers), as well as organizations making purchase and logistics decisions on behalf of operators. We also examine programs, such as those teaching eco-driving, that organizations can implement to change operator behaviors. Each program that we review changes individual behavior, however, in some cases, this happens within the context of a larger organization.

14.4.2.2 Efficient use by professional operators

As with personal transportation eco-driving programs, most professional eco-driving programs center around reduction of idling, harsh deceleration and acceleration, and overall speed reduction (Huang et al., 2018). These are encouraged through feedback (usually using onboard technology), training or incentives. The physical environment and punitive laws can also play an important role, but these strategies have already been discussed in the "personal transportation efficient use" section of this chapter. Efficient use programs are the most common approach used to change behavior for professional operators. Notably, professional operators may be less motivated to reduce fuel consumption than personal operators because they do not directly pay fuel costs. As such, many organizations provide financial incentives, contests, or recognition to spur eco-driving. In some cases, when an organization simply makes operators aware that they are being observed, the operators respond by reducing consumption (Gosnell et al., 2016).

14.2.2.2.1 Feedback and technology

Feedback has been tested and evaluated as a strategy for encouraging behavior change for bus drivers, freight transporters, and airline pilots.

Bus drivers in Lisbon reduced fuel consumption by approximately 5% when efficiency feedback devices were installed, relative to when they were removed (Rolim et al., 2017). Another evaluation found that bus drivers reduced their fuel consumption by 6.8%, relative to a control group, when the buses had feedback devices (with or without additional training; Strömberg and Karlsson, 2013). Truck drivers with devices that provided feedback and audible warnings achieved an average savings of 7.6% by encouraging slower acceleration and earlier upshift in gears (Vagg et al., 2013). Based on truck driving simulations, intermittent feedback may be safer, and just as effective for emissions reductions, as continuous feedback (Kircher et al., 2014).

One large-scale feedback study with airline captains found that providing feedback saved 266,000–704,000 kg of fuel across the airline over 8 months (Gosnell et al., 2016). Airplane pilot behaviors can significantly affect fuel consumption. For example, in Dallas/Fort Worth International Airport, idling, starting, and stopping during taxiing result in an estimated 18% of a flight's trips total fuel consumption (Nikoleris et al., 2011). The feedback program involved sending personalized letters to each airplane captain detailing their previous month's fuel consumption. The researchers found that the feedback helped reduce fuel consumption, and that some effect could also be achieved by simply informing pilots that their efficiency was being tracked (without feedback). However, a larger effect could be achieved by suggesting captains aim to achieve a specific goal (set at 25% better than their baseline efficiency). On board real-time energy management systems for airplanes also exist (e.g., Amelink et al., 2005; Catton et al., 2007), and they could be tested with a goal setting manipulation to see if it improves effectiveness. These programs make use of goal setting, observability/accountability, and feedback to influence behavior.

14.2.2.2.2 Training

Professional operator training can help encourage eco-driving and reduce fuel consumption. Operators of government fleet vehicles who received eco-driving training ($n = 15$) used less fuel by reducing idling posttraining by 4% for internal combustion engines and 10% for hybrids (Rutty et al., 2013). Truck drivers ($n = 4$) who underwent training in and out of the classroom on large double-trailer tanker trucks were able to reduce their fuel consumption by 24%–27%, when it was measured on a circuit up to 12 weeks later (Symmons and Rose, 2009). Bus driver training ($n = 2$)

reduced fuel consumption on actual bus routes by 4.35% after extensive training with an in-bus component (Zarkadoula et al., 2007). One interesting innovation that was found to improve bus driver eco-training was a social comparison tool. When drivers were trained to imitate the speed and revolutions per minute (RPMs) of the current most efficient drivers on same route, they drove significantly more efficiently (Huertas et al., 2018).

One example of bus driver training that we found particularly compelling involved the use of a bus simulator in conjunction with training. Drivers who completed full-day eco-driver training course with a simulator component ($n = 29$) saved 12% in fuel on their routes immediately after the training, and 17% 6 months later (Sullman et al., 2015). The control group, which completed the simulator ($n = 18$), but without training (only a first aid course instead), did not achieve savings.

14.2.2.2.3 Incentives and rewards

Busses and trucks can increase the effectiveness of eco-driver training with the use of incentives. One program found that bus drivers offered a monetary reward for driving efficiently decreased their fuel consumption by 10% (which lasted over time), and this reduction covered more than the cost of the monetary incentive (Lai, 2015). Similarly, a German logistics company achieved 3.5%–5% savings from nonmonetary and monetary incentives for eco-driving (Schall and Mohnen, 2017).

14.2.3 Purchasing vehicles for professional operation

Purchasing efficient fleet vehicles, trucks, or freight transportation options are a profoundly impactful behavior. Although these vehicles are few in number, they have a disproportionate impact on greenhouse gas emissions because they are driven further and more frequently than personal vehicles (e.g., Graus and Worrell, 2008; Environmental Protection Agency (EPA), 2018). Moreover, these vehicles are sometimes purchased in large quantities by single individuals or organizations. As such, the purchase decision by that single actor can have a very high impact on emissions for years into the future. Unlike most other behaviors addressed in this chapter, this behavior is frequently "organizational" (done by a business or other institution) as opposed to individual. Unfortunately, the purchase behavior of vehicles by organizations has not been studied from a behavioral science perspective, and the only intervention used with any regularity is purchase incentives or subsidies. Nevertheless, a small number of studies have examined potential motivators and barriers to purchase of efficient fleet vehicles or upgrades to current vehicles. These could form the basis of potential future behavioral interventions.

A variety of technologies (tires and aerodynamics upgrades) can be purchased for long-haul trucks that will make them more fuel-efficient. However, one barrier preventing these purchases is that American trucking firms discount the future heavily and are therefore unlikely to favor solutions with long payback periods (Guerrero, 2014). Additionally, approximately 23% of truck trailers are not owned by the people who use them and get paid for their fuel consumption (Vernon and Meier, 2012). As such, the trailer owners do not reap the benefits of investing in upgrades to aerodynamics or tire rolling resistance reduction (Vernon and Meier, 2012). For water-based shipping, the barriers to energy efficient transportation could be described as a lack of reliable information on cost and savings, difficultly in implementing energy efficiency measures (e.g., speed reduction, weather routing, waste heat recovery or air lubrication), lack of direct control over operations, and savings being ignored by decision-makers due to small impact (Rehmatulla and Smith, 2015). Behavioral interventions that address these barriers may prove successful for encouraging purchase of energy efficient fleet or freight transportation vehicles.

Encouraging the purchase of, or upgrade to, energy efficient taxis is often important to policymakers. Around the world, but especially in China, where the government has a goal of converting at least 50% of its taxis to BEVs, purchase subsidies are critically important for this goal (e.g., Gao and Kitirattragarn, 2008; Yang et al., 2018a,b). However, subsidizing public charging stations or natural gas upgrades could also help improve overall fuel economy (Yang et al., 2018a; Liu et al., 2012). In some cases, converting taxis to natural gas may be a more viable solution than incentivizing the purchase of new natural gas, or HEVs. Natural gas can reduce carbon emissions and fuel costs, but the cost of switching can be expensive and, therefore, in some cases, governments should consider helping to defer the costs of switching and incentivize the purchase of vehicles that can be converted (Liu et al., 2012). Directing the subsidies to young drivers of new taxi cabs might be most effective as those are the drivers that are most likely to invest in converting (Liu et al., 2012).

One group of Hong Kong researchers has suggested that, rather than rely entirely on government subsidies to encourage electric vehicle (EV) taxi purchase, taxi drivers can offer a "premium eTaxi service" that costs slightly more to customers (Yang et al., 2018b). The Tesla-operated airport taxi service in Amsterdam is an example of this approach (Netherlands Enterprise Agency, 2015). They argue that this higher quality service (e.g., guided by a ride-hailing app, using new high-end EVs, and with a driver permitting process) will command a higher fee and, therefore, will require a lower government subsidy. In Hong Kong, many taxis are rented and, therefore, are not owned by their drivers. Owners' purchase decisions would be based on the vehicle purchase price, potential rental income, and battery lifespan. The drivers' rental decision would be based on fare income, rental cost of the taxi, access time to chargers, and range per charge. This program has potential but has yet to be studied as a behavioral intervention.

14.2.4 Mode switching or overall travel reduction by professional operators

These types of behaviors are uncommon for professional operators, but several opportunities for mode switching or travel reduction do exist. For example, if delivery companies could convince shoppers to accept longer shipping times, then they may be able to use more efficient transportation options, such as trains or boats rather than airplanes. They could also ship more packages together and use fewer trucks to transport the same number of goods. Encouraging customers to shop locally or to buy goods at stores rather than online may, in some circumstances, reduce the number of vehicle trips required for deliveries.

Currently, mode switching to lighter smaller vehicles is already occurring to some degree for "last mile" distribution (Oliveira et al., 2017). Between 1990 and 2003 the number of light commercial vehicles in Europe increased 8.4% more than the number of commercial vehicles more than 3.5 t, suggesting that urban freight transport modes are changing (Schoemaker et al., 2005). Importantly, however, this switch to smaller vehicles will not reduce overall emissions if more vehicles are needed. The switch must be accompanied by increasing the efficiency of each vehicle, or switching to zero-emissions vehicles such as bicycles.

Delivery by bicycle or tricycle is becoming more common (Oliveira et al., 2017). For example, Coca-Cola delivers some of its products in Brazil by tricycle in order to avoid traffic congestion and restrictions (Ferreira, 2015). Behavioral interventions could help further this trend for mode switching, given that urban freight transportation can be quite inefficient. In London, for example, urban freight transportation, vehicle-occupancy rates were between 40% and 60% in 2005 (Schoemaker et al., 2005). Encouraging mode switching by shipping organizations is challenging and has yet to be investigated using a behavior science-based approach. Interventions using subsidies, social norms, laws, or structural environmental changes are several options for encouraging mode switching that could be investigated.

14.5 Guide for practitioners

Based on the work of several academics and behavioral insights teams around the world, we have developed a five-step process to "DRIVE" sustainable transportation behavior change. This process brings together the theories of behavior change with potential intervention strategies discussed in this chapter. It is supported by empirical research and follows the general approach of multiple other behavior research groups (Steg and Vlek, 2009; Geller, 2002; Hansen and Schmidt for the OECD, 2018; Ly et al., 2013; Bridgeable, 2017; Chan et al., 2017; Barrows et al., 2018; Alta Planning + Design and the Behavioural Insights Team, 2018). We depict the model in Fig. 14.1. Although we provide a linear overview of the behavioral design process, in practice it is typically more iterative in nature.

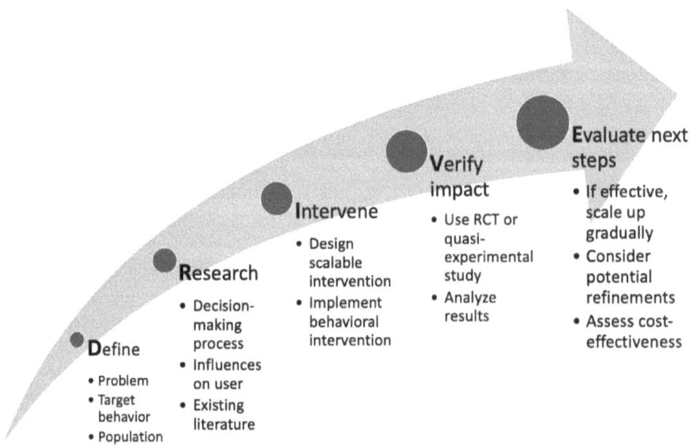

Figure 14.1 Five-step process to "DRIVE" behavior change. *RCT*, randomized control trial.

14.5.1 Step 1: define (problem, target behavior, and target population)

The first step is to define clearly the problem and the behavior that the intervention will change, as well as the population that is doing that behavior. The program implementer or developer (e.g., government agency, business, or nonprofit) should consider their capacity to implement a behavioral intervention, and then identify and reach out to potential partners and stakeholders as necessary. As described in Section 14.2, potential behaviors fall into two categories (personal or professional) and three domains (efficient use, purchase, or mode switching). These vary in their potential impact and feasibility depending on the target population. Choosing a behavior and specifically defining it is an important step that dictates the rest of the process.

14.5.2 Step 2: research

This step involves three subparts. With regard to the target population and behavior the program implementer should (1) explore the decision-making process as well as (2) the potential influences on the decision-making behavior. Finally, the program implementer should (3) consult existing evidence and others' successful approaches, such as those suggested in this chapter.

We suggest using this chapter as a reference for guiding research. After thinking about and mapping every step in the decision-making process for the target behavior, the program designer should survey or interview members of the target population (or several key informants) to learn more about what motivates (or demotivates) the target behavior. But what should the program designer ask? The questions should be constructed based on the TTM, TPB, and habit theories. For example, the survey might ask which stage of change the participant is in, using a question such as this:

Please select the answer that best represents you:

1. I currently do not do [target behavior] and do not intend to in the next 6 months.
2. I currently do not do [target behavior] but I am thinking about doing it in the next 6 months.
3. I am planning to do [target behavior] within the next month.
4. I started doing (or did) [target behavior] within the past 6 months.
5. I have been doing (or did) [target behavior] since 6 months ago.

The survey might also ask a series of questions about the target population's attitudes toward the behavior, subjective norms regarding the behavior, and perceived control over the behavior, as suggested by the TPB. One of the original TPB authors published a series of recommendations for constructing effective TPB survey items, and this document is available online.[20] Briefly, the survey could include questions such as:

1. Attitudes: How positive or negative are your overall attitudes toward [target behavior]? (rate from extremely negative [0] to extremely positive [10])
2. Attitudes: How does doing [target behavior] feel? (rate from extremely bad [0] to extremely good [10])
3. Subjective norms: Most people who are important to me do [target behavior] (rate from strongly disagree [0] to strongly agree [10])
4. Perceived behavioral control: I am confident that I can do [target behavior] (rate from strongly disagree [0] to strongly agree [10])

Additional measures and research may also help during this step. For example, if the target behavior is to encourage car owners to drive less, then recording baseline levels of driving is important. These baseline levels could be recorded by direct observation (e.g., watching intersections), data loggers (e.g., GPS monitors attached to cars), or self-reported behavior (e.g., asking drivers to maintain a diary or explain their driving over the past week in a survey). An in-depth interview with several drivers could also help uncover unknown barriers and motivators of the behavior that might not have been otherwise considered. In some cases, preliminary interviews can be used to help develop large-scale surveys. An overall goal of this stage is to learn where the target population is, in terms of changing their behavior, and what would motivate or demotivate them from changing (i.e., the barriers to and benefits of change).

In the third subpart the program designer can refer to sample interventions described in the "sample intervention programs for different behaviors" section for ideas. The designer can create a draft list of program ideas from those in this chapter that have previously been used to change the target behavior.

14.5.3 Step 3: intervene

This step has two subparts. The first is to design the intervention, which involves assessing which interventions can and cannot be used, and which

[20] https://people.umass.edu/aizen/pdf/tpb.measurement.pdf.

have the highest likelihood of success. The second is to implement the intervention, which involves finding a sufficient sample size and deciding how to create an intervention that can be fruitfully evaluated.

Data gathered during the previous step can be used to inform the strategy that is chosen for the intervention. For example, if members of the target group are primarily in the precontemplation stage and have generally negative attitudes toward the behavior, then a public awareness campaign might be effective. If the target group is in the action phase and has positive attitudes but low behavioral control, then a change to the physical environment might be enough to prompt shifts in behavior. Once barriers and benefits of change are known, Schultz (2014) offers some helpful suggestions for choosing which type strategy to employ.

In the second subpart of this step, program implementers will have to decide important design details, such as which outcome measures can be used to evaluate success and whether a randomized control trial (RCT) is possible (or whether another type of design and control group might be more feasible). This strategy for intervention allows testing strategies while implementing them. It is important for both small scale experimental studies and large intervention programs. We suggest consulting with a program evaluation company or academic institution during this step to discuss these details.

14.5.4 Step 4: verify effectiveness of intervention

Assuming that steps two and three were completed successfully, this step involves gathering data on outcome measures and conducting analyses on that data. This is also the step in which the program implementer interprets the results of the analyses and summarizes the strengths and weaknesses of the intervention.

The effectiveness of the intervention can be measured in several ways, beyond just the rate of behavior change; elements of the TTM or TPB may also be important. If the majority of the target population started the program in in the precontemplation stage of change, then the program may be unlikely to get them to the action stage (i.e., behavior change). Nevertheless, for this group, shifting from precontemplation to contemplation or planning may be a major success. Therefore, we recommend measuring both target behavioral outcomes (e.g., vehicle miles traveled, greenhouse gasses emitted, or number of EVs sold), as well as the TTM and TPB items that were measured during the research step. The TPB variables may act as mediators that explain the behavior change (or lack thereof, if no change is observed in the target behavior).

14.5.5 Step 5: evaluate next steps

In the final step of the program process the program implementer reviews the lessons learned from the program and decides how best to use them. Should the program implementer decide to scale up the program, he or she should calculate the potential costs and benefits of doing so, as well as whether any elements should be added, removed, or changed from the current implementation. Should he or she decide to terminate the program, the lessons learned should be retained and stored in a manner that is accessible the next time a similar program is planned.

14.6 Conclusion

Policymakers can reduce transportation greenhouse gas emissions using many different behavior change interventions. This review touches on several of those interventions and suggests theories and approaches for choosing the most appropriate option. A grounded understanding of human motivation and action, specifically given the context, population and target behavior of interest, can maximize the likelihood of success.

A systematic approach, such as DRIVE, should be used to develop transportation behavior change programs. DRIVE provides a framework and guide for walking program developers through the necessary steps for considering all behavioral science factors in effective programs. Importantly, it emphasizes a goal-focused approach that asks the developer to consider the desired end state and outcome, before deciding on the method to do so. Many program implementers and developers proceed in the reverse order, choosing the method before the goal. We believe a reverse-order strategy is less effective because outcomes are more important than the methods by which they are achieved.

In advocating for the DRIVE procedure, we also strongly suggest that program developers conduct extensive background research, both generally and in their target population. Theories such as the TTM and TPB, or theories of habit formation, can provide guidance for this preliminary research. For example, a preliminary survey of the target community may probe which stage of change they are in, or how they perceive social norms regarding the target behavior. This information can inform the choice of program that is implemented, and how that program is customized.

Notably, the DRIVE approach also describes the importance of evaluation. In our review of programs, few were tested using systematic behavioral science procedures. This should be improved, and program implementers are therefore advised to enlist the help of experts in evaluation to assist with program design and evaluation.

14.6.1 Overlooked behaviors to target and research

Our matrix of transportation domains and behaviors includes two transportation domains (personal and professional transportations) and three behaviors (efficient use, purchasing, and mode switching). However, far more research focuses on personal than professional transportation behavior changes and, within the category of personal transportation, almost no research addresses long-haul travel.

A plethora of personal transportation studies exist on efficient use, purchase behavior, and mode switching. Eco-driver training, eco-driver feedback devices and apps, PAYD insurance, vehicle trials, parking cash out, free bus passes, and monetary incentives are a few that have been studied in depth, but many others also hold promise. Monetary incentives are most commonly used for influencing vehicle purchase behaviors (which is important, given this is a difficult behavior to change), but mandatory vehicle labeling initiatives may also hold promise. Although labeling programs exist, field trials have not been designed to isolate the effects of labels and determine if they can affect behavior. Mode switching and efficient vehicle use can also be influenced by physical environmental changes such as provision of bike infrastructure and street layout, or by laws such as those restricting idling or charging congestion pricing.

One rich potential area for future behavior change interventions could be personal long-haul travel. Relatively, few travelers (higher income earners) are responsible for most emissions from this sector, and these emissions are significant on a per-trip basis. Personal long-haul travel emissions could be effectively reduced through programs encouraging local tourism or nonairplane travel, among others. However, preliminary surveys suggest that this behavior may be difficult to change because the most frequent flyers may be least willing to change. When presented with information about how their behavior affects the environment, they react by rationalizing and minimizing the effect. This poses an interesting challenge to behavior change researchers, but one that is important to tackle.

A second potential area for future behavior intervention is professional transportation. This domain also accounts for relatively high per capita emissions but is virtually unaddressed by the behavior change literature. Eco-driver training, feedback, and onboard devices have demonstrated some potential (most notably airline captain feedback), but no research has been conducted encouraging mode switching, and little has been done for encouraging purchasing of new vehicles or upgrades. These behavior interventions may need to target organizations as well as the individuals within them and, as such, should account for group decision-making and organizational behavior factors. Promising target behaviors may be mode switching to bicycle for last-mile distribution, encouraging purchase of electric taxis, or encouraging fuel-efficient truck purchases.

14.6.2 Looking to the future

The purpose of this review was to describe a landscape of successful transportation behavior change programs that have already been implemented. However, we recognize that this landscape may shift dramatically in the coming decades. Researchers and policymakers suggest that three revolutions in transportation may be on the horizon: automation, ride sharing, and electrification (Sperling, 2018). The widespread adoption of these innovations could drastically change all of the transportation behaviors described in this chapter. Behavior researchers in the future may look at questions of how to encourage the most efficient use of the available technologies. They may examine what factors influence choosing to share as opposed to traveling alone, or when people choose to charge their electric vehicles. It is also challenging to investigate whether new mobility technologies or social communications more effectively bring about a sustainable transport society. New mobility services may bring various changes in people's lives. For example, autonomous vehicles with full automation functions can allow people to perform activities inside vehicles without attending to the road. Autonomous vehicles can double as mobile home, office, or hotel spaces, thus providing a completely new activity space (Jiang and Zhang, 2018). Additionally, ride-sharing allows car owners to not only save money, but also reduce the resources needed for manufacturing new vehicles. And electrifying vehicles make them cheaper to operate than gasoline-powered vehicles. Hence, former drivers may use their financial savings for activities that consume new energy (Yu et al., 2013). Thus, sustainable transportation decisions will only lead to a sustainable future if they co-occur with other sustainable life choices that are interdependent with transportation (Zhang et al., 2017). We look forward to the next set of transportation behavior challenges, with the understanding that some insights may be applicable, others may be adaptable, and still others may be obsolete.

References

Aamaas, B., Borken-Kleefeld, J., Peters, G.P., 2013. The climate impact of travel behavior: a German case study with illustrative mitigation options. Environ. Sci. Policy 33, 273–282.

Abrahamse, W., Keall, M., 2012. Effectiveness of a web-based intervention to encourage carpooling to work: a case study of Wellington, New Zealand. Transp. Policy 21, 45–51.

Achten, W.M., Almeida, J., Muys, B., 2013. Carbon footprint of science: more than flying. Ecol. Indic. 34, 352–355.

Adkins, K.T., 2004. Two Essays in Applied Economics: Pay-As-You-Go Auto Insurance and Privatized Military Housing. Naval Postgraduate School, Monterey, CA.

Ahn, K., Rakha, H., 2009. A field evaluation case study of the environmental and energy impacts of traffic calming. Transp. Res., D: Transp. Environ. 14 (6), 411–424.

Ajzen, I., 1991. The theory of planned behavior. Organ. Behav. Hum. Decis. Process. 50, 179–211.

Alta Planning + Design and the Behavioural Insights Team, 2018. Applying behavioural insights to transportation demand management. Downloaded from: <https://altaplanning.com/resources/behavioral-insights-transportation-demand-management/>.

Amelink, M.H., Mulder, M., Van Paassen, M.M., Flach, J., 2005. Theoretical foundations for a total energy-based perspective flight-path display. Int. J. Aviat. Psychol. 15 (3), 205–231.

Anair, D., Mahmassani, A., 2012. State of Charge: Electric Vehicles' Global Warming Emissions and Fuel-Cost Savings Across the United States. United of Concerned Scientists. Available from: https://www.ucsusa.org/sites/default/files/legacy/assets/documents/clean_vehicles/electric-car-global-warming-emissions-report.pdf.

Andrieu, C., Saint Pierre, G., 2012. Comparing effects of eco-driving training and simple advices on driving behavior. Procedia—Soc. Behav. Sci. 54, 211–220.

Araña, J.E., León, C.J., Moreno-Gil, S., Zubiaurre, A.R., 2013. A comparison of tourists' valuation of climate change policy using different pricing frames. J. Travel Res 52 (1), 82–92.

Bachman, W., Katzev, R., 1982. The effects of non-contingent free bus tickets and personal commitment on urban bus ridership. Transp. Res., A: Gen. 16 (2), 103–108.

Balcombe, R., Mackett, R., Paulley, N., Preston, J., Shires, J., Titheridge, H., et al., 2004. The Demand for Public Transport: A Practical Guide.

Baldasano, J.M., Gonçalves, M., Soret, A., Jiménez-Guerrero, P., 2010. Air pollution impacts of speed limitation measures in large cities: the need for improving traffic data in a metropolitan area. Atmos. Environ. 44 (25), 2997–3006.

Bamberg, S., 2006. Is a residential relocation a good opportunity to change people's travel behavior? Results from a theory-driven intervention study. Environ. Behav. 38, 820–840. Available from: https://doi.org/10.1177/0013916505285091.

Barla, P., Gilbert-Gonthier, M., Castro, M.A.L., Miranda-Moreno, L., 2017. Eco-driving training and fuel consumption: impact, heterogeneity and sustainability. Energy Econ. 62, 187–194.

Barrows, A., Dabney, N., Hayes, J., R. Rosenberg for ideas42, 2018. Behavioural Design Teams: a model for integrating behavioral design in City Government. <http://www.ideas42.org/wp-content/uploads/2018/05/BDT_Playbook_FINAL-digital.pdf>.

Barth, M., Boriboonsomsin, K., 2009. Energy and emissions impacts of a freeway-based dynamic eco-driving system. Transp. Res., D: Transp. Environ. 14 (6), 400–410.

Beevers, S.D., Carslaw, D.C., 2005. The impact of congestion charging on vehicle emissions in London. Atmos. Environ. 39 (1), 1–5.

Bento, A.M., Hughes, J.E., Kaffine, D., 2013. Carpooling and driver responses to fuel price changes: evidence from traffic flows in Los Angeles. J. Urban Econ. 77, 41–56.

Beusen, B., Broekx, S., Denys, T., Beckx, C., Degraeuwe, B., Gijsbers, M., et al., 2009. Using on-board logging devices to study the longer-term impact of an eco-driving course. Transp. Res., D: Transp. Environ. 14 (7), 514–520.

Bjerkan, K.Y., Nørbech, T.E., Nordtømme, M.E., 2016. Incentives for promoting battery electric vehicle (BEV) adoption in Norway. Transp. Res., D: Transp. Environ. 43, 169–180.

Bigazzi, A.Y., Rouleau, M., 2017. Can traffic management strategies improve urban air quality? A review of the evidence. J. Transp. Health 7, 111–124.

Böhler, S., Grischkat, S., Haustein, S., Hunecke, M., 2006. Encouraging environmentally sustainable holiday travel. Transp. Res., A: Policy Pract. 40 (8), 652–670.

Bolderdijk, J.W., Knockaert, J., Steg, E.M., Verhoef, E.T., 2011. Effects of pay-as-you-drive vehicle insurance on young drivers' speed choice: results of a Dutch field experiment. Accid. Anal. Prev. 43 (3), 1181–1186.

Bonaccio, S., Dalal, R.S., 2006. Advice taking and decision-making: An integrative literature review, and implications for the organizational sciences. Organ. Behav. Hum. Dec. 101 (2), 127–151.

Bonham, J., Koth, B., 2010. Universities and the cycling culture. Transp. Res., D: Transp. Environ. 15 (2), 94–102.

Bordoff, J.E., Noel, P.J., 2008. The Impact of Pay-As-You-Drive Auto Insurance in California. Brookings Institution.

Boujbel, L., d'Astous, A., 2012. Voluntary simplicity and life satisfaction: exploring the mediating role of consumption desires. J. Consum. Behav. 11 (6), 487–494.

Bradley, M.J., Associates, 2014. Comparison of Energy Use & CO2 Emissions From Different Transportation Modes. Submitted for the American Bus Association. Available from: https://www.buses.org/assets/images/uploads/general/Report%20-%20Energy%20Use%20and%20Emissions.pdf.

Bridgeable, 2017. Designing for behaviour change toolkit. Downloaded from: <https://toolkit.bridgeable.com/>.

Buehler, R., 2012. Determinants of bicycle commuting in the Washington, DC region: the role of bicycle parking, cyclist showers, and free car parking at work. Transp. Res., D: Transp. Environ. 17 (7), 525–531.

Buehler, R., Pucher, J., 2012. Cycling to work in 90 large American cities: new evidence on the role of bike paths and lanes. Transportation 39 (2), 409–432.

Bühler, F., Cocron, P., Neumann, I., Franke, T., Krems, J.F., 2014. Is EV experience related to EV acceptance? Results from a German field study. Transp. Res., F: Traffic Psychol. Behav. 25, 34–49.

Buliung, R.N., Soltys, K., Bui, R., Habel, C., Lanyon, R., 2010. Catching a ride on the information super-highway: toward an understanding of internet-based carpool formation and use. Transportation 37 (6), 849–873. Available from: https://doi.org/10.1007/s11116-010-9266-0.

Burris, M., Alemazkoor, N., Benz, R., Wood, N.S., 2014. The impact of HOT lanes on carpools. Res. Transp. Econ. 44, 43–51.

Buxbaum, J.N., 2006. Mileage-based user fee demonstration project: pay-as-you-drive experimental findings. No. MN/RC−2006-39A.

Camilleri, A.R., Larrick, R.P., 2014. Metric and scale design as choice architecture tools. J. Public Policy Mark. 33 (1), 108−125.

Canning, P.E., Hughes, S.J., Hellawell, E.E., Gatersleben, B.C.M., Fairhead, C.J., 2010. Reasons for participating in formal employer-led carpool schemes as perceived by their users. Transp. Plan. Technol. 33 (8), 733−745.

Catton, L., Starr, A., Noyes, J.M., Fisher, A., Price, T., 2007. Designing energy display formats for civil aircraft: reply to Amelink, Mulder, van Paassen, and Flach. Int. J. Aviat. Psychol. 17 (1), 31−40.

Caulfield, B., Brazil, W., Fitzgerald, K.N., Morton, C., 2014. Measuring the success of reducing emissions using an on-board eco-driving feedback tool. Transp. Res., D: Transp. Environ. 32, 253−262.

Cervero, R., Griesenbeck, B., 1988. Factors influencing commuting choices in suburban labor markets: a case analysis of Pleasanton, California. Transp. Res., A: Gen. 22 (3), 151−161.

Chan, P. et al. for IRS, 2017. IRS behavioral insights toolkit. <https://www.irs.gov/pub/irs-soi/17rpirsbehavioralinsights.pdf>.

Chen, F., 2018. The Impact of High Occupancy Vehicle (HOV) Lane Access on Hybrid-Electric Vehicle Adoption: Evidence From US States (Doctoral dissertation). Georgetown University.

Chen, C., Chao, W., 2011. Habitual or reasoned? Using the theory of planned behavior, technology acceptance model, and habit to examine switching intentions toward public transit. Transp. Res., F: Traffic Psychol. Behav. 14, 128−137. Available from: https://doi.org/10.1016/j.trf.2010.11.006.

Chenoweth, J., 2009. Is tourism with a low impact on climate possible? Worldwide Hospitality Tourism Themes 1 (3), 274−287.

Cohen, S.A., Higham, J.E., Cavaliere, C.T., 2011. Binge flying: behavioural addiction and climate change. Ann. Tourism Res. 38 (3), 1070−1089.

Cristea, A., Hummels, D., Puzzello, L., Avetisyan, M., 2013. Trade and the greenhouse gas emissions from international freight transport. J. Environ. Econ. Manage. 65 (1), 153−173.

Degraeuwe, B., Beusen, B., 2013. Corrigendum on the paper "Using on-board data logging devices to study the longer-term impact of an eco-driving course". Transp. Res., D: Transp. Environ. 19, 48−49.

De Nazelle, A., 2016. Influencing people for healthier and more sustainable travel behaviours using smart phone technology. Front. Public Health 4. Available from: https://doi.org/10.3389/conf.FPUBH.2016.01.00011.

Den Tonkelaar, W.A.M., 1991. Speed limits, effects and benefits in terms of energy efficiency and reduction of emissions, Studies in Environmental Science, vol. 45. Elsevier, pp. 261−270.

Deslauriers, B.C., Everett, P.B., 1977. Effects of intermittent and continuous token reinforcement on bus ridership. J. Appl. Psychol. 62 (4), 369−375.

Diamond, D., 2008. Impact of high occupancy vehicle (HOV) lane incentives for hybrids in Virginia. J. Public Transp. 11 (4), 3.

Dijksterhuis, C., Lewis-Evans, B., Jelijs, B., de Waard, D., Brookhuis, K., Tucha, O., 2015. The impact of immediate or delayed feedback on driving behaviour in a simulated pay-as-you-drive system. Accid. Anal. Prev. 75, 93−104.

Dill, J., 2009. Bicycling for transportation and health: the role of infrastructure. J. Public Health Policy 30 (1), S95−S110.

Dolf, M., Teehan, P., 2015. Reducing the carbon footprint of spectator and team travel at the University of British Columbia's varsity sports events. Sport Manag. Rev. 18 (2), 244−255.

Edwards, P., Steinbach, R., Green, J., Petticrew, M., Goodman, A., Jones, A., et al., 2013. Health impacts of free bus travel for young people: evaluation of a natural experiment in London. J. Epidemiol. Community Health 67 (8), 641–647.

Egbue, O., Long, S., 2012. Barriers to widespread adoption of electric vehicles: an analysis of consumer attitudes and perceptions. Energy Policy 48, 717–729. Available from: https://doi.org/10.1016/j.enepol.2012.06.009.

Ehrhart, M.G., Naumann, S.E., 2004. Organizational citizenship behavior in work groups: a group norms approach. J. Appl. Psychol. 89 (6), 960.

Elvik, R., Vaa, T., Hoye, A., Sorensen, M. (Eds.), 2009. The Handbook of Road Safety Measures. Emerald Group Publishing.

Environmental Protection Agency (EPA), 2018. Fast facts: U.S. Transportation Sector Greenhouse Gas Emissions, 1990-2016. EPA-420-F-18-013, July 2018 [Fact Sheet]. Retrieved from: <https://nepis.epa.gov/Exe/ZyPDF.cgi?Dockey = P100USI5.pdf>.

EPA and DOT, 2010. Environmental Protection Agency Fuel Economy Label—Final Report. US Environmental Protection Agency and Department of Transportation, Washington, DCReport No. GS-23F-0364P. Available from: https://www.epa.gov/fueleconomy.

Eriksson, L., Forward, S.E., 2011. Is the intention to travel in a pro-environmental manner and the intention to use the car determined by different factors? Transp. Res., D: Transp. Environ. 16, 372–376. Available from: https://doi.org/10.1016/j.trd.2011.02.003.

Esser, J.K., 1998. Alive and well after 25 years: a review of groupthink research. Organ. Behav. Hum. Decis. Process. 73 (2–3), 116–141.

Farag, S., Lyons, G., 2010. Explaining public transport information use when a car is available: attitude theory empirically investigated. Transportation 37, 897–913. Available from: https://doi.org/10.1007/s11116-010-9265-1.

Ferguson, E., 1997. The rise and fall of the American carpool: 1970–1990. Transportation 24 (4), 349–376.

Ferreira, A., 2015. The use of tricycles in the physical distribution of beverages in Rio de Janeiro city: a qualitative approach of the operation. Rev. Engenharia Tecnol. 7, 136–151.

Forward, S.E., 2014. Exploring people's willingness to bike using a combination of the theory of planned behavioural and the transtheoretical model. Eur. Rev. Appl. Psychol. 64, 151–159. Available from: https://doi.org/10.1016/j.erap.2014.04.002.

Franke, T., Neumann, I., Bühler, F., Cocron, P., Krems, J.F., 2012. Experiencing range in an electric vehicle: understanding psychological barriers. Appl. Psychol. 61 (3), 368–391.

Franke, T., Günther, M., Trantow, M., Krems, J.F., Vilimek, R., Keinath, A., 2014. Examining user-range interaction in battery electric vehicles—a field study approach. Adv. Hum. Aspects Transp., II 334–344.

Franke, T., Rauh, N., Krems, J.F., 2016. Individual differences in BEV drivers' range stress during first encounter of a critical range situation. Appl. Ergon. 57, 28–35.

Franke, T., Günther, M., Trantow, M., Krems, J.F., 2017. Does this range suit me? Range satisfaction of battery electric vehicle users. Appl. Ergon. 65, 191–199.

Fraser, H., Soanes, K., Jones, S.A., Jones, C.S., Malishev, M., 2017. The value of virtual conferencing for ecology and conservation. Conserv Biol 31 (3), 540–546.

Friman, M., Huck, J., Olsson, L.E., 2017. Transtheoretical model of change during travel behavior interventions: an integrative review. Int. J. Environ. Res. Public Health 14. Available from: https://doi.org/10.3390/ijerph14060581.

Fujii, S., Kitamura, R., 2003. What does a one-month free bus ticket do to habitual drivers? An experimental analysis of habit and attitude change. Transportation 30 (1), 81–95.

Gao, H.O., Kitirattragarn, V., 2008. Taxi owners' buying preferences of hybrid-electric vehicles and their implications for emissions in New York City. Transp. Res., A: Policy Pract. 42 (8), 1064–1073.

Gardner, B., Abraham, C., 2007. What drives car use? A grounded theory analysis of commuters' reasons for driving. Transp. Res., F: Traffic Psychol. Behav. 10 (3), 187–200.

Gardner, G.T., Stern, P.C., 2008. The short list: the most effective actions US households can take to curb climate change. Environment: Sci. Policy Sustain. Dev. 50 (5), 12–25.

Geller, E.S., 2002. The challenge of increasing proenvironmental behavior. In: Churchman, A. (Ed.), Handbook of Environmental Psychology. John Wiley & Sons Inc, Hoboken, NJ, pp. 525–540.

Ghafghazi, G., Hatzopoulou, M., 2014. Simulating the environmental effects of isolated and area-wide traffic calming schemes using traffic simulation and microscopic emission modeling. Transportation 41 (3), 633–649.

Ghafghazi, G., Hatzopoulou, M., 2015. Simulating the air quality impacts of traffic calming schemes in a dense urban neighborhood. Transp. Res., D: Transp. Environ. 35, 11–22.

Gilbey, A., Perezgonzalez, J.D., Tani, K., 2011. The effect of information about the environmental impact of flying and people's desire to fly. Int. J. Appl. Aviat. Stud. 11, 59–65.

Gosnell, G.K., List, J.A., Metcalfe, R., 2016. A New Approach to An Age-Old Problem: Solving Externalities by Incenting Workers Directly. National Bureau of Economic Research (No. w22316).

Graus, W., Worrell, E., 2008. The principal–agent problem and transport energy use: Case study of company lease cars in the Netherlands. Energy Policy 36 (10), 3745–3753.

Guerrero, S.E., 2014. Modeling fuel saving investments and fleet management in the trucking industry: the impact of shipment performance on GHG emissions. Transp. Res., E: Logist. Transp. Rev. 68, 178–196.

Habib, K.M.N., Tian, Y., Zaman, H., 2011. Modelling commuting mode choice with explicit consideration of carpool in the choice set formation. Transportation 38 (4), 587–604.

Hansen, P.G., K. Schmidt for the OECD, 2018. BASIC toolkit and ethical guidelines for policy makers. <http://www.oecd.org/gov/regulatory-policy/BASIC-Toolkit-Draft-for-Consultation.pdf>.

Hardman, S., Chandan, A., Tal, G., Turrentine, T., 2017. The effectiveness of financial purchase incentives for battery electric vehicles—a review of the evidence. Renew. Sustain. Energy Rev. 80, 1100–1111.

Harvey, G., Deakin, E., 1998. Technical Methods for Analyzing Pricing Measures to Reduce Transportation Emissions. USEPA, *Report*.

Hartman, K., Dowd, E., 2015. State efforts to promote hybrid and electric vehicles. National conference of state legislatures 7, 2019. from: Retreived July https://www.lec.leg.mn/2014/NCSLHybridEVincentivesOct2014.pdf.

Heath, Y., Gifford, R., 2002. Extending the theory of planned behavior: predicting the use of public transportation. J. Appl. Soc. Psychol. 32, 2154–2189. Available from: https://doi.org/10.1111/j.1559-1816.2002.tb02068.x.

Heinen, E., Maat, K., Van Wee, B., 2011. The role of attitudes toward characteristics of bicycle commuting on the choice to cycle to work over various distances. Transp. Res., D: Transp. Environ. 16 (2), 102–109.

Hemingway, C.A., Maclagan, P.W., 2004. Managers' personal values as drivers of corporate social responsibility. J. Bus. Ethics 50 (1), 33–44.

Herzog, E., Bricka, S., Audette, L., Rockwell, J., 2006. Do employee commuter benefits reduce vehicle emissions and fuel consumption? Results of the fall 2004 best

workplaces for commuters survey. In: Transportation Research Board Annual Meeting, Washington, DC.
Hess, D.B., 2001. Effect of free parking on commuter mode choice: evidence from travel diary data. Transp. Res. Rec. 1753 (1), 35–42.
Higham, J.E., Cohen, S.A., Cavaliere, C.T., 2014. Climate change, discretionary air travel, and the "Flyers' Dilemma". J. Travel Res 53 (4), 462–475.
Ho, S.H., Wong, Y.D., Chang, V.W.C., 2015. What can eco-driving do for sustainable road transport? Perspectives from a city (Singapore) eco-driving programme. Sustain. Cities Soc. 14, 82–88.
Huang, Y., Ng, E.C., Zhou, J.L., Surawski, N.C., Chan, E.F., Hong, G., 2018. Eco-driving technology for sustainable road transport: a review. Renew. Sustain. Energy Rev. 93, 596–609.
Huertas, J.I., Díaz, J., Giraldo, M., Cordero, D., Tabares, L.M., 2018. Eco-driving by replicating best driving practices. Int. J. Sustain. Transp. 12 (2), 107–116.
Isaacs, C., 1980. Behavioral and ecological modifications that have increased bus ridership. Behav. Ther. 3 (5), 10–12.
Isenberg, D.J., 1986. Group polarization: a critical review and meta-analysis. J. Pers. Soc. Psychol. 50 (6), 1141.
James, A., Ravichandran, S., Chuang, N.K., Bolden III, E., 2017. Using lifestyle analysis to develop lodging packages for staycation travelers: An exploratory study. J. Qual. Assur. Hosp. Tour. 18 (4), 387–415.
Jamson, S.L., Hibberd, D.L., Jamson, A.H., 2015. Drivers' ability to learn eco-driving skills; effects on fuel efficient and safe driving behaviour. Transp. Res., C: Emerg. Technol. 58, 657–668.
Jeffreys, I., Graves, G., Roth, M., 2018. Evaluation of eco-driving training for vehicle fuel use and emission reduction: a case study in Australia. Transp. Res., D: Transp. Environ. 60, 85–91.
Jenn, A., Azevedo, I.L., Ferreira, P., 2013. The impact of federal incentives on the adoption of hybrid electric vehicles in the United States. Energy Econ. 40, 936–942.
Jiang, Y., Zhang, J., 2018. Capturing ownership behavior of autonomous vehicles in Japan based on a stated preference survey and a mixed logit model with repeated choices. Int. J. Sustain. Transp 13 (10). Available from: https://doi.org/10.1080/15568318.2018.1517841.
Katzev, R., Bachman, W., 1982. Effects of deferred payment and fare manipulations on urban bus ridership. J. Appl. Psychol. 67 (1), 83.
Kim, S.Y., Shin, D.J., Yoon, H.J., Bae, H.C., Kim, D.S., 2011. Development of Eco-Driving Guide System. SAE Technical Paper (No. 2011-28-0034).
Kircher, K., Fors, C., Ahlstrom, C., 2014. Continuous versus intermittent presentation of visual eco-driving advice. Transp. Res., F: Traffic Psychol. Behav. 24, 27–38.
Klæboe, R., Kolbenstvedt, M., Clench-Aas, J., Bartonova, A., 2000. Oslo traffic study—part 1: an integrated approach to assess the combined effects of noise and air pollution on annoyance. Atmos. Environ. 34 (27), 4727–4736.
Kormos, C., Gifford, R., 2014. The validity of self-report measures of pro-environmental behavior. J. Environ. Psychol. 40, 359–371. Available from: https://doi.org/10.1016/j.jenvp.2014.09.00.
Kormos, C., Sussman, R., 2018. Auto Buyers' Valuation of Fuel Economy: A Randomized Stated Choice Experiment. Consumers Union. <http://www.ConsumersUnion.org/MPGreport>.
Lai, W.T., 2015. The effects of eco-driving motivation, knowledge and reward intervention on fuel efficiency. Transp. Res., D: Transp. Environ. 34, 155–160.
Langbroek, J.H., Franklin, J.P., Susilo, Y.O., 2016. The effect of policy incentives on electric vehicle adoption. Energy Policy 94, 94–103.

Larue, G.S., Malik, H., Rakotonirainy, A., Demmel, S., 2014. Fuel consumption and gas emissions of an automatic transmission vehicle following simple eco-driving instructions on urban roads. IET Intel. Transport Syst. 8 (7), 590–597.

Lerner, J.S., Li, Y., Valdesolo, P., Kassam, K.S., 2015. Emotion and decision making. Annu. Rev. Psychol. 66, 799–823.

Litman, T., 2004. Pay-As-You-Drive Pricing for Insurance Affordability. Victoria Transport Policy Institute. <www.vtpi.org>.

Litman, T., 2005. Pay-as-you-drive pricing and insurance regulatory objectives. J. Insur. Regul. 23 (3).

Liu, B., Chen, C., Zhang, B., Bu, M., Bi, J., Yu, Y., 2012. Fuel use pattern and determinants of taxi drivers' fuel choice in Nanjing, China. J. Cleaner Prod. 33, 60–66.

Liu, Y., Sheng, H., Mundorf, N., Redding, C., Ye, Y., 2017. Integrating norm activation model and theory of planned behavior to understand sustainable transport behavior: evidence from China. Int. J. Environ. Res. Public Health 14. Available from: https://doi.org/10.3390/ijerph14121593.

Luoma, J., Sivak, M., 2012. Interactions of environmental and safety measures for sustainable road transportation. Eur. Transp. Res. Rev. 4 (4), 189–199.

Ly, K., Mažar, N., Zhao, M., Soman, D. for BEAR UofT, 2013. A Practitioner's Guide to Nudging.

Madireddy, M., De Coensel, B., Can, A., Degraeuwe, B., Beusen, B., De Vlieger, I., et al., 2011. Assessment of the impact of speed limit reduction and traffic signal coordination on vehicle emissions using an integrated approach. Transp. Res., D: Transp. Environ. 16 (7), 504–508.

Magaña, V.C., Muñoz-Organero, M., 2016. Artemisa: a personal driving assistant for fuel saving. IEEE Trans. Mob. Comput. 15 (10), 2437–2451.

Marshall, W.E., Garrick, N.W., 2012. Community design and how much we drive. J. Transp. Land Use 5 (2), 5–21.

Masson-Delmotte. V., Zhai. P., Pörtner H.O., Roberts D., Skea J., Shukla P.R., et al. (Eds.), 2018. Global Warming of 1.5°C. An IPCC Special Report on the Impacts of Global Warming of 1.5°C Above Pre-Industrial Levels and Related Global Greenhouse Gas Emission Pathways, in the Context of Strengthening the Global Response to the Threat of Climate Change, Sustainable Development, and Efforts to Eradicate Poverty. IPCC—Intergovernmental Panel on Climate Change. Retrieved from: <https://www.ipcc.ch/sr15/>.

McDonald, R.I., Chai, H.Y., Newell, B.R., 2015. Personal experience and the 'psychological distance' of climate change: an integrative review. J. Environ. Psychol. 44, 109–118.

McKercher, B., Prideaux, B., Cheung, C., Law, R., 2010. Achieving voluntary reductions in the carbon footprint of tourism and climate change. J. Sustain. Tourism 18 (3), 297–317.

Menczer, W.B., 2007. Guaranteed ride home programs: a study of program characteristics, utilization, and cost. J. Public Transp. 10 (4), 7.

Miller, G., 1991. The Impacts of Parking Prices on Commuter Travel. Metropolitan Washington Council of Governments.

Mokhtarian, P.L., 2002. Telecommunications and travel: the case for complementarity. J. Ind. Ecol. 6 (2), 43–57.

Mokhtarian, P., 2009. If telecommunication is such a good substitute for travel, why does congestion continue to get worse? Transp. Lett. 1 (1), 1–17.

Molan, A.M., Kordani, A.A., 2014. Optimization of speed hump profiles based on vehicle dynamic performance modeling. J. Transp. Eng. 140 (8), 04014035.

Mundorf, N., Redding, C.A., Paiva, A.L., 2018. Sustainable transportation attitudes and health behavior change: evaluation of a brief stage-targeted video intervention. Int. J. Environ. Res. Public Health 15. Available from: https://doi.org/10.3390/ijerph15010150.

National Household Travel Survey (2020), 2009. NHTS data. <https://nhts.ornl.gov/downloads> (accessed February 2019.).
Nelson/Nygaard Consulting Associates. 2005. Alameda County CMA Guaranteed Ride Home Program Evaluation: Executive Summary.
Neoh, J.G., Chipulu, M., Marshall, A., 2017. What encourages people to carpool? An evaluation of factors with meta-analysis. Transportation 44 (2), 423–447.
Netherlands Enterprise Agency, 2015. Electromobility in the Netherlands Highlights 2014. <https://www.rvo.nl/sites/default/files/2015/04/Electromobility%20in%20the%20Netherlands%20Highlights%202014.pdf/> (accessed 25.02.19.).
Nikoleris, T., Gupta, G., Kistler, M., 2011. Detailed estimation of fuel consumption and emissions during aircraft taxi operations at Dallas/Fort Worth International Airport. Transp. Res., D: Transp. Environ. 16 (4), 302–308.
O'Fallon, C., 2010. Bike Now: Exploring methods of building sustained participation in cycle commuting in New Zealand. Road Transp. Res. 19 (2), 77.
O'Mahony, M., Geraghty, D., Humphreys, I., 2000. Distance and time based road pricing trial in Dublin. Transportation 27 (3), 269–283.
Oliveira, C., Bandeira, R.A.M., Goes, G.V., Gonçalves, D.S., D'Agosto, M., 2017. Sustainable vehicles-based alternatives in last mile distribution of urban freight transport: a systematic literature review. Sustainability 9 (8), 1324.
Orfila, O., Saint Pierre, G., Messias, M., 2015. An android based ecodriving assistance system to improve safety and efficiency of internal combustion engine passenger cars. Transp. Res., C: Emerg. Technol. 58, 772–782.
Organization for Economic Co-operation and Development [OECD]/International Transport Forum [ITF], 2015. ITF Transport Outlook 2015. OECD Publishing, Paris. Available from: https://doi.org/10.1787/9789282107782-en.
Preston, I., White, V., Thumim, J., Bridgeman, T., 2013. Distribution of Carbon Emissions in the UK: Implications for Domestic Energy Policy. Joseph Rowntree Foundation, Retrieved on 29.03.19 from: https://www.jrf.org.uk/sites/default/files/jrf/migrated/files/carbon-emissions-energy-policy-full.pdf.
Prochaska, J.O., DiClemente, C.C., 1982. Transtheoretical therapy: toward a more integrative model of change. Psychother. Ther. Res. Pract. 19, 276–288. Available from: https://doi.org/10.1037/h0088437.
Prochaska, J.O., Velicer, W.F., 1997. The transtheoretical model of health behavior change. Am. J. Health Promot. 12, 38–48.
Rauh, N., Franke, T., Krems, J.F., 2015. Understanding the impact of electric vehicle driving experience on range anxiety. Hum. Factors 57 (1), 177–187.
Rauh, N., Franke, T., Krems, J.F., 2017. First-time experience of critical range situations in BEV use and the positive effect of coping information. Transp. Res., F: Traffic Psychol. Behav. 44, 30–41.
Rehmatulla, N., Smith, T., 2015. Barriers to energy efficient and low carbon shipping. Ocean Eng. 110, 102–112.
Rios-Torres, J., Sauras-Perez, P., Alfaro, R., Taiber, J., Pisu, P., 2015. Eco-driving system for energy efficient driving of an electric bus. SAE Int. J. Passenger Cars Electron. Electr. Syst. 8 (2015-01-0158), 79–89.
Rogers, E.M., 1982. Diffusion of Innovations, third ed. The Free Press, New York.
Rolim, C.C., Baptista, P.C., Duarte, G.O., Farias, T.L., 2014. Impacts of on-board devices and training on light duty vehicle driving behavior. Procedia—Soc. Behav. Sci. 111, 711–720.
Rolim, C., Baptista, P., Duarte, G., Farias, T., Pereira, J., 2017. Real-time feedback impacts on eco-driving behavior and influential variables in fuel consumption in a Lisbon urban bus operator. IEEE Trans. Intell. Transp. Syst. 18 (11), 3061–3071.

Rommerskirchen, C.P., Helmbrecht, M., Bengler, K.J., 2014. The impact of an anticipatory eco-driver assistant system in different complex driving situations on the driver behavior. IEEE Intell. Transp. Syst. Mag. 6 (2), 45–56.

Rose, G., Marfurt, H., 2007. Travel behaviour change impacts of a major ride to work day event. Transp. Res., A: Policy Pract. 41 (4), 351–364.

Rouzikhah, H., King, M., Rakotonirainy, A., 2013. Examining the effects of an eco-driving message on driver distraction. Accid. Anal. Prev. 50, 975–983.

Rutty, M., Matthews, L., Andrey, J., Del Matto, T., 2013. Eco-driver training within the City of Calgary's municipal fleet: monitoring the impact. Transp. Res., D: Transp. Environ. 24, 44–51.

Sangkapichai, M., Saphores, J.D., 2009. Why are Californians interested in hybrid cars? J. Environ. Plann. Manage. 52 (1), 79–96.

Schall, D.L., Mohnen, A., 2017. Incentivizing energy-efficient behavior at work: an empirical investigation using a natural field experiment on eco-driving. Appl. Energy 185, 1757–1768.

Schmalfuß, F., Mühl, K., Krems, J.F., 2017. Direct experience with battery electric vehicles (BEVs) matters when evaluating vehicle attributes, attitude and purchase intention. Transp. Res., F: Traffic Psychol. Behav. 46, 47–69.

Schoemaker, J., Allen, J., Huschebeck, M., Monigl, J., 2005. Quantification of Urban Freight Transport Effects Best Urban Freight Solutions II. Available from: <http://www.bestufs.net/download/BESTUFS_II/key_issuesII/BESTUF_Quantification_of_effects.pdf> (accessed 25.02.19.).

Schoner, J.E., Levinson, D.M., 2014. The missing link: bicycle infrastructure networks and ridership in 74 US cities. Transportation 41 (6), 1187–1204.

Schultz, P.W., 2014. Strategies for promoting proenvironmental behavior. Eur. Psychol. 19 (2), 107–117.

Shewmake, S., 2012. Can carpooling clear the road and clean the air? Evidence from the literature on the impact of HOV lanes on VMT and air pollution. J. Plann. Lit. 27 (4), 363–374.

Shewmake, S., Jarvis, L., 2014. Hybrid cars and HOV lanes. Transp. Res., A: Policy Pract. 67, 304–319.

Shoup, D.C., 1995. An opportunity to reduce minimum parking requirements. J. Am. Plann. Assoc. 61 (1), 14–28.

Sperling, D., 2018. Three Revolutions: Steering Automated, Shared, and Electric Vehicles to a Better Future. Island Press.

Spinellis, D., Louridas, P., 2013. The carbon footprint of conference papers. PLoS One 8 (6), e66508.

Steg, L., Vlek, C., 2009. Encouraging pro-environmental behaviour: an integrative review and research agenda. J. Environ. Psychol. 29, 309–317.

Stillwater, T., Kurani, K.S., Mokhtarian, P.L., 2017. The combined effects of driver attitudes and in-vehicle feedback on fuel economy. Transp. Res., D: Transp. Environ. 52, 277–288.

Strömberg, H.K., Karlsson, I.M., 2013. Comparative effects of eco-driving initiatives aimed at urban bus drivers–Results from a field trial. Transp. Res., D: Transp. Environ. 22, 28–33.

Strömberg, H., Karlsson, I.M., Rexfelt, O., 2015. Eco-driving: drivers' understanding of the concept and implications for future interventions. Transp. Policy 39, 48–54.

Sullman, M.J., Dorn, L., Niemi, P., 2015. Eco-driving training of professional bus drivers—does it work? Transp. Res., C: Emerg. Technol. 58, 749–759.

Symmons, M.A., Rose, G., 2009. Ecodrive training delivers substantial fuel savings for heavy vehicle drivers. Proceedings of the Fifth International Driving Symposium on

Human Factors in Driver Assessment, Training and Vehicle Design. Public Policy Center, University of Iowa, Iowa City, IA, pp. 46–53.

Teal, R.F., 1987. Carpooling: who, how and why. Transp. Res., A: Gen. 21 (3), 203–214.

Therrien, S., Brauer, M., Fuller, D., Gauvin, L., Teschke, K., Winters, M., 2014. Identifying the leaders: applying diffusion of innovation theory to use of a public bike-share system in Vancouver, Canada. Transp. Res. Rec. 2468, 74–83. Available from: https://doi.org/10.3141/2468-09.

Thogersen, J., Moller, B., 2008. Breaking car use habits: the effectiveness of a free one-month travel card. Transportation 35, 329–345. Available from: https://doi.org/10.1007/s11116-008-9160-1.

Toth, C., 2015. Carpooling in Hungary: can it reduce the GHG emissions of personal transport? Eur. Transp.—Trasporti Europei 58, 4.

Tyng, C.M., Amin, H.U., Saad, M., Malik, A.S., 2017. The Influences of Emotion on Learning and Memory. Front. Psychol. 8, 1454.

Ubbels, B.J., Verhoef, E.T., 2006. Behavioural responses to road pricing: empirical results from a survey among Dutch car owners. Eur. Transp.—Trasporti Europei 31, 101–117.

Ungemach, C., Camilleri, A.R., Johnson, E.J., Larrick, R.P., Weber, E.U., 2017. Translated attributes as choice architecture: aligning objectives and choices through decision signposts. Management Science 64 (5), 2445–2459.

U.S. Department of Transportation, 2017. Average Age of Automobiles and Trucks in Operation in the United States. Bureau of Transportation Statistics. Retrieved from: <https://www.bts.gov/content/average-age-automobiles-and-trucks-operation-united-states>. (accessed 13.03.19.).

Vagg, C., Brace, C., Hari, D., Akehurst, S., Ash, L., 2013. A Driver Advisory Tool to Reduce Fuel Consumption. SAE Technical Paper (No. 2012-01-2087).

Van Acker, V., Mokhtarian, P.L., Witlox, F., 2011. Going soft: on how subjective variables explain modal choices for leisure travel. Eur. J. Transp. Infrastruct. Res.—Delft 11 (2), 115–146.

Van der Loop, H., Willigers, J., Haaijer, R., 2019. Empirical estimation of effects of flexible working on mobility and congestion in the Netherlands 2000-2016. In: Transportation Research Annual Meeting, Washington, DC.

Van der Voort, M., Dougherty, M.S., van Maarseveen, M., 2001. A prototype fuel-efficiency support tool. Transp. Res., C: Emerg. Technol. 9 (4), 279–296.

Vanoutrive, T., Van De Vijver, E., Van Malderen, L., Jourquin, B., Thomas, I., Verhetsel, A., et al., 2012. What determines carpooling to workplaces in Belgium: location, organisation, or promotion? J. Transp. Geogr. 22, 77–86.

Vantage Marketing Research, 2003. Commuter transportation survey 2003. Denver Regional Council of Governments RideArrangers, Denver, CO.

Van Vugt, M., Van Lange, P.A., Meertens, R.M., Joireman, J.A., 1996. How a structural solution to a real-world social dilemma failed: A field experiment on the first carpool lane in Europe. Soc. Psychol. Q. 364–374.

Várhelyi, A., 2002. The effects of small roundabouts on emissions and fuel consumption: a case study. Transp. Res., D: Transp. Environ. 7 (1), 65–71.

Vernon, D., Meier, A., 2012. Identification and quantification of principal–agent problems affecting energy efficiency investments and use decisions in the trucking industry. Energy Policy 49, 266–273.

Verplanken, B., Aarts, H., 1999. Habit, attitude, and planned behaviour: is habit an empty construct or an interesting case of goal-directed automaticity? Eur. Rev. Soc. Psychol. 10, 101–134. Available from: https://doi.org/10.1080/14792779943000035.

Verplanken, B., Roy, D., 2016. Empowering interventions to promote sustainable lifestyles: testing the habit discontinuity hypothesis in a field experiment. J. Environ. Psychol. 45, 127–134.

Waechter, S., Sütterlin, B., Siegrist, M., 2015. Desired and undesired effects of energy labels—An eye-tracking study. PLoS One 10 (7), e0134132.
Walton, R., January 2017. Georgia electric vehicle sales shrink 80% in wake of tax credit repeal. Utility Dive. Retrieved March 26, 2019 from: <https://www.utilitydive.com/news/georgia-electric-vehicle-sales-shrink-80-in-wake-of-tax-credit-repeal/434092/>.
Wang, M., Daamen, W., Hoogendoorn, S., Van Arem, B., 2011. Estimating acceleration, fuel consumption, and emissions from macroscopic traffic flow data. Transp. Res. Rec. 2260 (1), 123–132.
Wang, X., Liu, C., Kostyniuk, L., Shen, Q., Bao, S., 2014. The influence of street environments on fuel efficiency: insights from naturalistic driving. Int. J. Environ. Sci. Technol. 11 (8), 2291–2306.
Wang, S., Fan, J., Zhao, D., Yang, S., Fu, Y., 2016. Predicting consumers' intention to adopt hybrid electric vehicles: using an extended version of the theory of planned behavior model. Transportation 43, 123–143.
Wee, S., Coffman, M., La Croix, S., 2018. Do electric vehicle incentives matter? Evidence from the 50 US states. Res. Policy 47 (9), 1601–1610.
Weinberger, R., Kaehny, J., Rufo, M., 2010. US parking policies: an overview of management strategies. In: Presentation at Transportation Research Board Annual Meeting.
Wilson, R.W., 1992. Estimating the travel and parking demand effects of employer-paid parking. Reg. Sci. Urban Econ. 22 (1), 133–145.
Wu, Y., Zhao, X., Rong, J., Zhang, Y., 2017. How eco-driving training course influences driver behavior and comprehensibility: a driving simulator study. Cognit. Technol. Work 19 (4), 731–742.
Xu, Y., Liu, H., Rodgers, M.O., Guin, A., Hunter, M., Sheikh, A., et al., 2017. Understanding the emission impacts of high-occupancy vehicle (HOV) to high-occupancy toll (HOT) lane conversions: experience from Atlanta, Georgia. J. Air Waste Manage. Assoc. 67 (8), 910–922.
Yang, J., Dong, J., Hu, L., 2018a. Design government incentive schemes for promoting electric taxis in China. Energy Policy 115, 1–11.
Yang, W.H., Wong, R.C.P., Szeto, W.Y., 2018b. Modeling the acceptance of taxi owners and drivers to operate premium electric taxis: policy insights into improving taxi service quality and reducing air pollution. Transp. Res., A: Policy Pract. 118, 581–593.
Yu, B., Zhang, J., Fujiwara, A., 2013. Rebound effects caused by the improvement of vehicle energy efficiency: an analysis based on an SP-off-RP survey. Transp. Res., D: Transp. Environ. 24, 62–68.
Zailani, S., Iranmanesh, M., Masron, T.A., Chan, T., 2016. Is the intention to use public transport for different travel purposes determined by different factors? Transp. Res., D: Transp. Environ. 49, 18–24. Available from: https://doi.org/10.1016/j.trd.2016.08.038.
Zarkadoula, M., Zoidis, G., Tritopoulou, E., 2007. Training urban bus drivers to promote smart driving: a note on a Greek eco-driving pilot program. Transp. Res., D: Transp. Environ. 12 (6), 449–451.
Zhang, J., Xiong, Y., Jiang, Y., Tanaka, N., Ohmori, N., Taniguchi, A., 2017. Behavioral changes in migration associated with jobs, residences, and family life. In: Zhang, J. (Ed.), Life-oriented Behavioral Research for Urban Policy. Springer, pp. 479–505.
Zhao, X., Wu, Y., Rong, J., Zhang, Y., 2015. Development of a driving simulator based eco-driving support system. Transp. Res., C: Emerg. Technol. 58, 631–641.
Zhou, Y., Wang, M., Hao, H., Johnson, L., Wang, H., 2015. Plug-in electric vehicle market penetration and incentives: a global review. Mitig. Adapt. Strateg. Global Change 20 (5), 777–795.

CHAPTER 15

Life-oriented household energy consumption research

Weiyan Zong[1], Junyi Zhang[1,*] and Ying Jiang[2]
[1]Hiroshima University, Higashi Hiroshima, Japan
[2]Hiroshima University, Hiroshima, Japan

15.1 Introduction

Faced with current energy issues, encouraging household behavioral changes toward an energy-saving life is of significance for sustainable development. However, it has been difficult to reduce energy consumption in the domestic and passenger transport sectors, mainly involving households. Various energy-saving technologies have been deployed in the market; however, the energy consumption in these two sectors is still growing. As energy consumption needs to spend money, nobody would reject saving money, if they could. Why people have to consume energy is simply because energy is needed for meeting their needs in life. Therefore, for encouraging people to reduce energy consumption, it is necessary to make sure whether their needs in life could be met without consuming more energy caused by, for example, frequent use of a car due to unavailable convenient public transport or staying more time at home for caring of elderly members. Related to this argument, lifestyle has been identified to affect household energy consumption (Bin and Dowlatabadi, 2005; Anable et al., 2012; Yu and Zhang, 2017). There are various definitions of lifestyle, where life choices (but only a limited number of life choices, e.g., time use, consumption, activities, and ownership of durable goods) are commonly included. On the other hand, decisions on life choices are complex and may be interdependent with each other, probably because a change in one life domain may cause changes in other life domains due to the influences of limited time and monetary resources and/or intrahousehold interactions, etc. Such interdependencies make the evaluation of effects of the various life choices on household energy consumption difficult. Another difficulty in reducing household energy

* Corresponding author.

DOI: https://doi.org/10.1016/B978-0-12-815965-1.00015-6

consumption is that, for example, even though people can actually use a public transport because it is very close to their home and also very convenient, they still use a car simply because they like using a car for all activities. Thus encouraging people to change their lifestyles is very challenging. Due to the influences of lifestyles, improving energy efficiency does not necessarily result in a decline of energy consumption, because people may spend the money saved from using energy-saving end uses to use them more, that is, the existence of rebound effects (Yu et al., 2011, 2013a,b).

Surprisingly, the influences of life choices on household energy consumption have remained unclear, because of limited studies. There are surely some existing studies, but only a limited set of life choices are examined. This has motivated us to investigate household energy consumption based on the life-oriented approach, which argues that life choices are not independent of each other and such interdependencies should be captured within a unified and consistent research framework (Zhang, 2017). The importance of the life-oriented approach has been argued and evidenced in various research fields, including not only transport and energy but also heath, time use, tourism, employment, migration, expenditure, information and communication technologies, and general lifestyle studies. For more details, refer to Zhang (2017).

This study attempts to give a comprehensive investigation about the influences of various life choices on household energy consumption based on the life-oriented approach, by using a data collected in 2014 from 890 households living in different areas of Japan. The data contain household monthly consumption in both the domestic (i.e., electricity, gas, kerosene) and transport sectors (i.e., gasoline, diesel) and about 90 life choices in 10 domains (i.e., residence, family budget, health, neighborhood, education, job, family life, leisure, tourism, and transport). Because this is the first attempt in the literature to jointly examine such a large set of life choices, there is no a priori knowledge about how to select life choices for bettering the representation of household energy consumption. To avoid the influence of any arbitrary judgment in advance as much as possible, all available life choice variables are included in the analysis based on a random forest approach (Breiman, 2001).

The remainder of this chapter comprises five parts: Section 15.2 reviews the existing studies. Section 15.3 provides a brief aggregate analysis of survey data. Section 15.4 introduces the random forest approach.

Section 15.5 explains and discusses modeling results. The last section summarizes main findings and future research directions.

15.2 Literature review

The transport and residential sectors account for 27% and 33% of global energy consumption, ranking the first and third largest energy consumers (Nejat et al., 2015). Thus research on household energy consumption has attracted many researchers' attention. Swan and Ugursal (2009) provided a systematic review on how to model residential energy use. Ding et al. (2018) reviewed studies on influential factors of urban residents' low-carbon consumption behavior, including individual, family, and situational factors. Liu et al. (2015) revealed the influences of sociodemographic attributes (i.e., age, educational background, and income level) on household energy consumption behaviors.

A number of studies are in-line with the life-oriented approach. Yu et al. (2011) examined influential factors to both in-home and out-of-home energy consumption, and Yu et al. (2012a) further added residential location choices into energy consumption research. Yu et al. (2012b) jointly modeled households time use and energy consumption by incorporating not only interactivity interactions but also intrahousehold interactions. Wiedenhofer et al. (2013) revealed that households in urban areas had lower direct energy use and higher indirect and total energy consumption, and households living in suburban and rural areas consumed more energy on transport than urban households did. Yu et al. (2013a,b) evaluated various direct and indirect rebound effects of household energy consumption, where indirect effects are measured in terms of interactions across both the in-home and out-of-home end uses and in association with choices in the transport and residential domains. Wahlström and Hårsman (2015) explored the impacts of housing attributes on energy use for heating/cooling and electricity consumption. Poruschi and Ambrey (2016) examined the relationship between residential energy consumption and income, dwelling type, tenure type, and city living, in Australia, and found that living in the capital city and living in a detached house led to higher energy consumption. Thus only a limited number of life choices have been investigated in the literature in the context of household energy consumption.

This study attempts to fill the abovementioned research gaps by focusing on a much larger set of life choices in association with household energy consumption in the context of Japan. Using such a larger set of life choices is expected to avoid the influences of omitted variables on household energy consumption. Statistically, for example, including an explanatory variable in a regression model and excluding, it may not only affect the modeling accuracy but also result in different parameter estimations of other explanatory variables. Behaviorally, omitting life choice A in explaining life choice B may cause not only statistical biases but also structural/systematic errors.

15.3 Survey and data

To explore the relationships between household energy consumption and life choices in other life domains, an online questionnaire survey was conducted in 2014, and as a result, 890 households in different cities/towns of Japan provided valid data. According to a whitepaper published by Japanese Agency for Natural Resources and Energy, electricity, gas, and kerosene accounted for 50.9%, 32.0%, and 16.2% of Japanese household domestic energy consumption, respectively, in 2014.[1] The corresponding expenditure shares are 58.5%, 29.7%, and 11.9% in our survey, which are roughly consistent with the national data. The differences may be due to the fact that in the questionnaire survey, there were more respondents aged 60 years old or above.

First of all, the ratios of expenditures for different energy sources in spring/fall, summer, and winter are summarized in Fig. 15.1. Household expenditures for domestic energy use occupied 76.0%, 73.5%, and 81.5% of the total energy expenditure in spring/fall, summer, and winter, respectively. These percentages are larger than those in the transport sector. Within the domestic sector, electricity played the major role, and it was gasoline that occupied a large part in expenditure of transport energy use. Apparently, energy expenditure patterns varied across seasons: compared with spring/fall, the share of transport-related energy expenditure was larger in summer, and households allocated more money on electricity and less on gas and kerosene in the domestic sector. Differently, in winter, expenditures for gas and kerosene consumptions increased, while expenditures for electricity, gasoline, and diesel consumptions decreased.

[1] The whitepaper is accessible at <https://www.enecho.meti.go.jp/about/whitepaper/> [in Japanese; accessed 10.05.19.].

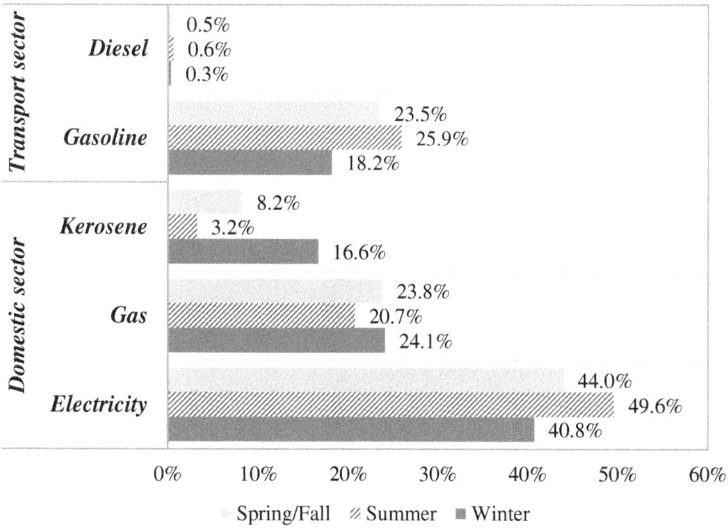

Figure 15.1 Ratios of expenditures for different energy types in spring/fall, summer, and winter.

Descriptive statistics of household seasonal consumption for various energy types are displayed in Table 15.1. It is observed that electricity, gas in the domestic sector, and gasoline in the transport sector were consumed more than kerosene and diesel. Moreover, as mentioned earlier, energy consumption varied across seasons, especially in the domestic sector as households spent most in winter, then in spring/fall, and the least in summer. As expected, such seasonal variations were most obvious in kerosene expenditure that is mainly used for heating in Japan. And as stated in the aforementioned whitepaper, energy consumption for heating accounted for 22.9% of the total domestic energy consumption in 2014, which is a considerably high proportion. On the other hand, the use of transport-related energy consumption was much more stable over seasons regarding to the proportions of households using such energy type and average expenditures.

To get more details on such seasonal variations in energy consumption, cumulative curves of consumptions for different types of energy in different seasons are illustrated in Fig. 15.2. Similarly, the differences of kerosene consumption across seasons were the largest, while the curves for gasoline and diesel consumption are almost overlapped. Compared with in spring/fall and summer, the share of households spending more than 9000 JPY on electricity in winter was larger. The cumulative curves

Table 15.1 Descriptive statistics of household energy consumption.

		Electricity	Gas	Kerosene	Gasoline	Diesel
Spring/fall	Proportion of households using (%)	94.16	76.85	28.99	66.63	3.03
	Average expenditure (JPY)	10,494.30	6,217.95	10,264.67	10,081.74	5,995.06
Summer	Proportion of households using (%)	92.92	73.82	14.83	65.62	3.37
	Average expenditure (JPY)	10,622.48	5,087.85	6,305.94	10,130.90	5,373.28
Winter	Proportion of households using (%)	93.71	76.52	54.72	65.73	2.36
	Average expenditure (JPY)	12,576.17	7,969.88	11,989.05	10,014.22	5,580.74

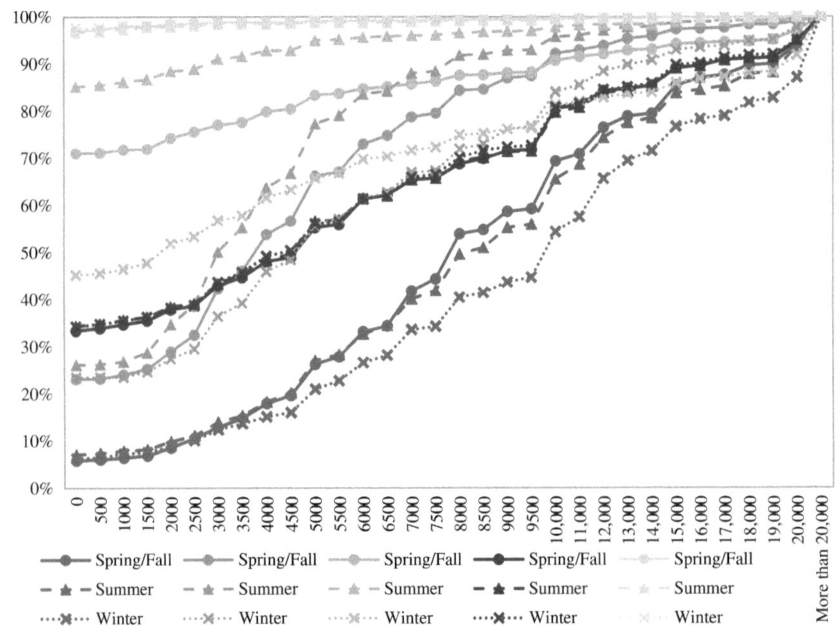

Figure 15.2 Cumulative curves of consumptions for different energy types (unit: JPY).

of gas expenditure grow faster before 8000 JPY than other types of energy, indicating that a large number of households spent less on gas use.

15.4 Methodology

15.4.1 Random forest approach

Random forest approach, a machine learning technique, was first proposed by Breiman (2001) by combining classification and regression tree (Breiman, 1984) and bagging (Breiman, 1996). Briefly, in a random forest, prediction is obtained by averaging the results of classification and regression trees that are grown on bootstrap samples. Thus, when growing a tree, training data are divided into a bootstrap sample data and out-of-bag (OOB) data, and cross validation is possible in random forest by using the OOB data. In addition, at each node, variables are randomly selected from all predictor variables, and one of them would be selected as the splitting criteria.

One important feature of random forest approach is the ability to rank predictors according to internal measure of variable importance (Verikas et al., 2011; Chen and Ishwaran, 2012). Permutation importance is a frequently used variable importance measure in random forest, which is calculated by the difference of the prediction errors before and after permuting a certain predictor variable in the OOB data. In this study the permutation importance measure is adopted and calculated to identify important variables. A predictor is regarded to be highly predictive when its importance score is larger than 0.001 (Yatsunenko et al., 2012). The R package *randomForest* was implemented to grow random forests for household energy consumption.

15.4.2 Predictors for household energy consumption

In random forests, response variables are household monthly expenditure for domestic energy use (including electricity, gas, and kerosene) and transport energy use (including gasoline and diesel) in different seasons of the year 2014. To enhance calculation efficiency, all response variables are categorized into 32 categories: (1) equal to 0, 1 category; (2) (0, 10,000 JPY), 20 categories, evenly spaced by 500 JPY; (3) (10,000 and 20,000 JPY), 10 categories, evenly spaced by 1000 JPY; (4) more than 20,000 JPY, 1 category.

As shown in Table 15.2, altogether, 87 predictor variables were selected, comprising life choices in 10 domains (i.e., residence, family budget, health, neighborhood, education, job, family life, leisure, tourism, and transport).

Table 15.2 Response variables and predictor variables.

Life domain		List of life choices [identification number of each life choice variable]
Energy	15	Consumptions of electricity, gas, kerosene, gasoline, and diesel in spring/fall, summer, and winter
Residence	5	Residence province (47 categories) [1]; residence city (22 categories) [2]; duration of residence [3]; house type I (owned house, public rental house, private rental house, company domitory, etc.) [4]; house type II (detached house, semidetached house, collective house, etc.) [5]
Family budget	11	Annual income [1]; percentages of expenditure for food [2], housing [3], energy [4], furniture [5], cloth [6], health [7], transport [8], education [9], leisure [10], and savings [11]
Health	8	Health status [1]; daily sleeping time [2]; frequency [3] and duration [4] of contact sports; frequency [5] and duration [6] of noncontact sports; frequency [7] and duration [8] of gentle sports
Neighborhood	6	Frequency of participation in neighborhood activity [1]; time [2] and cost [3] for participation in neighborhood activity; whether participating in community activity [4]; time [5] and cost [6] of participation in community activity
Education	12	Frequency [1], time [2], and travel mode [3] of language-related education; frequency [4], time [5], and travel mode [6] of knowledge-related education; frequency [7], time [8], and travel mode [9] of hobby-related education; frequency [10], time [11], and travel mode [12] of job-related education
Job	5	Commuting mode [1]; commuting time [2]; job type [3]; daily working hours [4]; monthly workdays [5]
Family life	6	Indoor time use [1] and outdoor time use [2] on weekday; indoor time use [3] and outdoor time use [4] on weekend/holiday; weekly frequency of family meal [5]; frequency of contact with relatives [6]
Leisure	17	Leisure time use on weekday [1] and weekend/holiday [2]; frequency [3], time [4], and travel mode [5] of activity at cinema/theater; frequency [6], time [7], and travel mode [8] of sports-related activity; frequency [9], time [10], and travel mode [11] of activity at amusement parks; frequency [12], time [13], and travel mode [14] of entertainment activity; frequency [15], time [16], and travel mode [17] of racing-related activity

(Continued)

Table 15.2 (Continued)

Life domain		List of life choices [identification number of each life choice variable]
Tourism	10	Frequency [1] and cost [2] of leisure tour; frequency [3] and cost [4] of domestic tour; frequency [5] and cost [6] of oversea tour; frequency [7] and cost [8] of domestic visit; frequency [9] and cost [10] of oversea visit
Transport	7	Ownership [1] and annual mileage [2] of light car; ownership [3] and annual mileage [4] of compact car; ownership [5] and annual mileage [6] of ordinary car; main travel mode [7]

Note: The second column indicates the number of life choices and the random forest approach can directly use a categorical data without transforming it into dummy variables.

15.5 Empirical results

15.5.1 Predictors to household energy consumption

In total, 15 random forests were generated with respect to 15 response variables of household monthly consumption of electricity, gas, kerosene, gasoline, and diesel in spring/fall, summer, and winter. The numbers of highly predictive variables are summarized in Table 15.3, and all highly predictive predictors are listed in Table 15.4. Apparently, compared with other three energy types, gasoline (19–19–20) and kerosene (12–5–15) possess much more predictors with high predictive ability. Electricity and diesel only have a few influential factors in several seasons, indicating their weak relationships with choices in other life domains. Regarding predictors, choices in the transport domain (2.4–1.8–2.6) are most influential, followed by the residence domain (1.6–1.6–1.6) and the family life domain (1.4–1.0–1.4). In contrast, the health domain has not any highly predictive predictor, and only one highly predictor is found in the life domains of family budget and tourism, meaning that these three life domains are barely associated with household energy consumption.

15.5.1.1 Electricity

There are four highly predictive variables for electricity consumption in spring/fall, including indoor time use on weekday, house type, residence province, and frequency of contact with relatives. As displayed in Fig. 15.2, household electricity consumption is more evenly distributed than other energy types with least zero values. For electricity consumption in summer and winter, no highly predictive variables are observed.

Table 15.3 Numbers of highly predictive variables to household energy consumption.

	Electricity	Gas	Kerosene	Gasoline	Diesel	Average number
Residence	2–0–0	2–2–2	2–3–4	2–3–2	0	1.6–1.6–1.6
Family budget	0	0	0–0–1	0	0	0–0–0.2
Health	0	0	0	0	0	0
Neighborhood	0	0	2–0–1	1–0–1	0	0.6–0–0.4
Education	0	0	0–0–1	1–0–1	0	0.2–0–0.4
Job	0	0–1–0	3–1–0	1–1–1	1–1–0	1.0–0.8–0.2
Family life	2–0–0	2–1–0	0–0–3	3–4–4	0	1.4–1.0–1.4
Leisure	0	0	0	4–4–4	0	0.8–0.8–0.8
Tourism	0	0	1–0–0	0	0	0.2–0–0
Transport	0	1–1–1	4–1–5	7–7–7	0	2.4–1.8–2.6
Total	4–0–0	5–5–3	12–5–15	19–19–20	1–1–0	8.2–6.0–7.6

Note: All numbers are listed in the order of spring/fall, summer, and winter, for example, "2–3–4" in the second row and fourth column means that there are 2, 3, and 4 highly predictive variables in residence domain for household kerosene consumption in spring/fall, summer, and winter, respectively. And "0–0–0" is represented by "0" for simplification.

Table 15.4 Influential factors to energy consumption responses: ranked in a descending order of variable importance.

		Electricity	Gas	Kerosene	Gasoline	Diesel
Residence	*S/F*	[5][1]	[1][5]	[1][5]	[1][5]	–
	S	–	[1][5]	[1][5][2]	[1][5][2]	–
	W	–	[1][5]	[1][5][3][2]	[5][1]	–
Family budget	*S/F*	–	–	–	–	–
	S	–	–	–	–	–
	W	–	–	[3]	–	–
Health	*A*	–	–	–	–	–
Neighborhood	*S/F*	–	–	[6][5]	[5]	–
	S	–	–	–	–	–
	W	–	–	[5]	[6]	–
Education	*S/F*	–	–	–	[9]	–
	S	–	–	–	–	–
	W	–	–	[9]	[9]	–
Job	*S/F*	–	–	[1][4][5]	[1]	[3]
	S	–	[2]	[2]	[1]	[3]
	W	–	–	–	[1]	–
Family life	*S/F*	[1][6]	[6][3]	–	[5][3][1]	–
	S	–	[3]	–	[5][1][6][3]	–
	W	–	–	[5][3][6]	[5][3][1][6]	–
Leisure	*S/F*	–	–	–	[5][8][11][14]	–
	S	–	–	–	[5][11][8][14]	–
	W	–	–	–	[5][11][8][14]	–
Tourism	*S/F*	–	–	[2]	–	–
	S	–	–	–	–	–
	W	–	–	–	–	–
Transport	*S/F*	–	[7]	[7][2][1][4]	[7][3][4][1][2][6][5]	–
	S	–	[7]	[7]	[7][3][4][1][2][6][5]	–
	W	–	[7]	[7][2][4][1][6]	[7][3][4][1][2][6][5]	–

Note: "*S/F*" denotes spring/fall; "*S*" denotes summer; "*W*" denotes winter; "*A*" denotes all seasons. The number in [ ] refers to the variable identification number in each type of life choices (predictors) (see Table 15.2).

15.5.1.2 Gas

Residence province and house type in the residence domain and main travel mode in the transport domain are of high predictive ability for gas expenditures in all seasons. For gas consumption in spring/fall, frequency of contact with relatives and indoor time use on weekend/holiday are also influential. Commuting time and frequency of contact with relatives are highly predictive to summer gas consumption.

15.5.1.3 Kerosene

Differences of the numbers of highly predictive variables between seasons are obvious for kerosene consumption, with 12 in spring/fall, 5 in summer, and 15 in winter. The possible reason is that there are few households using kerosene in summer than in other seasons. There are three common influential factors shared by kerosene consumption in spring/fall, summer, and winter, including residence province, house type, and main travel mode. For spring/fall, nine more predictors are highly predictive: annual mileage of light car, ownership of light car, annual mileage of compact car, cost for participation in community activity, commuting mode, daily working hours, cost of leisure tour, time of participation in community activity, and monthly workdays. Commuting time and residence city are influential to kerosene consumption in summer. As indicated in Table 15.1, the proportion of households using kerosene and the average expenditure in winter are higher than that in other seasons, and the number of highly predictive variables is also larger. In addition to three common predictors, these 12 factors also show high predictive ability: annual mileage of light car, duration of residence, annual mileage of compact car, residence city, time of participation in community activity, percentages of expenditure for housing, weekly frequency of family meal, indoor time use on weekend/holiday, ownership of light car, annual mileage of ordinary car, travel mode of hobby-related education, and frequency of contact with relatives.

15.5.1.4 Gasoline

Gasoline consumption has the largest number of highly predictive predictors among all response variables. Household gasoline consumptions in spring/fall, summer, and winter share 17 common highly predictive predictors: residence province, house type in residence domain; commuting mode in job domain; weekly frequency of family meal, indoor time use on weekend/holiday, indoor time use on weekday in family life domain; travel mode of activity at cinema/theater, travel mode of activity at amusement parks, travel mode of sports-related activity, travel mode of entertainment activity in leisure domain; and all predictors in the transport domain. Evidently, many influential factors from other domains are travel modes for different activities. Moreover, for gasoline consumption in spring/fall, time of participation in community activity and travel mode of hobby-related education are also influential; for summer, residence city and frequency of contact with relatives are influential; for winter, cost of participation in community activity, travel mode of hobby-related education, and frequency of contact with relatives are influential.

15.5.1.5 Diesel

Only one predictor, job type, is influential to household diesel consumption in spring/fall and summer. For winter, none of the predictors is highly predictive.

15.5.2 Relationships between household energy consumption and other life choices

As mentioned earlier, highly predictive predictors are mainly from the transport, residence, and family life domains, followed by the job and leisure domains. In comparison, choices related to family budget, health, neighborhood, education, and tourism are less associated with household energy consumption.

15.5.2.1 *Residence*

Among all predictors in the residence domain, residence province and house type are influential to consumption for electricity, gas, kerosene, and gasoline. As geographical conditions and energy suppliers vary across provinces in Japan, energy prices too vary, which affects household energy consumption and expenditure. The predictor "house type" contains four categories: detached house, semidetached house, public house, and others. According to the analysis results of survey data in Fig. 15.3, electricity consumption in spring/fall of households living in detached houses is much larger. In the case of winter gas consumption, public house stands out of all house types, followed by detached house. However, households living in public house consume less kerosene than in other house types, which may be caused by limited housing conditions. As for gasoline, households living in detached house consume more than households living in other types do.

15.5.2.2 *Job*

Commuting mode is influential to household gasoline consumption regardless of seasons. This is not surprising because commuting is an important part of household daily traveling. Commuting by driving results in high gasoline consumption, while commuting by walking, bicycle, railway, and bus is associated with lower gasoline consumption. Furthermore, commuting time is highly predictive for gas and kerosene consumption in summer. Job type is the only highly predictive predictor for household diesel consumption, and residents in agricultural, forestry, and aquatic industries consume more diesel in spring/fall and summer. The possible reason is that most of the machineries are powered by diesel and regularly used during these seasons.

15.5.2.3 *Family life*

According to the numbers of highly predictive predictors, compared with outdoor time use, indoor time use is more influential to household energy consumption. With the increase in indoor time use on weekday,

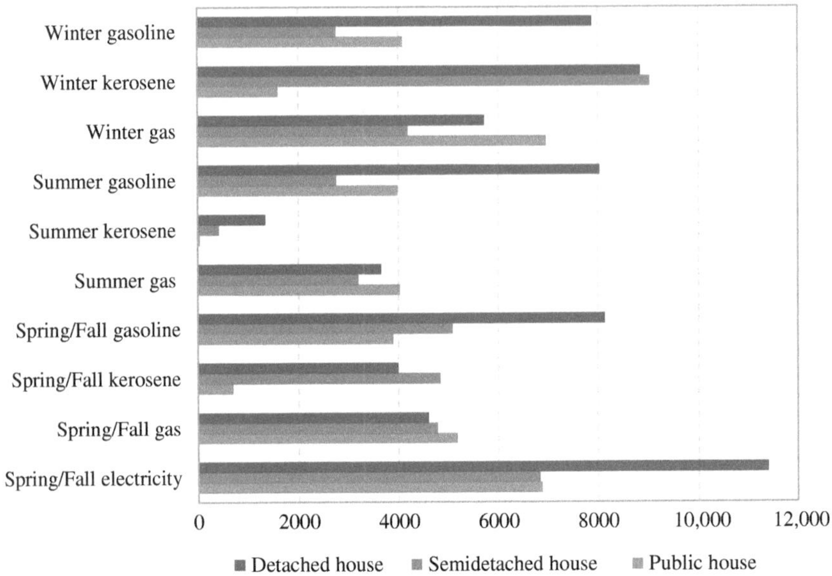

Figure 15.3 Average energy expenditure of households living different types of houses.

household electricity consumption in spring/fall increases. And households spending 1–8 hours on indoor activities use more gas in spring/fall and summer and more gasoline in all seasons. Frequency of family meal and frequency of contact with relatives are also influential. As shown in Fig. 15.4, the trends of average gasoline consumption and average kerosene consumption (winter) are totally different, indicating the competitive relationship between these two energy types.

15.5.2.4 Leisure

Predictors in the leisure domain only appear in the list of highly predictive predictors for gasoline consumption, containing travel mode for various activities. Driving to participate in activities at cinema/theater, sports facilities, amusement parks, and entertainment facilities is associated with higher gasoline consumption regardless of seasons.

15.5.2.5 Transport

Main travel mode is highly predictive for consumption of gas, kerosene, and gasoline. The shares of households using walking, bicycle, car, and railway as main travel mode are 13.37%, 16.52%, 47.75%, and 11.24%, respectively. It is as expected that households traveling mainly by car use

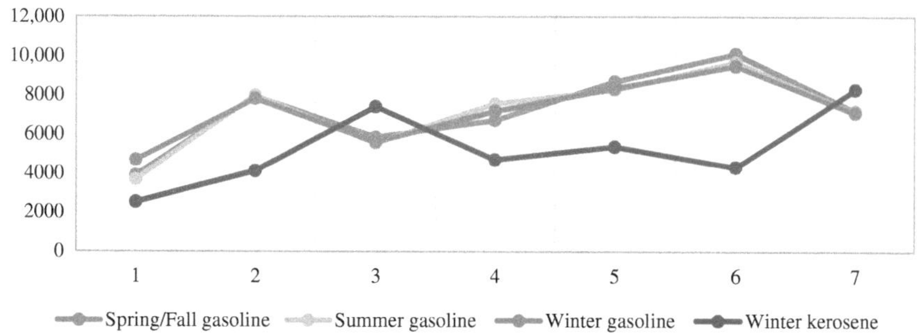

Figure 15.4 Average energy expenditure of households with different family meal frequencies.

more gasoline (see Fig. 15.5). Noticeably, in the domestic sector, households traveling mainly by car also use more kerosene but less gas than households traveling mainly by walking, bicycle, and railway. This may be because gasoline and kerosene are sold at gas stations in Japan, and kerosene is more accessible for households with a driver than for households without a driver.

Moreover, kerosene consumption in spring/fall and winter is associated with annual mileage of (light and compact) cars. As excepted, concerning gasoline consumption, all predictors in transport domain are of high predictive ability.

15.6 Conclusions and future research directions

This chapter has clarified the influences of various life choices on energy consumption in domestic (i.e., electricity, gas, kerosene) and transport sectors (i.e., gasoline, diesel). There are only few studies on exploring the influences of a limited number of life choices on household energy consumption. Focusing on such a large set of life choices (87 variables), this is the first study in the literature to investigate household energy consumption based on the life-oriented approach.

The random forest approach is a powerful tool to jointly capture the influences of such a large number of predictors, without any pregiven assumption, even though it cannot provide any theoretical explanation about the influential factors obtained from the analysis. To tackle such an issue caused by the random forest approach, behavior-oriented models should be applied/developed by reflecting different decision-making mechanisms with respect to different life domains.

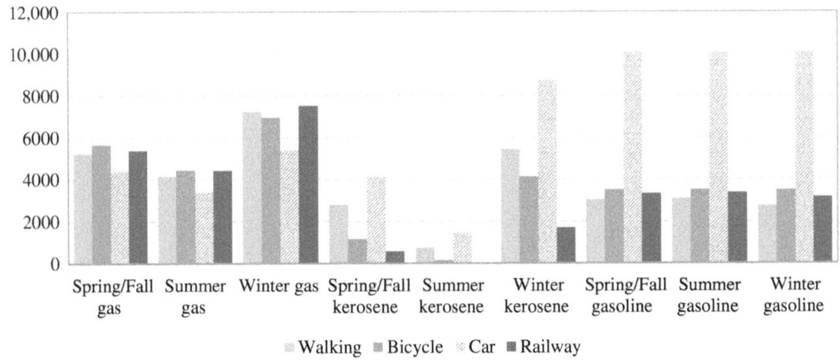

Figure 15.5 Average energy expenditure of households with different main travel modes.

Even though about 90 life choice variables are investigated in this study, there are more life choice variables, in reality. Life choices in all domains may have mixed effects on household energy consumption. Here, it is first discussed how choices in each life domain should be associated with different types of energy consumption, and then future behavioral research issues are summarized.

As examined in Chapters 4, 7, 10, and 11, the residence and transport domains are directly associated with household energy consumption via use of in-home and out-of-home end uses that are connected with other choices in these two domains. Detailed discussions about decision-making mechanisms related to energy consumption in the residence and transport domains refer to the abovementioned four chapters.

In the job domain, working outside home means less use of energy at home but longer commuting time to workplace may lead to more energy consumption from commuting. On the other hand, working also indicates that a higher income can be expected, which may induce more energy consumption via use of in-home and out-of-home end uses. If proenvironmental attitude appears more in high-income people, because of the potential education effects, then the income increase to a certain level may lead to less energy consumption. Thus choices in the job domain may have various types of mixed effects on household energy consumption. To properly capture the effects of choices in the job domain, choices of workplace location, working hours and frequency, and commuting mode choices, as well as the activities performed before and after work, and so on should be modeled jointly by incorporating various interdependencies in these discrete/continuous choices.

Education choices may also have mixed effects on household energy consumption in terms of schooling distance/time and travel mode, proenvironmental attitude, learning activities outside home, etc.

Chapter 11, Japanese young people's car ownership and usage: research from an expenditure perspective, has already examined how transport and energy expenditures are affected by the various other expenditures and discussed relevant behavioral research issues in the future. Here, no further discussion will be given about how choices in the family budget domain affect energy consumption.

Health is an important lifestyle factor. Even though no health factors are identified to be highly predictive to energy consumption in this case study, the use of behaviorally oriented models may derive different findings. For example, if more frequent use of nonmotorized travel modes (e.g., walk, man-powered bicycle, kickboard, and other nonmotorized personal mobility) can reduce the dependence on car, car-related energy consumption will be reduced. Sleeping time may have both positive and negative effects on energy consumption. More sleeping time under a comfort weather condition may lead to less use of an air conditioner at home, while more sleeping time under a very hot weather condition may result in excessive use of an air conditioner. If the liking of more sports is associated with more use of nonmotorized travel modes, then sport activities may reduce car-related energy consumption. On the other hand, if jogging is often performed, then the use of shower at home may become more frequent, inducing more use of energy.

As for the neighborhood communication and more general social networking, they are usually associated with more social activities. More social activities outside home means less time staying at home, which results in a decline of in-home energy consumption.

The family life domain includes time use at home: more time spent at home may result in more energy use at home, if energy efficiency remains unchanged. On the other hand, for example, use of passive cooling reduces energy use, and use of energy-saving end uses may lead to less use of energy at home, if rebound effects do not exist.

Leisure and tourism indicate more time use outside home. If the frequency of leisure and tourism is very high, they may lead to an obvious decline of in-home energy use; otherwise, they may be not a significant factor affecting in-home energy use.

Note that, even though out-of-home activities are linked with less use of energy at home, it does not necessarily mean that total energy consumption from a household will decrease, because some out-of-home activities also need energy consumption, such as eating at a restaurant and shopping at a department store.

Thus future research issues are very clear, that is, to clarify the above unclear and mixed effects. However, various challenges are waiting. First, it is necessary to develop integrated choice models that can fully capture all decisions associated with household energy consumption, for each life domain. Next, it is necessary to develop cross-domain integrated choice models. As suggested by the life-oriented approach (Zhang, 2017), ideally, it is important to develop integrated choices models with both within-domain and cross-domain decision-making mechanisms. Either of the above integrated choice models is more complicated than the models dealing with a single-faceted choice and therefore needs more data.

References

Anable, J., Brand, C., Tran, M., Eyre, N., 2012. Modelling transport energy demand: a socio-technical approach. Energy Policy 41, 125–138.

Bin, S., Dowlatabadi, H., 2005. Consumer lifestyle approach to US energy use and the related CO_2 emissions. Energy Policy 33, 197–208.

Breiman, L., 1984. Classification and Regression Trees. Routledge, New York.

Breiman, L., 1996. Bagging predictors. Mach. Learn. 24 (2), 123–140.

Breiman, L., 2001. Random forests. Mach. Learn. 45 (1), 5–32.

Chen, X., Ishwaran, H., 2012. Random forests for genomic data analysis. Genomics 99 (6), 323–329.

Ding, Z., Jiang, X., Liu, Z., Long, R., Xu, Z., Cao, Q., 2018. Factors affecting low-carbon consumption behavior of urban residents: a comprehensive review. Resour. Conserv. Recycl. 132, 3–15.

Liu, L.C., Wu, G., Zhang, Y.J., 2015. Investigating the residential energy consumption behaviors in Beijing: a survey study. Nat. Hazards 75 (1), 243–263.

Nejat, P., Jomehzadeh, F., Taheri, M.M., Gohari, M., Majid, M.Z.A., 2015. A global review of energy consumption, CO_2 emissions and policy in the residential sector (with an overview of the top ten CO_2 emitting countries). Renew. Sustain. Energy Rev. 43, 843–862.

Poruschi, L., Ambrey, C.L., 2016. On the confluence of city living, energy saving behaviours and direct residential energy consumption. Environ. Sci. Policy 66, 334–343.

Swan, L.G., Ugursal, V.I., 2009. Modeling of end-use energy consumption in the residential sector: a review of modeling techniques. Renew. Sustain. Energy Rev. 13 (8), 1819–1835.

Verikas, A., Gelzinis, A., Bacauskiene, M., 2011. Mining data with random forests: a survey and results of new tests. Pattern Recognit. 44 (2), 330–349.

Wahlström, M.H., Hårsman, B., 2015. Residential energy consumption and conservation. Energy Build. 102, 58–66.

Wiedenhofer, D., Lenzen, M., Steinberger, J.K., 2013. Energy requirements of consumption: urban form, climatic and socio-economic factors, rebounds and their policy implications. Energy Policy 63, 696–707.

Yatsunenko, T., Rey, F.E., Manary, M.J., Trehan, I., Dominguez-Bello, M.G., Contreras, M., et al., 2012. Human gut microbiome viewed across age and geography. Nature 486 (7402), 222–228.
Yu, B., Zhang, J., 2017. Household energy consumption behavior. In: Zhang, J. (Ed.), Life-Oriented Behavioral Research for Urban Policy. Springer, pp. 123–148.
Yu, B., Zhang, J., Fujiwara, A., 2011. Representing in-home and out-of-home energy consumption behavior in Beijing. Energy Policy 39 (7), 4168–4177.
Yu, B., Zhang, J., Fujiwara, A., 2012a. Analysis of the residential location choice and household energy consumption behavior by incorporating multiple self-selection effects. Energy Policy 46, 319–334.
Yu, B., Zhang, J., Fujiwara, A., 2012b. A household time use and energy consumption model with multiple behavioral interactions and zero-consumption. Environ. Plann. C 40 (2), 330–349.
Yu, B., Zhang, J., Fujiwara, A., 2013a. Evaluating the direct and indirect rebound effects in household energy consumption behavior: a case study of Beijing. Energy Policy 57, 441–453.
Yu, B., Zhang, J., Fujiwara, A., 2013b. Rebound effects caused by the improvement of vehicle energy efficiency: an analysis based on an SP-off-RP survey. Transp. Res., D 24, 62–68.
Zhang, J., 2017. Life-Oriented Behavioral Research for Urban Policy. Springer, Tokyo.

CHAPTER 16

To connect the past with the future

Junyi Zhang*
Hiroshima University, Higashi Hiroshima, Japan

16.1 What can be learnt from this book

After reading the previous chapters, readers should have learnt a lot. For helping readers to have a much better understanding of this book, here, findings and insights as well as future research issues from Chapters 2–15 are described and discussed in a comparative way.

16.1.1 Integrated global assessment model is useful but needs more improvements

The Paris Agreement ambition to achieve a neutral balance of emissions by the middle of this century requires a paradigm shift in the transport system. By building a global transport model that can incorporate behavioral and technological factors for being integrated into an integrated assessment model, Chapter 2, The role of the transport sector in energy transition and climate change mitigation: insights from an integrated assessment model, estimated that stringent penetration of electric vehicles (EVs) together with carbon pricing projects a significant transition of energy mix and the highest potential for emission reduction. CO_2 emissions from transport can be reduced dramatically by technological transformations and lifestyle changes. The economic cost associated with pricing carbon can be lowered by transport electrification and shared mobility. The contribution of shared mobility to reducing mitigation cost is more significant than transport electrification. Balanced technological and lifestyle innovations deserve consideration for the decarbonization of transport sector.

The coupling of transport-economic-climate model enables the integrated assessment model to capture the cross-sectoral interactions, where

* Corresponding author.

Transport and Energy Research
DOI: https://doi.org/10.1016/B978-0-12-815965-1.00016-8

the transport model is a multinomial logit (MNL) model. To develop such a global-level integrated model, it is surely operational to use the MNL model. However, the MNL model suffers from an issue of the IIA (independence of irrelevant alternatives) property. To overcome the IIA issue and better represent the decision-making mechanism related to transport-mode-choice decisions, non-IIA choice models should be adopted, such as nested logit model, nested paired combinatorial logit model, and mixed logit model. In addition, users' responses to EVs, shared mobility, and lifestyle changes could be better reflected by investigating their stated preferences (SPs) based on experimental surveys. Furthermore, transport choice behaviors can also be integrated with energy-consumption behaviors within a unified choice modeling framework. With the above improvements the integrated global-assessment model could be used to better identify the role of transport in energy transition and climate-change mitigation.

16.1.2 Policy instruments for reducing energy differ across regions

Chapter 3, Effects of green policies on household transport and energy consumption in Europe: a literature review, reviewed and evaluated the effectiveness of the EU green policies from transportation and home-energy-consumption/production perspectives by categorizing relevant literatures into household long-term transport decisions and household in-home behavior. Aiming to examine the overall strategic plan of the European Union and to understand variations between countries, this chapter also summarized the EU green policies that have a direct and indirect relationship with household energy consumption. As a result, this chapter excavated consistent findings with respect to the effectiveness of the EU household—energy-consumption policies or interventions and discussed the strengths and shortcomings of current policies and interventions. It is found that using nonprice-based policy instruments such as regulation and educational outreach will make green policies more effective. Enhancing households' energy-related knowledge and the ability to make energy-efficient investments is important, from the supply perspective. Educating consumers about their energy consumption and how to identify the most efficient appliances need to be continued from the demand perspective. However, the above measures are not necessarily applicable to other countries. For example, in the

United States, technological improvements and energy-efficiency policies are still dominating[1], and in the context of Asia, mitigation policies (e.g., urban design, information and communication systems, and behavioral measures) are more effective to reduce CO_2 emissions (Shukla and Dhar, 2015). Therefore when designing policies for reducing transport energy consumption, national heterogeneities should be properly reflected in association with different economic and social development stages.

16.1.3 Travel choices: fundamentals to address energy issues in the transport sector

16.1.3.1 More efforts can be made from travel choices and preferences

As shown in Chapter 5, Travel choices, preferences, and energy implications in the United States, understanding transportation behavior and preferences can help develop effective transportation and other related policies, and to forecast changes in future travel behavior, energy use, and CO_2 emissions.

Reducing the frequency of driving can be just as important as shifting mode choice in reducing energy consumption. The frequency of driving is closely related to the frequency of participating in out-of-home activities. In this regard, teleworking can reduce out-of-home activity participation related to commuting and work business, even considering possible extra out-of-home requests, such as ordering foods from a restaurant via the Internet. In recent years, more and more people have started relying on online services, which, however, have to rely on certain transport mode(s) to deliver the ordered goods. Thus more dependence on online services does not necessarily reduce energy consumption. This, of course, needs more careful investigation. Related to the frequency of driving, reducing the travel distance is also important. For this, promoting compact city development is useful. However, it is quite a challenge whether people's car-dependent lifestyles would be altered or not by compact city development. If people could not enjoy a much better quality of life, they would not live under a compact city environment.

Although travel cost is important to most individuals when considering their commute choices, it is not always the sole determinant. More

[1] <http://needtoknow.nas.edu/energy/energy-use/transportation/> (accessed 27.06.19.).

precisely, moderate changes in transportation pricing may not achieve long-term mode shifts for commuters who are regular drivers without feasible travel alternatives. Similarly, for those who prefer to use public transportation shifting to driving will not occur under circumstances when driving becomes less costly. Transit incentives were found to be attractive; however, they may only be attractive for current public transportation users, and improvements in services would have to occur in order to attract more users. This implies that transportation policies that are targeted at increasing transit subsidies may not be effective when existing transit services are unsatisfactory.

Personal preferences and perceptions based on past experiences and current needs, commute constraints caused by dependents and residential location are all influential to mode choices. The impact of transportation policies on transportation demand will thus be region specific and individual specific. As long as people's lives have to rely on transport, influences of life choices on transportation demand cannot be ignored, and car dependence should be therefore reduced by efforts from not only the transport sector but also other sectors supporting nontransport life activities/choices.

Having different sets of transportation policies or programs that are targeted at specific groups of drivers will be more effective than having one generic program. For example, since "regular but flexible" drivers are fairly independent, programs that support the improvement of public transportation services in the form of convenience or shorter travel time would be more attractive. Any future changes in transportation and parking policies have to address the travel needs to encourage substantial shifts in mode choices. Drivers in the "regular but flexible" category tend to be more likely to change their travel behavior and to drive less than "regular drivers."

The impacts of land use, public transportation, and pricing policies on mode choices may not be very effective in influencing current regular drivers. Similar to the influence of travel cost, moderate changes in these policies may not work well to reduce current car ownership and usage, suggesting the necessity and importance of dramatic changes in these policies. For example, land use planning policies or measures that can increase public transportation supply, change the travel distance, and increase public transportation demand may thus achieve the goal of significant carbon reductions.

16.1.3.2 Children's school travel matters to energy consumption

Delivering and picking up children to/from schools are common in many countries. The main factors are parental concern about traffic safety and time pressure of parents. Chapter 6, Children's travel behavior and implication to transport energy consumption of household: a case study of three Australian cities, examined how extra energy is used in association with children being driven to school, based on the interview survey in Sydney, Melbourne, and Canberra. Parents are aware of extra energy consumption and understand that their children could walk to school but priority on time-saving outweighs other considerations. The complexity of everyday schedules is interrelated with barriers to using alternative transport modes.

A more integrated approach to secure traffic safety of children is critical in saving energy use of family, suggesting the importance of making cross-sectoral policies. In Japan, for example, primary school students must walk to/from schools, and students of secondary and high schools can use a bicycle for going to schools. This has been made possible because of various safe and secure traffic measures taken by the government and the involvement of local communities. Even though no relevant research can be found from Japan, safe and secure traffic to/from schools has surely convinced parents to allow their children to walk and take a bicycle to/from schools, which consequently reduced energy consumption from parental delivery and pick up.

Subsidizing families who choose low-energy cars would be effective. Support for families who wish to live in the inner area, with the trade-off of limited car use, is also recommended. Such financial supports need further theoretical justification and more empirical evidence on the one hand, while promoting effective collaborations between relevant stakeholders/sectors are surely crucial to reduce energy consumption from school traffic on the other.

Children learn many things from their parents (i.e., intergenerational interactions exist), including ownership and usage of vehicles. However, this study has not explored how parental ownership and usage (both behavior and preference) of vehicles affects their children's vehicle ownership and usage. For doing so, panel data on intergenerational interactions is required; however, it is costly and time-consuming. As an alternative survey method, a retrospective survey (especially, a life history survey) may be more operational in the sense that it allows respondents to recall what happened in the past and report how their current vehicle ownership and usage had been affected by their parents.

16.1.4 Household choices matter: many difficulties and diverse angles in energy reduction

16.1.4.1 Comparison of factors affecting household energy consumption in Asian megacities

From a household perspective in the context of three Asian megacities (Chapter 4: Energy consumption in the transport and domestic sectors: a household-level comparison between capital cities of Japan, China, and Indonesia), an multiple discrete-continuous extreme value (MDCEV) model estimated that the most influential factors to energy consumption vary with energy types (electricity, gas, and gasoline) in Beijing and Jakarta. In contrast, residential attributes are the most influential factors to all three types of energy consumption in Tokyo.

Transport-related attributes explain about 40% of the total variance of gasoline consumption, indicating that it is the most influential factor to gasoline consumption, except in Tokyo, which gasoline consumption is mostly influenced by residential attributes. For reducing gasoline consumption, it is necessary to reduce the numbers of car driving licenses and cars as well as the commuting time in Tokyo, Beijing, and Jakarta. For Jakarta, it is further necessary to reduce the number of motorcycles. Transport-related attributes mostly affect the gas consumption in Beijing, followed by that in Jakarta and in Tokyo. In all three cities, transport-related attributes do not affect electricity consumption at all, except electricity consumption in winter in Beijing.

Residential behavior is positively related to the use of vehicle, but if the spending on residence is larger than a certain threshold, an inverse relationship occurs, even though its influence on gasoline consumption is very limited.

Life cycle stage factors show a larger influence on the electricity consumption in Tokyo and all three types of energy consumption in Beijing; however, its influence is minor on energy consumption in Jakarta. The effects of life cycle stage variables vary across cities. In Tokyo, low income does not affect any energy consumption, but high income is associated with less gasoline consumption. On the contrary, in Beijing and Jakarta, high income is not influential at all. Low-income households in Beijing use more gas; however, low-income households in Jakarta use less gas and gasoline. Full-time workers consume less gas and gasoline in Tokyo, more gas in Beijing and Jakarta, and more gasoline in Jakarta. Thus Tokyo and Jakarta show a completely different picture about the influence of full-time employment.

Tokyo, Beijing, and Jakarta are typical megacities in Asia and the capital cities of three countries with 1.8 billion populations. Even though the generality of the above findings should be carefully examined in the future, the findings have important implications into energy policymaking. First, it is suggested that household energy consumption should be investigated with respect to both in-home and out-of-home energy consumption, ignoring either would lead to biased estimation of the rebound effects of energy-efficiency policies. Second, it is revealed that residential behavior and attributes play significant roles in affecting household energy consumption. Third, energy policymaking targeting households should take nation-level and city-level heterogeneities into account, because of the diverse influences of life cycle stages and income.

16.1.4.2 Energy policies need to be made by considering interpersonal interactions

In Chapter 7, Residential location and commuting mode choices: intrahousehold interaction modeling and its implications for energy policy, residential location choice is jointly analyzed with commuting mode choice at the household level. Both choices are closely related to energy consumption. Such choices are usually intrahousehold interactions. To capture intrahousehold interactions involved in future choices of residential location and commuting mode, it is suggested to use an SP survey, where both individual members' independent choice outcomes and their joint choice outcomes are investigated. Such an SP survey is very flexible and powerful to explore complicated decision-making mechanisms, and it has been widely applied in many research fields. Unfortunately, it has not attracted sufficient attention from energy researchers. For example, energy policymakers have been bothered by the rebound effect of using energy-saving technologies. This rebound effect has not well understood, and as a result, energy-saving technologies have not contributed that much to energy reduction, especially at the household level. For this, SP surveys can be used to understand why the rebound effect occurs and how to reduce it via behavioral and policy interventions. In Chapter 7, Residential location and commuting mode choices: intrahousehold interaction modeling and its implications for energy policy, the household head's commuting mode choice is targeted. In other words, husband and wife were asked to negotiate with each other about what kind of commuting mode should be used by the household head, as a household matter. As argued in Chapter 14, Behavioral interventions for sustainable transportation: an overview of programs and guide for practitioners, behavioral interventions are of importance to save energy in the transport sector. The existence of intrahousehold interaction further suggests that behavioral interventions should target right members in a household. Similarly, in the case of an organization, interpersonal interactions also exist and affect energy consumption. It is therefore necessary to target right persons in organizations for reducing energy consumption via effective behavioral interventions.

16.1.4.3 Expenditure data is useful to understand young people's less car dependence

The phenomenon of young people's less car dependence has attracted more and more attention from both researchers and practitioners, mainly in developed countries. In Japan the target country of Chapter 11, Japanese young people's car ownership and usage: research from an expenditure perspective, this phenomenon has been observed after the end of the bubble economy (1990s). This study provided additional evidence on this phenomenon in the context of Japan and further explored influential factors from an expenditure perspective by building a household expenditure allocation model based on the concept of multilinear utility. Especially, interactions between car-related, public transport, residence, energy, and other expenditures are explicitly modeled. Based on a national household expenditure survey data in 2009, this study confirmed the significant influence of interexpenditure interactions (especially, the interactions between car-related and other types of expenditures) on household expenditure decisions for different life domains. This result suggests that understanding young people's car ownership and usage from the expenditure perspective cannot ignore how other types of expenditures are associated with car-related expenditure. Comparing the attributes of household, residential location, and vehicle, the results of modeling analysis showed that household size is the most influential factor, followed by the number of cars owned by household and the number of household members aged under 18 years old. The major profiles of young people with less ownership and usage of car are those low-income people, living in a megacity and in an urban area.

16.1.4.4 Survey and data collection/disclosure methods should be further improved

In Chapter 4, Energy consumption in the transport and domestic sectors: a household-level comparison between capital cities of Japan, China, and Indonesia, a questionnaire survey was conducted in each of three Asian megacities, where respondents were asked to report their energy consumption according to energy type by referring to the receipts. The good feature of such a survey is to allow respondents to report what type(s) of end use(s) to own and how much to use for each owned end use by season. As a result, a comprehensive set of household energy-consumption data can be easily collected. Such a comprehensive data set is necessary to simultaneously capture household decisions on different types of end uses, because a household needs to make trade-offs between different end uses under its budget constraint. However, especially for the web survey, it is difficult to check whether each respondent actually referred to as the receipts, as requested. In this sense, there is a survey quality issue. For this, if household-level data (Big Data) stored in energy supply companies could be accessible, the survey quality issue would be resolved. In the future, it is further necessary to combine household-level questionnaire survey data and Big Data from energy supply companies for deriving more general scientific insights into energy-saving policymaking.

Existing studies have found that the built environment affect transport and energy-consumption behaviors. Unfortunately, the national household expenditure data used in Chapter 11, Japanese young people's car ownership and usage: research from an expenditure perspective, only includes the residential location information at the prefecture level, because of the national policy of privacy protection. Such prefecture-level location information cannot properly capture the influence of the built environment. Actually, there are various ways of privacy protection when opening such national data to the public. The Japanese national data disclosure policy should be improved. Such data disclosure issues in Japan have also been pointed out with respect to other types of government-owned data for years. Significant improvements have been made in recent years; however, further improvements are surely required for promoting scientific research at the global level.

In Chapter 15, Life-oriented household energy-consumption research, a nation-wide online questionnaire survey was implemented in Japan, which included a huge number of question items based on the life-oriented approach (Zhang, 2017). As it was conducted via a web survey system, it was possible to select those respondents who provided valid answers to all question items. It is unclear how answering such a huge number of question items affected the survey quality due to the influence of answering burdens, and therefore it is worth investigating such survey quality issues and exploring how to improve the life-oriented behavioral survey method in the future.

16.1.4.5 Methodological improvements are crucial to address household energy-consumption issues

1.4.5.1 Multiple discrete-continuous choice models are useful

Households need to decide what type(s) of end use(s) to own and how much to use for each owned end use. To represent such household decisions, multiple discrete-continuous choice models are applicable, which are built under the random utility-maximizing principle. In Chapter 4, Energy consumption in the transport and domestic sectors: a household-level comparison between capital cities of Japan, China, and Indonesia, an MDCEV model is used, which is flexible to accommodate varied utility forms. However, MDCEV accommodates interchoice interactions in an implicit and unobservable way. This shortcoming can be overcome, for example, by using the concept of multilinear utility, which can directly use observed information to capture the interchoice interactions. In the future, it may be worth developing a much better model that can make use of the advantages of both MDCEV and the multilinear utility model.

1.4.5.2 Intrahousehold interaction modeling

Intrahousehold interactions have not been modeled in a satisfactory way in the literature. Chapter 7, Residential location and commuting mode choices: intrahousehold interaction modeling and its implications for energy policy, presents two new ways to incorporate intrahousehold interactions into household's joint choice of residential location and commuting mode: one assumes that joint decisions occur at the attribute level and the other assumes that joint decisions are made by negotiating individual members' independent choice outcomes. As it is not easy to observe individual member's independent choice outcome before joint decision, it is more feasible to make use of the first type of intrahousehold models. Such intrahousehold interactions can also be observed with other energy-related decisions, such as time use, expenditure, and ownership and usage of various (in-home and out-of-home) energy-consuming end uses. The attribute-based intrahousehold interaction modeling approach would be applicable to other household joint choices related to energy consumption.

1.4.5.3 Life-oriented approach is crucial to understand household energy consumption

Household energy consumption is actually associated with various life choices, including residence, family budget, health, neighborhood, education, job, family life, leisure, tourism, and transport. In reality, only a limited set of life choices have been targeted in the existing studies, consequently leading to various biased findings/insights.

There are various reasons causing the above research gap, among which lack of cross-domain behavioral research methodologies is a major reason. For this, Chapter 15, Life-oriented household energy-consumption research, showed how the life-oriented approach can be used to accommodate all possible life choices at different life domains within a unified and consistent framework.

Empirical evidence on the influences of the above life choices suggests that, to reduce household energy consumption, a cross-sectoral energy policymaking is required by connecting energy policy with not only transport policy but also policies about residence, family budget, health, neighborhood, education, job, family life, leisure, and tourism. In reality, because of difficulties in negotiation between various stakeholders, such cross-sectoral policies are very difficult to be made. For this, institutional designs for realizing cross-sectoral policies are essential, which shape/regulate the patterns of cross-stakeholder interactions and provide confidence to various stakeholders that the regulatory process is credible, legitimate, and predictable. For making institutional designs possible, leaderships and public participation become important, among various other factors. On the other hand, without scientifically sophisticated and practically operational interdisciplinary methodologies, cross-sectoral policymaking would not be possible.

A random forest approach is applied to simultaneously capture how about 90 life choice variables affect household energy consumption in an objective way. This approach is flexible to deal with the influences of a large number of explanatory variables on a dependent variable; however, it cannot provide a behaviorally sound explanation about its estimation results. This shortcoming is not unique to this random forest approach, but it is a common issue for all data mining or machine learning approaches, which have been widely applied in those fields with Big

Data. Unfortunately, such a shortcoming has not well been recognized by data scientists. To better make use of the advantages of such approaches, they should be better integrated with models incorporating behavioral decision-making mechanisms.

The life-oriented approach has shown various possibilities in capturing people's decisions in various life domains in association with different public policies. It presents a brand new research stream for interdisciplinary behavioral research and for supporting a long-lasting paradigm shift in public policymaking.

16.1.5 Opportunities and challenges in promoting new mobility in China

16.1.5.1 Electric vehicles in China

Facing the challenges of energy dependence, environmental pollution, and the development opportunity of automobile industry, it is significant for China to popularize EVs that are a representative vehicle type of new-energy vehicles. For this a series of policies has been published from national level to local government level to ensure the implementation of national strategic targets. Among policies at different administrative levels, because of the highly centralized policymaking system in China, the nation-level policies play the most important role in the sense that the local-level policies must be in-line with the national policies; otherwise, they would not work.

Even though the current number of EVs in China is just about 2.0 million vehicles, but there is a prediction by the automobile industry saying that the number will reach 100 million by 2030[2]. There is no way for us to verify the prediction, but it is true that the number of EVs in China is growing very rapidly. Such a rapid growth of EV is a unique Chinese phenomenon. In China, currently, electric bicycles already reached 250 million vehicles (as of June 2019)[3], and the total number of automobiles reached 240 million (as of the end of 2018)[4]. The current vehicle ownership rate is just about 20% in China. However, if young people in China are asked whether a vehicle is required for marriage, most of them would

[2] <http://auto.people.com.cn/n1/2019/0114/c1005–30525866.html> (in Chinese; accessed 25.06.19.).

[3] <https://tech.sina.com.cn/it/2019-06-14/doc-ihvhiqay5524207.shtml> (in Chinese; accessed 25.06.19.).

[4] <http://auto.ce.cn/auto/gundong/201901/15/t20190115_31260415.shtml> (in Chinese; accessed 25.06.19.).

say "yes." In reality, various constraints have hindered young people to purchase a car, but such a strong desire surely exists. This may be a part of the reasons why the above prediction of 100 million EVs was made. The research in this chapter should be further expanded by investigating more surveys with more factors in more Chinese cities. Both quantitative and qualitative surveys should be cleverly combined for figuring out true factors (not only prices and policies but also more life-oriented and sociopsychological factors) pushing such a strong growth of automobiles. The deployment of shared mobility (see Chapter 8: Battery electric vehicles in China: ownership and usage) will be surely influential to the ownership and usage of EVs. It is therefore important to investigate the influences of various factors by properly considering structural changes in the Chinese transport sector. Currently, China dominates the EV supply chain[5]. It is not clear how the US–China trade war will affect the global EV market, how China will adjust its automobile industry structure, and whether and how the Chinese customers will alter their attitudes toward and decisions on EV ownership and usage. All these need to be addressed in the future.

16.1.5.2 Shared mobility in China

China's transport sector has experienced the fastest growth of shared mobility in the world in the most unexpected ways, largely because Chinese people are much more willing to try new digital products and services, especially cashless services. However, the emergence of such a new mobility market is not aimed at reducing negative impacts of road traffic, but meeting people's mobility needs, considering the fact that shared mobility has been operated by the private sector from its birth. In China the transport sector is a very big industry and it is leading the development of digital technologies. In recent years the Chinese government has made various efforts to take the advantage of such new mobility services to mitigate various negative impacts of road traffic, including excessive use of energy and the resulting environmental emissions.

Shared mobility has an evident influence on passengers' current travel model choices and their future attitudes toward purchasing new vehicles or replacing the old vehicles. Promoting ridesharing among the passengers in China will substantially mitigate environmental pressures in the transport sector. However, travel mode shifts to ride-sourcing travel caused

[5] <https://www.scmp.com/economy/china-economy/article/3004953/united-states-sets-sights-china-new-electric-vehicle-push> (accessed 25.06.19.).

additional direct energy consumption and CO_2 emissions. Therefore joint measures of promoting the use of EVs, switching to clean electricity generation, shortening pick up time, increasing vehicle occupancy, and enhancing the ride-sourcing usage would make the major cities in China achieve nearly 50% of energy-savings and emissions reduction, based on a Big Data analysis.

The use of Big Data in the field of transportation (both research and practice) in China has become more and more popular. Such popularity is mainly because of the rapid growth of new mobility services. The difficulty (in terms of both time and cost as well as quality) of collecting questionnaire data from individuals further makes the use of Big Data more popular. Strongly influenced by the popularity of data science, statistical and mathematical models have been dominating in the literature and practice. More and more methodological improvements in terms of data handling are expected on the one hand, while lack of behavioral decision-making mechanisms may hinder the theoretical breakthroughs of transportation research in China on the other.

16.1.6 Transport-related energy policymaking needs to address social issues

Rural migrant workers' issues are not unique in China, the target country of Chapter 10, Rural migrant workers' energy consumption in passenger transport and domestic sectors of china: case study from a social perspective. But China's issue is special because of the existence of *hukou*, a household registration system (people with a rural *hukou* cannot have equal accesses to public services as those with an urban *hukou*). The *hukou* system has been relaxed gradually, allowing more and more migrant workers to permanently reside in cities. Considering the huge number of rural migrant workers in Chinese cities, the future improvements of social issues bothering rural migrants may induce more energy consumption.

The case study in this chapter empirically showed that such a social issue is surely influential to migrant workers' current energy consumption, even though the influence is not very large. This finding is very important, considering that urbanization in many developing countries is mainly driven by rural-to-urban migration. Rural migrant workers suffer from various social exclusion issues caused by low education level and the resulting low-income level. The above finding suggests that energy consumption (closely related to the quality of life) of this special population group depends on what kinds of daily life difficulties they experience.

This kind of association implies that transport and energy policymakers need to prepare for the rising energy consumption once social exclusion issues are resolved. Because of the income constraint, rural migrant workers attempt to save energy as much as possible. This is not because they concern about environmental issues, but about the limited income. However, such an energy-saving habit is very important. If such a habit could remain even after their income increases via effective interventions (as suggested in Chapter 14: Behavioral interventions for sustainable transportation: an overview of programs and guide for practitioners), transport and energy policymakers' worries would disappear. Considering the history of environmental battles, this is clearly not an easy task. What's important is to make rural migrant workers behave in an environment-friendly way and to help them understand how they could benefit from energy and environmental policies. Considering the widely observed rebound effects of energy-efficiency improvements (especially in developed countries), technological and economic measures may not be the top-priority policy. Behavioral interventions should be paid more attention in a careful way, considering the heterogeneity of different population groups.

16.1.7 Limited behavioral insights from aviation transport and energy research

Chapter 12, Behavioral research on transport and energy in the context of aviation, presents a review examining behavioral impacts of aviation transport on energy consumption and emissions from different stakeholders' perspectives. It is found that there are not many behavioral studies on aviation energy consumption, even though air travelers' behavioral changes are of great need.

Choice models are most widely used tool for investigating air travelers' energy-related behavior. The low degree of transparency of airlines has influenced consumers' actual behavior. Mandatory policies involving energy saving are necessary in some cases. Psychological impetus of behavioral change is more effective than infrastructural changes in air travelers' environmental consciousness; however, efforts should be made to eliminate attitude–behavior gap.

Airlines' energy-related behavior involves choice of fuel usage, carbon emissions, and aircraft size, which are mainly investigated by regression models. Equity is important for developing economic instruments, and carbon price is the most important factor to energy consumption in aviation logistics.

Moral decision-making theories, and choice models with social interactions and sociopsychological factors, should be paid enough attention in future research.

16.1.8 Many unclear points in tourism transport and energy research

The rapid growth of tourism industry has increased the energy consumption in visiting destinations. With the further growth of global economy (even though various uncertainties exist), this trend is expected to be further strengthened. The variety-seeking property of tourist behavior suggests that with the increase of income, people may be more likely to visit more faraway destinations, because those faraway destinations may have different attractions that can meet people's unmet tourism needs. At the same time, quality-tourism experiences have been paid more attention by tourists, and activity participations during tourism have, therefore, become more and more important, in comparison with destination visitation-focused tourism (i.e., activity participation is not the major focus). Moreover, the aging phenomenon in many countries indicates that more and more elderly people will become a customer of the tourism industry. Health concerns during tourism may discourage elderly people to travel too far from their home countries. Thus to understand how tourism transportation in the future affects energy consumption, it is important to clarify how the travel distance is influential to tourism decisions. If promoting attractive activities at a destination closer to home could attract tourists, promoting such shorter distance tourism may reduce tourism-related energy consumption. Unfortunately, the reality is more complicated, because a reduction in shorter distance tourism may be canceled out by the energy consumption from the transportation and activity participation by too many tourists. All the above issues need to be carefully examined in the future.

Tourism energy consumption is a valid parameter to understand tourism and environment, and it exists in the whole process of tourism, stretching across original generating regions, transit routes, and visiting destinations. Such a process-based research is important to properly capture the effects of tourism on energy consumption in a seamless and consistent way. Tourism energy-related behavior is not only related to tourists but also various stakeholders, such as tourism suppliers and local governments. Measuring tourism energy consumption is possible to better assess sustainability in a quantitative and objective way. The studies on energy consumption should be connected to tourism and destination development within sustainable goals.

Research on energy consumption and sustainability can be bettered from a behavioral perspective and an integrated view of tourism development. Unfortunately, relevant research is still very limited in the literature. This is partially due to the lack of available data. Therefore more research efforts should be made to collect more data at individual level. It may be worth exploring a standardized tourist behavior data collection system covering origin, destinations, major travel modes, activity participation, and time use as well as monetary consumption at facility level, based on questionnaire surveys. Such a data system may include counting data at facility level (hotel, attraction, station, restaurant, store, etc.), log data with time and location information of smartphone usage, as well as other Big Data that are expected to be accumulated gradually.

16.1.9 Behavioral interventions should be a "must" in transport and energy policy menus

Proper interventions on travel behavior help improve the transportation sustainability. Such interventions recognize that behavioral change takes time, and it is therefore important to focus on decision-making processes. In this regard, Chapter 14, Behavioral interventions for sustainable transportation: an overview of programs and guide for practitioners, presents three main established models of behavior change (Transtheoretical Model, Diffusion of Innovation, and Theory of Planned Behavior), as well as the concept of habits for research pertaining to sustainable transportation behaviors. Related to these models, there are actually more than 90 evidence-based behavior change techniques from nudge psychology[6] (goals & planning: 9; feedback & monitoring: 7; social support: 3; shaping knowledge: 4; natural consequences: 6; comparison of behavior: 3; associations: 8; repetition & substitution: 7; comparison of outcomes: 3; reward & threat: 11; regulation: 4; antecedents: 6; identity: 5; scheduled consequences: 10; self-belief: 4; covert learning: 3). Applying right nudge techniques to right persons will be a key. To this end, various

[6] https://digitalwellbeing.org/nudge-psychology-all-93-behavior-change-techniques-listed-and-summarised-free-app/ (accessed June 23, 2019).

segmentation methods should be applied to identify robust segments of people for different nudge techniques.

The target transportation behavior includes two main domains (personal and professional transportations) and three types (efficient use, purchasing, and mode switching), and a bigger picture of activities and lifestyle. Although a plethora of personal transportation studies exist in literature, there still remain neglected potential areas, such as the issues of monetary incentive, mode switching, and personal long-haul travel. The domain of professional transportation is unaddressed by behavior change literature and requires more in-depth studies. A five-step process named "DRIVE" (define, research, intervene, verify impact, and evaluate next steps) is developed to examine sustainable transportation behavior change. A goal-focused approach, such as DRIVE and background research, helps with the grounded understanding of human motivation and behavior change, can maximize the likelihood of success.

The landscape of transportation will shift dramatically in the coming decades, and future research may look at the questions of how to encourage more effective use of available technologies. In this regard, various smart technologies, equipped with artificial intelligence (AI), internet of things (IoT), and sensors, etc., have been developed and deployed in practice. Autonomous vehicles and smartphone-based travel-related apps are two major examples of smart technologies in the field of transportation. Effects of these smart transportation technologies on energy consumption have remained unknown, by considering their connections with various activities (both in-home and out-of-home) and trips. It is not clear, either, how to make use of these smart technologies to encourage people's behavioral changes toward a sustainable transport society.

16.2 What should be done for an energy-saving transport society

Even though there are various unresolved issues about behavioral research on transport and energy, some "what-should-be-done" for reducing energy consumption from the future transport sector, based on existing studies and practices, can be summarized. The following are some examples, among others, which are illustrated with behavioral statements.

1. Policies should be packaged.

   Transport and energy-consumption behaviors involve multifaceted decisions, which are further associated with other life choices.

Such multidimensional and interrelated behaviors suggest that there will be no single policy suitable to change/modify different behaviors in a consistent way toward energy reduction. It is necessary to package various context-sensitive policies with both "candy and whip" instruments.

2. Behavioral interventions should be taken in a continuous way.

   Technological developments, pricing and regulation policies, have limited effects on reducing energy consumption. The roles of these measures will continue to be important; however, to reduce energy consumption from the transport sector to the required level, behavioral interventions should be taken. Considering human behaviors usually change over time, such interventions should be done in a continuous way.

3. Transport and energy policymaking should address heterogeneous responses of different actors.

   Assuming average responses of different actors or different population groups to a policy will no longer work. Heterogeneous responses of different actors to various policies have been well revealed in various research fields: no exception in the transportation field. In other words, segment-sensitive policies should be made.

4. Environmental pricing policy should be made.

   People should pay for the environmental emissions from their energy consumption, because energy is (consciously or unconsciously) consumed more than actually needed. It is also because of the existence of the rebound effects of technological improvements.

5. Technologies for personal usage should be developed at a proper level.

   This is because not everyone can afford for the most advanced technologies that do not necessarily contribute to energy reduction. To reduce energy consumption by a majority of population, energy-saving technologies for personal usage should be developed at a proper level. The private sector of technological developments should pay more attention to not only how to attract more customers but also how much their high-tech products contribute to energy reduction in the real world.

6. Land use should be supported by suitable transportation.

   Land use generates activities that usually involve transport. If high-density land use is only accessible by car, traffic congestion and the resulting unnecessary energy consumption could not be avoided.

In such a case, public transport with proper capacity is more suitable, and if there is no proper public transport, people have to reply on car.

7. Reducing car dependence should consider how it affects people's daily life.

   It has been unsuccessful for many transport policies of reducing car, because they cannot reflect the role of car in supporting many needs in life. Considering that cars are usually parked for the majority of the whole day, owning a car is actually determined not only by travel behavior itself but also by various life considerations.

8. Shared mobility should be prompted.

   Cars are usually parked for the majority of the whole day. This does not only waste the money spent on parking but also the land used for parking and the resources used for making vehicles. For avoiding these wastes, shared mobility should be promoted.

9. City boundary should be properly controlled.

   Improvements of transportation systems can reduce travel time, consequently reducing people's resistance to live farther from the city center. Faster transport allows people to have more choices about their residences and activity destinations, making a city to become bigger and bigger. More and more smart technologies have been developed and deployed in the real world, which may allow people to perform various activities without having more spatial constraints. For example, autonomous vehicles may encourage people to live farther from the city center because of the reduced driving burdens. If city growth could not be properly controlled, lengthened and induced trips from improved transportation systems and smart technologies would lead to more energy consumption.

10. A city should be walkable and walking environment should be safe and comfortable.

    Tourists usually walk a lot at destinations. This is simply because there are attractions for them, which generate more utility than the negative utility derived from walking. In other words, walking environment allowing people to experience good feelings will encourage people to walk more. Cities have housed more and more people, who have a high tendency to use a car. Making cities more compact with safe and comfortable walkable environment is crucial to reduce car dependence.

11. A neighborhood should meet its residents' most daily necessities.

Nobody likes to buy his/her daily necessities at a far place, unless he/she wants to buy them together with other shopping. Allowing the purchase of most daily necessities at a neighborhood will reduce trips made for the purchase at a farther place and the resulting energy consumption.

12. Transportation system should be comprehensive and affordable.

For allowing a majority of people to have more travel choices, transportation system should be comprehensive and affordable. "Comprehensive" does not mean that all kinds of travel modes should be made available, but a suitable set of travel modes should be available in consideration of area-specific conditions. "Affordable" mainly refers to affordable public transportation systems in terms of fares and access/egress costs.

13. Public transport should be prioritized and seamless connectivity of different public transport modes should be guaranteed.

Public transport is not as convenient as private transport, and therefore, in essence, people like to use private transport more than public transport to support their daily lives, if possible. If public transport is not to be prioritized in transportation systems, there is no reason for people to put it as the first choice target for making a trip. Furthermore, public transport should be safe, environment-friendly, convenient, and comfortable; otherwise, car users would never use it. Public transport service providers should understand that users need a well-connected public transport networks. Seamless connectivity between different public transport modes is a "must" for people to use public transport in a continuous way.

14. Transit-oriented development with affordable houses should be promoted.

Transit-oriented development usually results in a rise in land prices of the areas surrounding transit stations. If there are not enough affordable houses nearby transit stations, the higher housing price will discourage these "loyal" users of public transport to live far from transit stations, which may eventually push them to move toward a car-dependent lifestyle.

15. Transportation facilities, vehicles, and equipment should be environment-friendly through the whole lifecycles.

Travel choices are influenced by the attributes of transportation alternatives and trip makers' environmental attitudes, where the latter is not independent from the former. Environment-friendly transport services bring a "green" image to transport users and are more likely to encourage them to travel in an environment-friendly way.

16. Travel demand should be better managed.

People usually make travel choices without properly considering the influence of limited supply and travel demand changes over time. On the other hand, travel supply cannot be overconstructed due to the influence of budget limitation and the consideration of waste of limited resources. As a result, transport supply cannot always meet the demand, resulting in various negative impacts (including too much environmental emissions due to unnecessary use of energy). Better travel demand management will allow an effective use of existing transportation infrastructure and services.

17. Transport-related energy-consumption issues should be resolved under a cross-sectoral scheme.

Transport is a derived demand from human activities, which are performed at a certain location and supported by certain service(s). Therefore to resolve transport-related energy-consumption issues, policies should be made by involving not only transport-related but also activity-related stakeholders. In other words, the policies should be made under a cross-sectoral scheme.

18. Accountable transport and energy policymaking should be supported by models with behavioral mechanisms.

The target of transport and energy policymaking is trip makers. Without understanding their behaviors, it is impossible to make effective policies, because effects of policies are actually context-dependent. One policy being effective in a certain city does not necessarily work well in the other cities.

19. After transport and energy planning/policy are made, it should be properly monitored continuously.

Transport and energy planning/policy are made for meeting some goals. It is implemented because it is predicted that the goals can be achieved. Because various uncertainties are involved in planning and policymaking and people change their behaviors in a way that cannot be fully understood by policymakers and researchers, it is unclear whether the expected goals can be actually achieved or not. It is therefore important to monitor the progress of the planning and policy implementation continuously.

20. Stakeholders' behaviors should be better understood.

It is difficult to solely rely on private sectors' and households' own efforts to reduce energy consumption. Governments will still be a key stakeholder to regulate firms' and households' energy-consumption behaviors. However, little is known about policymakers' behaviors in the literature, even though their behaviors are crucial to transport and energy policymaking in practice. Especially, it is important to know what motivates them to choose or not to choose a certain policy, and what factors encourage them to make policies to reduce energy usage in the transport sector. On the other hand, energy reduction in organization needs close cooperation of individuals in organizations; however, relevant research has been rare. Furthermore, citizen participation is extremely important to realize sustainable transportation. For making full use of citizens' power, massive participation is required, for which incentive designs become important. Nevertheless, behavioral insights about citizen participation are also very limited.

21. Advanced technologies should be developed for use in all countries.

Reducing energy consumption in the transport sector needs advanced technologies, for example, existing ITS (intelligent transport systems) and other smart technologies supported by AI and IoT, and so on. These technologies are needed in not only developed countries but also developing countries. Because of the influence of globalization and the Internet, people in developing countries are also very adaptive to advanced technologies, which means that the advancement of technologies in developed countries is also likely to be applicable to developing countries. Advanced technologies should be developed and used not only for developed countries but also for developing countries. Advanced technologies developed in developed countries should not be used for hindering the quality development of developing countries.

16.3 Epilog

Behavioral change is crucial for realizing an environmentally sustainable transport society. However, both the researchers and policymakers have not well understood behaviors of transport service providers and users. Behaviors of policymakers and other stakeholders have not been well understood by researchers, either.

Cross-sectoral policies are important to reduce transport-related energy; however, in reality, conflicts of transport and energy sectors as well as other related sectors are still a big barrier to make cross-sectoral policies. For this, institutional design is important; however, limited behavioral knowledge about different stakeholders has made institutional design difficult and ineffective. One more barrier is lack of scientifically sophisticated cross-disciplinary methodologies. Collaborations between different research fields by involving both the public and private sectors should be further promoted via more innovative funding schemes.

Even though recent years have observed a rapid growth of various Big Data, behavioral research still needs direct observation of people's and organizational behaviors. This is because Big Data has rich information of behavior outputs, but it does not include sufficient information about the factors resulting in the behavior outputs. However, in recent years, it has become more and more difficult to implement surveys to people and organizations, because of the ever increasing concern about privacy issues. This is not a good signal for evidence-based policymaking. Citizens and organizations should still be welcomed and encouraged to actively provide their information for deriving scientifically sound insights into policies supporting citizens' quality of life and organizations' values.

Energy consumption in the transport sector of developing countries has increased faster than that in developed countries, as evidenced by the ever increasing pollutions. However, transport and energy research in developing countries is still very limited. Such a phenomenon is common in a majority of research fields. Behavioral research on transport and energy in the context of developing countries should be given a proper position in more high-impact international journals. Researchers in developed countries should show more interest in transport and energy research in developing countries. They should be further encouraged to have more collaborations with researchers in developing countries for resolving transport-related energy issues in developing countries.

References

Shukla, P.R., Dhar, S., 2015. Energy policies for low carbon sustainable transport in Asia. Energy Policy 81, 170–175.

Zhang, J., 2017. Life-Oriented Behavioral Research for Urban Policy. Springer, Tokyo.

Index

Note: Page numbers followed by "*f*" and "*t*" refer to figures and tables, respectively.

C

D

N

O

P

Q

R

U

V

W

Y

Kolibri 360 GmbH
Ferdinand-Jühlke-Straße 7
99095 Erfurt, Deutschland
produktsicherheit@kolibri360.de